NEUROMETHODS

For further volumes:
http://www.springer.com/series/7657

Neuromethods publishes cutting-edge methods and protocols in all areas of neuroscience as well as translational neurological and mental research. Each volume in the series offers tested laboratory protocols, step-by-step methods for reproducible lab experiments and addresses methodological controversies and pitfalls in order to aid neuroscientists in experimentation. *Neuromethods* focuses on traditional and emerging topics with wide-ranging implications to brain function, such as electrophysiology, neuroimaging, behavioral analysis, genomics, neurodegeneration, translational research and clinical trials. *Neuromethods* provides investigators and trainees with highly useful compendiums of key strategies and approaches for successful research in animal and human brain function including translational "bench to bedside" approaches to mental and neurological diseases.

Behavioral Neurogenetics

Edited by

Daisuke Yamamoto

National Institute of Information and Communications Technology, Kobe, Hyogo, Japan

Humana Press

Editor
Daisuke Yamamoto
National Institute of Information and
Communications Technology
Kobe, Hyogo, Japan

ISSN 0893-2336 ISSN 1940-6045 (electronic)
Neuromethods
ISBN 978-1-0716-2320-6 ISBN 978-1-0716-2321-3 (eBook)
https://doi.org/10.1007/978-1-0716-2321-3

This Humana imprint is published by the registered company Springer Science+Business Media, LLC, part of Springer Nature.
The registered company address is: 1 New York Plaza, New York, NY 10004, U.S.A.

Preface to the Series

Experimental life sciences have two basic foundations: concepts and tools. The Neuromethods series focuses on the tools and techniques unique to the investigation of the nervous system and excitable cells. It will not, however, shortchange the concept side of things as care has been taken to integrate these tools within the context of the concepts and questions under investigation. In this way, the series is unique in that it not only collects protocols but also includes theoretical background information and critiques which led to the methods and their development. Thus, it gives the reader a better understanding of the origin of the techniques and their potential future development. The Neuromethods publishing program strikes a balance between recent and exciting developments like those concerning new animal models of disease, imaging, in vivo methods, and more established techniques, including, for example, immunocytochemistry and electrophysiological technologies. New trainees in neurosciences still need a sound footing in these older methods in order to apply a critical approach to their results.

Under the guidance of its founders, Alan Boulton and Glen Baker, the Neuromethods series has been a success since its first volume published through Humana Press in 1985. The series continues to flourish through many changes over the years. It is now published under the umbrella of Springer Protocols. While methods involving brain research have changed a lot since the series started, the publishing environment and technology have changed even more radically. Neuromethods has the distinct layout and style of the Springer Protocols program, designed specifically for readability and ease of reference in a laboratory setting.

The careful application of methods is potentially the most important step in the process of scientific inquiry. In the past, new methodologies led the way in developing new disciplines in the biological and medical sciences. For example, physiology emerged out of anatomy in the nineteenth century by harnessing new methods based on the newly discovered phenomenon of electricity. Nowadays, the relationships between disciplines and methods are more complex. Methods are now widely shared between disciplines and research areas. New developments in electronic publishing make it possible for scientists that encounter new methods to quickly find sources of information electronically. The design of individual volumes and chapters in this series takes this new access technology into account. Springer Protocols makes it possible to download single protocols separately. In addition, Springer makes its print-on-demand technology available globally. A print copy can therefore be acquired quickly and for a competitive price anywhere in the world.

Saskatoon, SK, Canada ***Wolfgang Walz***

Preface

Behavioral neurogenetics is a rapidly growing research field that seeks to elucidate how animal behavior is generated at the organismal, cellular, and molecular levels with the aid of a variety of synthetic and analytical techniques in genetics and neurosciences. Traditionally, behavioral neurogenetics has focused on several key questions. How are individual motions composing a fixed action pattern generated by the circuit and how do composite neurons accomplish their dedicated roles in the circuit? How are the wiring patterns and mode of circuit actions modulated for the acquisition of context-dependent changes in behavior? How is the circuit for a behavior organized in development? What genes are involved in the formation and operation of the neural circuit underlying behavior? Although much ground has been broken, these mysteries also remain central to contemporary behavioral neurogenetics. This book gathers achievements in addressing these questions to illustrate the framework in which different scientific disciplines will be integrated to guide future studies in behavioral neurogenetics. This book further endeavors to illuminate emerging new trends in behavioral neurogenetics, which are characterized by data-driven and/or exhaustive comparative approaches toward a complete understanding of the system controlling behavior across the animal kingdom. For instance, novel histological techniques are filling the gap between light and electron microscopic images of neural networks in higher organisms from flies to mammals, unveiling the global web of neural connections throughout the nervous system. Such success in connectomics will revolutionize neuroscience by making a deterministic approach feasible, in addition to a conventional probabilistic approach, in elucidating the causal relationship between the molecular/cellular events and behavioral outcomes. This trend is being fueled by rapid advances in next-generation sequencing technology, leading to transcriptome analysis at the single neuron level. Combining transcriptome approaches with connectomics for behavioral analysis in living organisms will permit us to identify molecular events occurring in single neurons that underlie behavior changes at the organismal level. Recent studies are even extending our view of the organism by unraveling the complex interactions between the host and inhabiting microorganisms, which are often decisive in the psychobehavioral traits of the host. Lastly, invention of the CRISPR/Cas9 technique has opened up an avenue for neurogenetic studies in non-model organisms, providing the means to dissect the genetic basis for behavioral diversification. This book covers all these important aspects of current behavioral neurogenetics, thereby illuminating the future direction of the field.

Kobe, Hyogo, Japan *Daisuke Yamamoto*

Contents

Contributors

KAI AMINO • *Laboratory of Applied Entomology, Department of Agricultural and Environmental Biology, The University of Tokyo, Tokyo, Japan*

MURAT ARTAN • *Institute of Science and Technology (IST) Austria, Klosterneuburg, Austria*

JEAN-CHRISTOPHE BILLETER • *Groningen Institute for Evolutionary Life Sciences, University of Groningen, Groningen, The Netherlands*

BHIM B. BISWA • *Mouse Genomics Resource Laboratory, National Institute of Genetics, Shizuoka, Japan; Department of Genetics, SOKENDAI, Shizuoka, Japan*

YA-HUI CHOU • *Institute of Cellular and Organismic Biology, Academia Sinica, Taipei, Taiwan*

MASAHIRO DAIMON • *Graduate School of Life Sciences, Tohoku University, Sendai, Miyagi, Japan*

MARIO DE BONO • *Institute of Science and Technology (IST) Austria, Klosterneuburg, Austria*

YUN DING • *Department of Biology, University of Pennsylvania, Philadelphia, PA, USA*

JULIA C. DUCKHORN • *Department of Biology, Villanova University, Villanova, PA, USA*

KHALED GHANDOUR • *Department of Biochemistry, Graduate School of Medicine and Pharmaceutical Sciences, University of Toyama, Toyama, Japan; Research Center for Idling Brain Science, University of Toyama, Toyama, Japan; Department of Biochemistry, Faculty of Pharmacy, Cairo University, Cairo, Egypt*

RALPH J. GREENSPAN • *Kavli Institute for Brain and Mind, University of California-San Diego, La Jolla, CA, USA; Division of Biological Sciences, University of California-San Diego, La Jolla, CA, USA; Department of Cognitive Science, University of California-San Diego, La Jolla, CA, USA*

DHRUV GROVER • *Kavli Institute for Brain and Mind, University of California-San Diego, La Jolla, CA, USA*

AKI HORI • *Faculty of Pharmacy, Institute of Medical, Pharmaceutical and Health Sciences, Kanazawa University, Kanazawa, Ishikawa, Japan*

LONGWEN HUANG • *State Key Lab of Brain and Cognitive Science, Institute of Biophysics, Chinese Academy of Sciences, Beijing, China*

YASUKO ISOE • *Department of Molecular and Cellular Biology, Harvard University, Cambridge, MA, USA*

NAOTO JUNI • *Department of Synthetic Chemistry and Biological Chemistry, Graduate School of Engineering, Kyoto University, Kyoto, Japan*

IAN P. JUNKER • *Department of Biology, University of Pennsylvania, Philadelphia, PA, USA*

HIBIKI KADOGUCHI • *Faculty of Pharmacy, Institute of Medical, Pharmaceutical and Health Sciences, Kanazawa University, Kanazawa, Ishikawa, Japan*

MASAHIRO KAI • *Department of Synthetic Chemistry and Biological Chemistry, Graduate School of Engineering, Kyoto University, Kyoto, Japan*

TAKEO KATSUKI • *Kavli Institute for Brain and Mind, University of California-San Diego, La Jolla, CA, USA*

SOH KOHATSU • *National Institute of Information and Communications Technology, Kobe, Japan*

PHILIP KOHLMEIER • *Groningen Institute for Evolutionary Life Sciences, University of Groningen, Groningen, The Netherlands*

TSUYOSHI KOIDE • *Mouse Genomics Resource Laboratory, National Institute of Genetics, Shizuoka, Japan; Department of Genetics, SOKENDAI, Shizuoka, Japan*
TAKAYUKI KURAISHI • *Faculty of Pharmacy, Institute of Medical, Pharmaceutical and Health Sciences, Kanazawa University, Kanazawa, Ishikawa, Japan; AMED-PRIME, Japan Agency for Medical Research and Development, Chiyoda-ku, Tokyo, Japan; JST-FOREST, Japan Science and Technology Agency, Tokyo, Japan*
TAKASHI MATSUO • *Laboratory of Applied Entomology, Department of Agricultural and Environmental Biology, The University of Tokyo, Tokyo, Japan*
TAKAAKI MIYAZAKI • *IIIS, University of Tsukuba, Ibaraki, Japan*
KOHJIRO NAGAO • *Department of Synthetic Chemistry and Biological Chemistry, Graduate School of Engineering, Kyoto University, Kyoto, Japan*
RYUSUKE NIWA • *Life Science Center for Survival Dynamics, University of Tsukuba, Ibaraki, Japan*
NORIAKI OHKAWA • *Division for Memory and Cognitive Function, Research Center for Advanced Medical Science, Comprehensive Research Facilities for Advanced Medical Science, Dokkyo Medical University, Tochigi, Japan*
TERUHIRO OKUYAMA • *Institute for Quantitative Biosciences, The University of Tokyo, Tokyo, Japan*
TROY R. SHIRANGI • *Department of Biology, University of Pennsylvania, Philadelphia, PA, USA*
TAKAAKI SOKABE • *Division of Cell Signaling, National Institute for Physiological Sciences, Okazaki, Aichi, Japan*
TAKUTO SUITO • *Division of Cell Signaling, National Institute for Physiological Sciences, Okazaki, Aichi, Japan*
HIDEAKI TAKEUCHI • *Graduate School of Life Sciences, Tohoku University, Sendai, Miyagi, Japan*
HIROFUMI TODA • *IIIS, University of Tsukuba, Ibaraki, Japan*
KUO-TING TSAI • *Institute of Cellular and Organismic Biology, Academia Sinica, Taipei, Taiwan*
MASATO UMEDA • *Department of Synthetic Chemistry and Biological Chemistry, Graduate School of Engineering, Kyoto University, Kyoto, Japan; Research Institute for Mechanobiology, Organization for Research Initiatives and Development, Doshisha University, Kyoto, Japan*
THOMAS A. VERSCHUT • *Groningen Institute for Evolutionary Life Sciences, University of Groningen, Groningen, The Netherlands; Department of Zoology, Stockholm University, Stockholm, Sweden*
MU-YUN WANG • *Institute for Quantitative Biosciences, The University of Tokyo, Tokyo, Japan*
JULIE A. WILLIAMS • *Department of Neuroscience, Smilow Research Center, Perelman School of Medicine, Chronobiology & Sleep Institute, University of Pennsylvania, Smilow Research Center, Philadelphia, PA, USA*
DAISUKE YAMAMOTO • *National Institute of Information and Communications Technology, Kobe, Japan*
SAORI YOKOI • *Faculty of Pharmacological Sciences, Hokkaido University, Sapporo, Hokkaido, Japan*
YUTO YOSHINARI • *Life Science Center for Survival Dynamics, University of Tsukuba, Ibaraki, Japan*

Chapter 1

Combined in Situ Hybridization Chain Reaction and Immunostaining to Visualize Gene Expression in Whole-Mount *Drosophila* Central Nervous Systems

Julia C. Duckhorn, Ian P. Junker, Yun Ding, and Troy R. Shirangi

Abstract

Methods to visualize gene expression in the *Drosophila* central nervous system are important in fly neurogenetic studies. In this chapter, we describe a detailed protocol that sequentially combines in situ hybridization chain reaction (HCR) and immunostaining to detect mRNA and protein expression in whole-mount *Drosophila* larval and adult central nervous systems. We demonstrate the application of in situ HCR in comparisons of nervous system gene expression between *Drosophila* species and in the validation of single-cell RNA-Seq results in the fly nervous system. Our protocol provides a simple, robust, multiplexable, and relatively affordable means to quantitatively visualize gene expression in the nervous system of flies, facilitating its general use in fly neurogenetic studies.

Key words *Drosophila*, In situ hybridization chain reaction, Immunohistochemistry, Whole-mount CNS tissues

1 Introduction

Neurogenetic studies of *Drosophila* often require methods to visualize and quantify gene transcription in the larval or adult central nervous system. Fluorescence in situ hybridization of mRNAs, or mRNA-FISH, is widely used across animal systems to detect gene expression in a variety of whole-mount tissue types. However, the application of mRNA-FISH on whole-mount *Drosophila* central nervous systems has faced long-standing challenges, including weak probe penetration and poor sample preservation, especially in adult nervous systems.

In situ hybridization chain reaction (HCR) is an enzyme-free method to detect one or more sequence-specific nucleic acids in cell, tissue, and whole-mount samples via detection and

Authors Julia C. Duckhorn and Ian P. Junker have contributed equally to this chapter.

Daisuke Yamamoto (ed.), *Behavioral Neurogenetics*, Neuromethods, vol. 181, https://doi.org/10.1007/978-1-0716-2321-3_1,

amplification stages [1–4]. During detection, an initiator sequence assembles on target mRNA molecules using adjacent, paired DNA probes. The initiator sequence subsequently triggers the self-assembly of a fluorescently tagged DNA polymer from a mixture of two amplifier hairpins. The use of short DNA probe-sets greatly enhances tissue penetration and signal detection. The detection and amplification reactions are done under mild experimental conditions, thus improving sample integrity and permitting the application of immunostaining protocols. Moreover, the use of split-paired probes suppresses off-target amplification, thereby facilitating high signal-to-noise ratios [4].

In recent years, two whole-mount mRNA-FISH protocols have been developed for use on *Drosophila* central nervous systems [5, 6]. Both protocols detect mRNA expression with single-molecule resolution using fluorescently labeled probe-sets but differ in sample preparation and imaging protocols. In comparison to these protocols, an HCR-based approach is, in principle, more sensitive and robust, allowing it to detect transcripts across different expression levels. It is also more cost-effective and does not require the experimenter to design or create probe-sets, which are provided by Molecular Instruments™. HCR also offers greater flexibility to mix and match color channels for multiplexing and post-HCR immunostaining, since the amplifier hairpins and not the probes themselves are labeled with a fluorochrome. Although in situ HCR-based FISH has been successfully implemented using whole-mount *Drosophila* CNS tissues [7, 8], the procedure is not widely used in the field.

In this chapter, we report a detailed in situ HCR protocol that provides a multiplexable, straightforward, affordable, rapid, sensitive, and robust means to detect mRNA expression in the *Drosophila* larval and adult central nervous system using confocal light microscopy. Our protocol can be used with standard immunohistochemistry protocols to simultaneously detect protein expression. We demonstrate the utility of in situ HCR with two important applications. First, we visualize the mRNA expression of multiple orthologous genes in the adult and larval central nervous systems of two *Drosophila* species. Second, we validate recently published single-cell RNA-Seq data [9] on neurons derived from a common neuronal stem cell in the fly ventral nerve cord.

2 Materials

2.1 Split-Initiator Probes and Amplification Hairpins

Single-stranded DNA HCR probe-sets and fluorescently tagged DNA HCR amplifier hairpins were purchased from Molecular Instruments™. Each probe- and hairpin-set are matched by the identity of the HCR initiator. For example, if a particular probe-set utilizes a B1 HCR initiator, a B1 complementary hairpin set is required for successful polymerization. We used a probe-set size of 20 for all in situ HCR experiments. Probes are complementary DNA sequences of the desired endogenous mRNA target. The sequences of each probe within a probe-set are available upon request from Molecular Instruments™. The lot numbers for the probe-sets used in this study are PRH303 (*Antp*), PRE237 (*Ubx*), PRF810 (*Abd-B*), and PR1675 (*Sox21B*). For long-term storage, probes and hairpins were aliquoted and frozen at −20 °C. During use, aliquots were defrosted and stored short term at 4 °C. All fluorescently tagged hairpins were protected from light during storage.

2.2 Reagents and Buffers for HCR and Immunostaining

Central nervous systems were dissected in 1X phosphate buffered saline (PBS; Fisher Cat. # BP2438-4) and fixed in 4% paraformaldehyde (PFA; Electron Microscopy Sciences, Cat. # 15710) buffered in 1X PBS. 4% PFA was stored at 4 °C for no more than 2 weeks or as single-use aliquot at −20 °C. Probe hybridization buffer (stored at −20 °C), probe wash buffer (stored at −20 °C), and amplification buffer (stored at 4 °C) were purchased from Molecular Instruments™ but can be easily prepared in the lab [4]. 5X SSCT (UltraPure 20X SSC, ThermoFisher Scientific, Cat. # 15557044 plus Tween 20 diluted to 0.1%) was made fresh the day of usage. Nervous tissue was washed in PBTx (1X PBS with 1% Triton X-100) stored at room temperature or 4 °C. Nervous systems were blocked in 5% heat-inactivated normal goat serum (NGS; Jackson Immunochemicals, Cat. # 005-000-121) in PBTx. The following primary antibodies were used in this study: mouse anti-Antp 8C11 (Developmental Studies Hybridoma Bank, diluted 1:20 in PBTx/5% NGS), mouse anti-Ubx FP3.38 (DSHB, diluted 1:20), and mouse anti-Abd-B 1A2E9 (DSHB, diluted 1:10). All primary antibodies were used with a goat anti-mouse Alexa Fluor® 568 secondary diluted in PBTx/5% NGS (Invitrogen, Cat. # A11004, diluted 1:500).

2.3 Reagents for Mounting

We mounted larval and adult nervous systems in VECTASHIELD (Cat. # H-1000) or ProLong Gold Antifade with DAPI (Invitrogen, Cat. # P36931). If users prefer to clear with xylenes before mounting, we recommend passing the nervous systems through an ascending ethanol series (30%, 50%, 70%, 95%, 100%, 100%) for 5 min each, then placing the coverslip-adhered (Corning; Cat. # 2845-22) nervous systems in xylenes (Sigma-Aldrich, Cat. # 247642-500ML) twice for 5 min. The coverslip with nervous systems adhered to it can be mounted on a slide using Dibutylphthalate Polystyrene Xylene (DPX; Sigma-Aldrich, Cat. # 06522-500ML).

2.4 Drosophila Strains

Fly stocks were maintained at room temperature. The following genotypes were used: *D. melanogaster* Canton S and *D. yakuba* Ivory Coast (UCSD Stock #14021-0261.02).

2.5 Confocal Imaging

Central nervous systems were imaged on Leica SP8 (larval tissues) and Leica DMi8 (adult tissues) confocal microscopes at 40X and 63X with optical sections at 0.3 or 1 μm intervals.

3 Methods

3.1 Multiplexable HCR

To demonstrate the use of in situ HCR on the whole-mount central nervous system of *Drosophila melanogaster* (Canton S), we simultaneously probed the mRNA expression of three *Hox* genes in larvae and adults using the protocol described below: *Antennapedia* (*Antp*), *Ultrabithorax* (*Ubx*), and *Abdominal-B* (*Abd-B*). The expression of these *Hox* genes has been well-characterized in the fly nervous system [9], and good antibodies are available against proteins of all three genes. Probe-sets and corresponding amplification hairpins were designed and supplied by Molecular Instruments™ (Antp-B1/B1-Alexa Fluor® 488, Ubx-B2/B2-Alexa Fluor® 594, and Abd-B-B3/B3-Alexa Fluor® 647). The HCR protocol takes 3 days to complete. An overview of the protocol workflow is illustrated in Fig. 1.1. Exemplary images of resulting larval and adult nervous systems are shown in Fig. 1.2.

Day 1:

1. Dissect nervous systems from late-stage 3rd instar larvae or adult female flies aged 4–7 days in cold 1X PBS and transfer them to a glass embryo dish (Electron Microscopy Sciences; Cat. # 70543–30) on ice containing 1X PBS. All subsequent steps are performed in the dish.
2. Fix the nervous systems in 4% PFA fixation buffer for 35 minutes at room temperature. Following fixation, quickly rinse samples three times with PBTx, and then wash the samples with PBTx four times for 5 min each at room temperature.
3. Pre-hybridize samples in 200 μl prewarmed probe hybridization buffer for 30 min at 37 °C. In case the tissues are floating on the surface of the buffer, gently prod the tissues downward so they become immersed in the buffer. During pre-hybridization, prepare the probe solution by adding 0.8 pmol of each probe-set to 200 μl of prewarmed probe hybridization buffer.
4. Replace the pre-hybridization solution with the probe solution, and incubate overnight (12–16 h) at 37 °C. The embryo dish should be covered and sealed tightly with parafilm to prevent

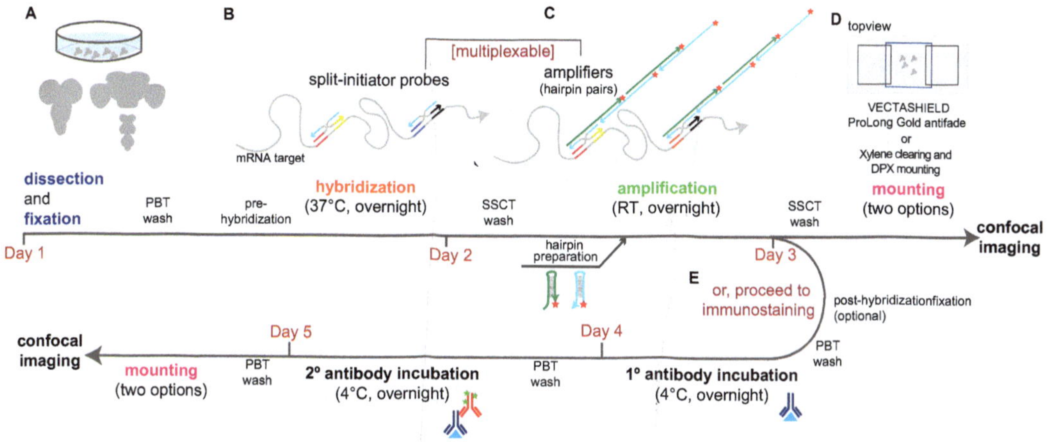

Fig. 1.1 Overview of the in situ HCR protocol combined with immunostaining. Summary of the main steps of in situ HCR and the mechanisms of detection and amplification are provided along a timeline. (**a**) Dissection and fixation. All steps are performed in a glass dish. (**b**) HCR hybridization. mRNA targets are bounded by split-initiator probe-sets (we used a set of 20 pairs in this study). (**c**) HCR amplification. mRNA targets are illuminated in the amplification stage, where initiators assembled by individual probe pairs trigger the self-assembly of hairpin amplifiers conjugated with fluorophores (denoted by stars). HCR is multiplexable by matching probe-sets designed for different transcripts with hairpin pairs conjugated with different fluorophores. The timeline is the same regardless of the number of targeted mRNA. (**d**) Mounting. Samples can be mounted directly with generic mounting media such as VECTASHIELD or ProLong Gold Antifade or be subjected to xylene clearing and DPX mounting for potentially optimized signal-to-noise ratio. (**e**) Immunostaining. Users may perform this step after HCR and an optional post-HCR PFA fixation (for better preservation of HCR signals). Some antibodies are sensitive to this additional fixation step, and users can skip it or adjust the conditions for a more desirable outcome. Users may also try immunostaining before HCR, if the antibody staining is consistently problematic

the probe solution from evaporating overnight. All following steps should be completed with protection from light.

Day 2:

5. Preheat the probe wash buffer for 15 min at 37 °C. Remove probe solution, and wash the nervous systems with 200 μl of preheated probe wash buffer four times, 15 min each at 37 °C. Next, remove the probe wash buffer, and wash the tissues with 5X SSCT three times, 5 min at room temperature. 5X SSCT should be prepared fresh the day of use.
6. Pre-amplify the samples in 200 μl of amplification buffer for 30 min at room temperature. Prewarm the amplification buffer to room temperature before use.
7. During pre-amplification, prepare the hairpin solution:

   Each sample receives 12 pmol of each hairpin in a total of 200 μl of amplification buffer. Add 12 pmol of hairpin h1 to one Eppendorf tube and 12 pmol of hairpin h2 to another Eppendorf tube (i.e., if the stock solution of each hairpin is 3 μM or 3 pmol/μl, the experimenter should add 4 μl of the

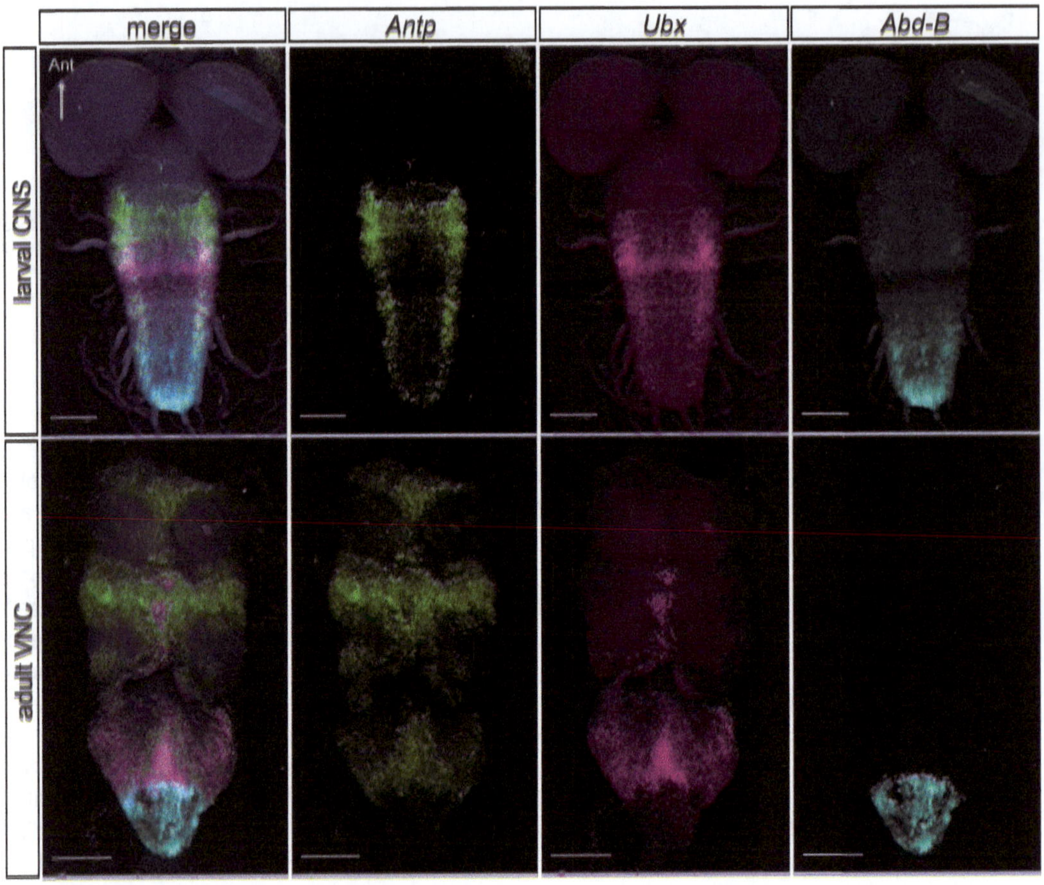

Fig. 1.2 Detection of *Hox* gene mRNA expression in the larval and adult central nervous system of *D. melanogaster* by in situ HCR. Antp-B1, Ubx-B2, and Abd-B-B3 probe-sets were used with B1-Alexa Fluor® 488, B2-Alexa Fluor® 594, and B3-Alexa Fluor® 647 amplifiers in multiplexed in situ HCR. mRNAs of *Antp* and *Ubx* are expressed primarily in cells of the thoracic neuromeres; *Abd-B* is expressed in cells of the abdominal ganglion. *Hox* gene expression occurs in a pattern consistent with that previously published (Allen et al. 2020). Scale bar = 50 μm

hairpin to the tube). Snap-cool by placing both tubes in a preheated heat-block set at 95 °C for 90 s, and then allow the hairpins to cool to room temperature for 30 min in a dark drawer. Then add 100 μl of amplification buffer to each hairpin, mix gently, and combine the two hairpins for a total of 200 μl.

8. Replace pre-amplification solution with the hairpin solution, and incubate overnight (12–16 h) at room temperature. The dish should be covered and sealed with parafilm.

Day 3:

9. Remove the hairpin solution, and wash samples with 5X SSCT at room temperature using the following series: two times, 5 min each; two times, 30 min each; and one time for 5 min.

10. Mount the nervous systems. Users may mount tissues using standard mounting procedures with VECTASHIELD or ProLong Gold Antifade. Alternatively, users may choose to mount with DPX after clearing the nervous systems with xylenes. Users should note that compared to tissues that pass through immunohistochemistry-only protocols, post-HCR tissues often do not stick well to poly-L-lysine-coated coverslips. This can lead to sample loss in the subsequent ethanol dehydration steps. We therefore recommend that ethanol dehydration be performed in an Eppendorf tube instead of on a poly-L-lysine-coated coverslip. Upon completion of the series, transfer the nervous systems onto a poly-L-lysine-coated coverslip before clearing with xylene, which makes the tissues transparent and difficult to see and handle.

3.2 Combined HCR and IHC

Users may wish to detect mRNA expression while probing protein expression by immunohistochemistry (IHC). To demonstrate the use of in situ HCR with IHC, we probed *Antp*, *Ubx*, and *Abd-B* mRNA expression individually with antibodies directed to the proteins of each gene. When combining in situ HCR with IHC, the HCR protocol is performed as described above, but instead of mounting, users proceed with the IHC protocol described below. The combined HCR/IHC protocol takes 5 days to complete. An overview of the protocol workflow is illustrated in Fig. 1.1. The results of these experiments are shown with exemplary images presented in Fig. 1.3. Although we apply the IHC protocol after the in situ HCR module, users may try performing IHC before in situ HCR, in case immunostaining quality is compromised from the HCR procedure.

Day 3:

1. A post-HCR PFA fixation may be applied to stabilize HCR signals. Note that post-HCR fixation may affect IHC signals depending on the antibody. Fixation conditions, including PFA concentration and fixation time, can be adjusted to determine the conditions to obtain optimal HCR and IHC results. Here, we perform 2% PFA fixation for 35 min at RT for adult CNS and no fixation for larval CNS. The nervous systems are rinsed three times and washed for 20 min in PBTx before applying the fixative or moving onto the following steps.
2. Remove the fixative (if applied), and rinse samples three times in PBTx at RT. Incubate the samples in a blocking solution (e.g., 5% NGS in PBTx) for 1.5 h at room temperature.
3. Replace the blocking solution with a primary antibody prepared in blocking solution, cover and seal the dish with parafilm, and incubate overnight at 4 °C.

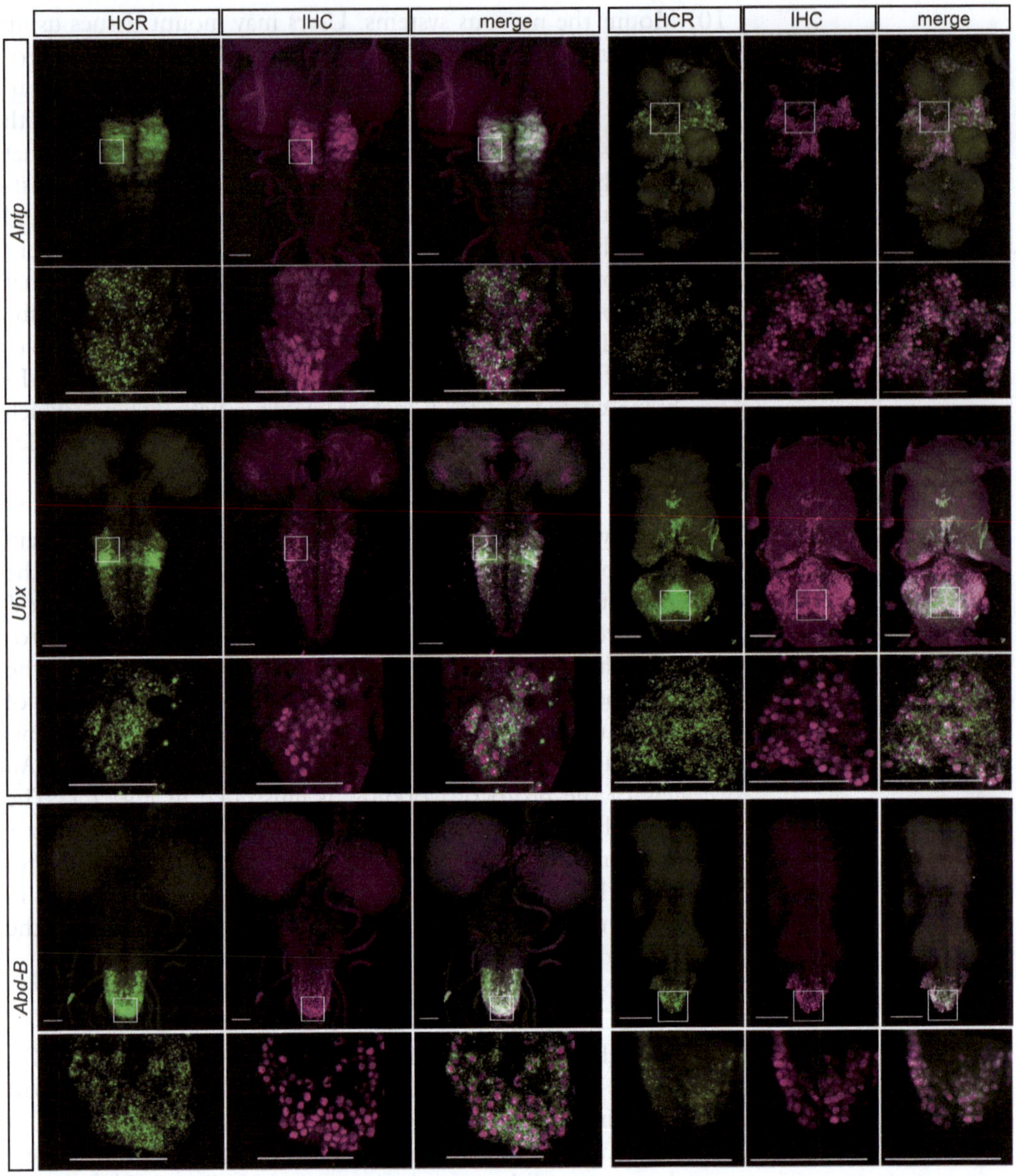

Fig. 1.3 Combined in situ HCR and immunostaining to detect *Hox* mRNA and protein expression in the *D. melanogaster* larval and adult central nervous system. Antp-B1, Ubx-B2, and Abd-B-B3 probe-sets were used individually with Alexa Fluor® 488-based amplifiers to detect *Hox* mRNA expression. The in situ HCR protocol was followed with IHC using antibodies against individual Hox proteins. As expected, mRNA and protein expression of each *Hox* gene occurs in largely the same cells in the larval and adult central nervous system. While Hox protein expression is nuclear, mRNA expression is largely cytoplasmic. Scale bar = 50 μm

Day 4:

4. Remove the primary antibody solution, and rinse the nervous systems three times in PBTx. Then wash three times, 30 min each, with PBTx.
5. Replace the final wash with the secondary antibody prepared in blocking solution, cover and seal the dish with parafilm, and incubate overnight at 4 °C.

Day 5:

6. Remove the secondary antibody solution, and rinse the nervous systems three times in PBTx. Then wash three times, 30 min each, with PBTx.
7. Proceed with the mounting protocol.

3.3 Applications: Comparing Gene Expression in the Central Nervous System of Drosophila Species

Like other animals, species in the genus *Drosophila* exhibit a rich diversity of behaviors, which likely evolved through changes in the expression of genes that influence the morphology, physiology, or molecular properties of discrete neurons in the nervous system. The protocol we describe in this chapter provides an effective way to visualize gene expression in the central nervous system across the *Drosophila* phylogeny and identify gene expression changes that are relevant to the evolution of nervous systems and behaviors.

To demonstrate this application, we visualized the mRNA expression of *Antp*, *Ubx*, and *Abd-B* in the *D. yakuba* nervous system, a closely related species that diverged from *D. melanogaster* ~ 12.8 million years ago [10]. Using the protocol described above with the initiators and amplifier hairpins we applied to *D. melanogaster* nervous systems, we probed the expression of *Antp*, *Ubx*, and *Abd-B* in the *D. yakuba* larval and adult ventral nerve cords. The expression of all three genes in the *D. yakuba* larval and adult nervous system (Fig. 1.4) was detected in a pattern that was similar to that observed in the *D. melanogaster* nervous system (Fig. 1.2). The sensitivity of detection using *D. yakuba* nervous systems was comparable to that obtained using *D. melanogaster* nervous systems. This suggests that when qualitatively comparing mRNA expression of well-conserved orthologous genes in the nervous systems across *Drosophila* species, users can potentially apply a single common probe-set.

3.4 Applications: Validation of Single-Cell RNA-Seq Results

Recent advances in single-cell mRNA sequencing technologies (scRNA-Seq) have been transformative in generating atlases of gene expression with cellular resolution and identifying neuronal cell types defined by expression of unique molecular markers [11]. Fluorescence in situ hybridization provides an effective means to visualize the expression of genes identified in scRNA-

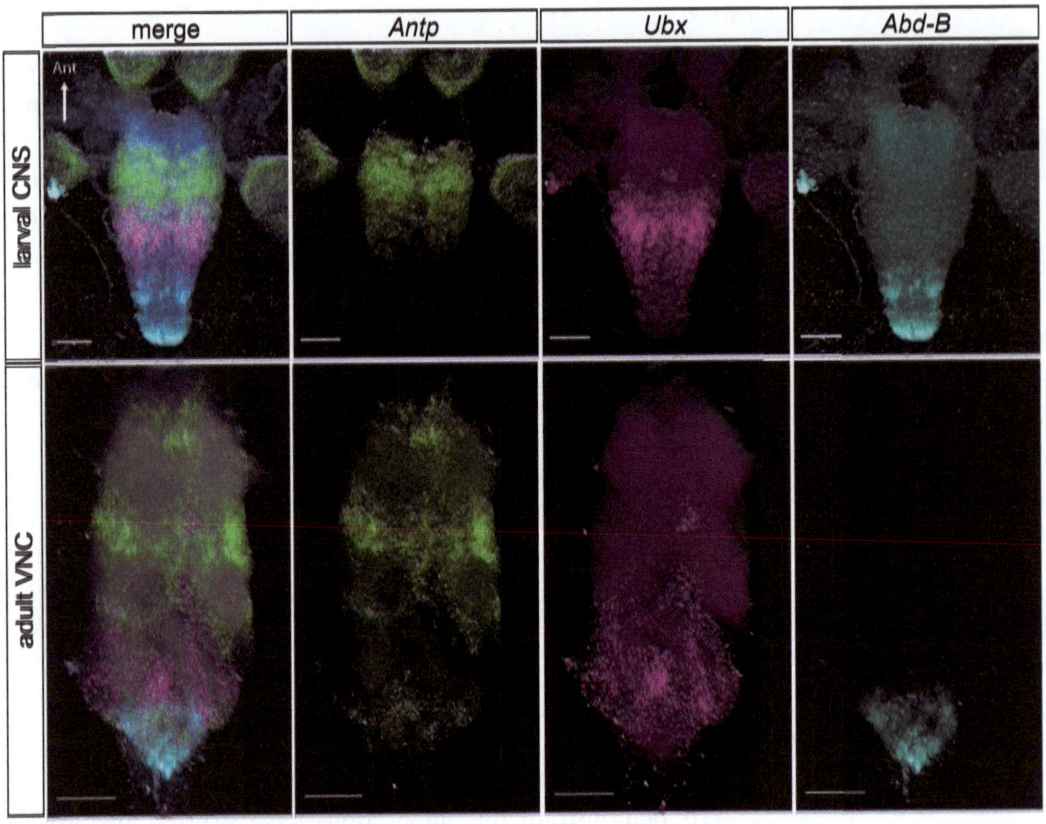

Fig. 1.4 Detection of *Hox* gene mRNA expression in the larval and adult central nervous system of *D. yakuba* by in situ HCR. Using the same probe-sets and amplifiers as those used in the experiment shown in Fig. 1.1, *Hox* mRNA expression was detected in the larval and adult central nervous system of *D. yakuba*. *Antp*, *Ubx*, and *Abd-B* expression in the *D. yakuba* nervous system is similar to that observed in *D. melanogaster*. Scale bar = 50 μm

Seq experiments, thus providing a powerful way to validate identified molecular markers. Identify homologous populations of neurons across conditions and genotypes, and guide subsequent studies to determine if and how a given candidate gene influences neural development, physiology, and behavior.

Most neurons in the adult *Drosophila* central nervous system are born during larval life from a set of segmentally repeating, bilateral neural stem cells called neuroblasts [12]. A given neuroblast will often give rise to two lineages, i.e., hemilineages, of neurons. Neurons of a given hemilineage may share molecular, anatomical, or functional properties [12–15]. A recent study used a scRNA-Seq approach to identify new molecular markers that label neurons derived from one or more hemilineages in the *Drosophila* ventral nerve cord [9]. One marker, *Sox21B*, was found to label ventral nerve cord neurons specifically from hemilineage 13B, but no antibody against SOX21B protein is available for validation. To demonstrate the application of our protocol in validating scRNA-

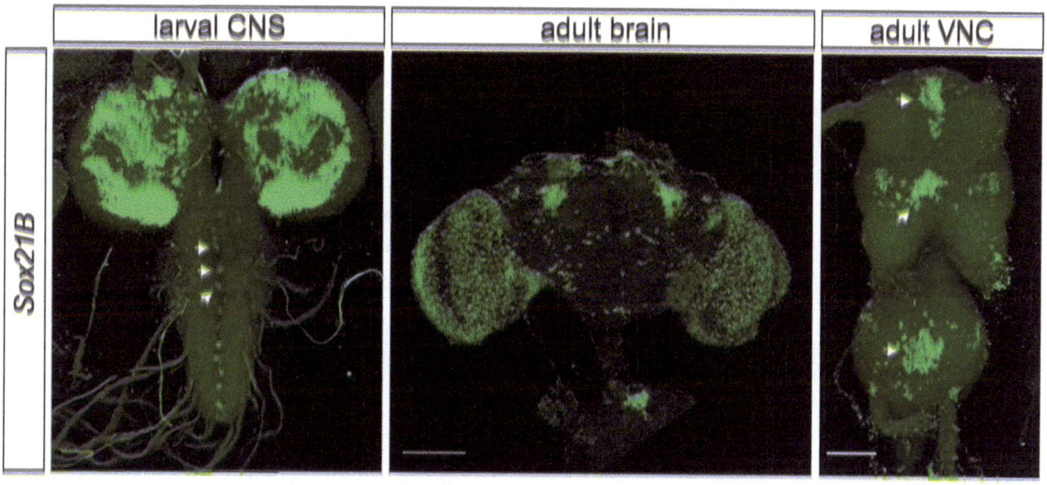

Fig. 1.5 Validation of scRNA-Seq results using in situ HCR. *Sox21B* expression in the larval and adult central nervous system of *D. melanogaster* was detected by in situ HCR. In a previous study (Allen et al. 2020), expression of *Sox21B* was found in neurons derived from hemilineage 13B using scRNA-Seq. Expression of *Sox21B* in hemilineage 13B was visualized by in situ HCR using a B1 probe-set against *Sox21B* and a B1-Alexa Fluor® 488 amplifier. Neurons of hemilineage 13B occur as segmentally repeating ventromedial clusters through the thoracic neuromeres of the ventral nerve cord. *Sox21B* expression occurs in clusters of cells where hemilineage 13B normally resides in the adult and larval ventral nerve cord (arrowheads). *Sox21B* is also expressed in cells of the larval and adult brain. Scale bar = 50 μm

Seq data, we used in situ HCR to visualize the expression of *Sox21B* mRNA in the adult and larval central nervous system. Consistent with published RNA-Seq data, *Sox21B* expression was detected in a cluster of segmentally repeating cells located ventromedially in the three thoracic neuromeres of the adult ventral nerve cord (Fig. 1.5). These clusters are situated where neurons derived from hemilineage 13B are normally found [12]. *Sox21B* expression was also detected in segmentally repeating ventromedial groups of cells in the larval nerve cord, which may correspond to neurons of hemilineage 13B (Fig. 1.5). In addition, *Sox21B* was widely expressed in the adult brain, consistent with published scRNA-Seq data of the *Drosophila* brain [16]. In the larval and adult nervous systems, *Sox21B* was found in several bilateral clusters of cells outside of hemilineage 13B (Fig. 1.5).

4 Notes

The protocols presented in this chapter provide an effective, relatively affordable and straightforward means to visualize mRNA and protein expression in whole-mount adult and larval central nervous systems. The protocol is especially useful for applications where antibodies are often not available or ideal such as cross-species gene expression comparisons and validation of RNA-Seq results in the fly nervous system. Although our protocol focuses on quantitative mRNA detection with subcellular resolution, in situ HCR is also compatible with single-molecule imaging. For such applications, Molecular Instruments™ recommends the use of relatively large probe-sets (e.g., probe-sets of ≥30) for maximal signal-to-noise fidelity and shorter amplification times (e.g., 45–90 min) for short amplification polymers that allow individual mRNAs to be resolved as diffraction-limited dots [2].

Based on our experiences, we share a few notes below for users to further optimize the protocol according to their specific needs.

4.1 Sample Treatment: Proteinase K Digestion and Post-Hybridization Paraformaldehyde Fixation

Proteinase K digestion is commonly used in in situ protocols as a way to increase the permeability of thick whole-mount tissues and enhance in situ probe penetration. Using the experimental conditions described above, the addition of a proteinase K step did not offer any obvious benefit for adult or larval nervous systems. Applying proteinase K at a concentration of 4 μg/ml for 7 min after fixation and before hybridization had no noticeable effect on the signal-to-noise ratio of the mRNAs we detected (data not shown). Nevertheless, users should keep in mind that optimized proteinase K treatment can, in some circumstances, improve in situ results, whether due to better tissue penetration or to removal of RNA-binding proteins that thwart probe hybridization (H. Choi, personal communication).

Prior to immunostaining, a post-hybridization paraformaldehyde fixation step may be applied to stabilize the amplification polymers and HCR signals. We found that this treatment appeared to compromise immunostaining quality in an antibody-specific manner. For instance, some antibodies, such as rabbit anti-GFP (ThermoFisher, A-11122), were largely insensitive to post-HCR fixation; however, other antibodies, such as anti-N-cadherin (DSHB, BP104), were highly sensitive to it (data not shown). We have tested three post-HCR fixation conditions (i.e., no fix, 2% paraformaldehyde, or 4% paraformaldehyde each for 35 min) and found that the absence of a post-HCR fixation step performs relatively better for larval nervous tissues, while a 2% post-HCR fixation performs better for the adult (data not shown). We recommend that users test different post-HCR fixation conditions to achieve their desirable balance between in situ and immunostaining signals. Furthermore, users can try IHC before in situ HCR, if post-HCR immunostaining is consistently problematic.

4.2 Detection Sensitivity: Probe Concentration, Probe-Set Size, and Amplification Conditions

In situ HCR is in principle well suited to detect low abundance transcripts. In other mRNA-FISH approaches for whole-mount *Drosophila* nervous systems [5, 6], the fluorophores are directly conjugated to each individual probe; e.g., a probe-set of 20 will lead to the assembly of, at most, 20 fluorophores per transcript. It is estimated that most amplification polymers in in situ HCR will each contain ~200 fluorophores [3]. Remarkably, this means that a set of 20 probes could, in theory, lead to the construction of upwards of 4000 fluorophores on each transcript.

If users are probing the expression of a low abundance target gene, there are some "tweaks" that may improve detection sensitivity. First, users can try adjusting the concentration of the probe. The standard probe concentration, as recommended by Molecular Instruments™, is 4 nM, but increasing probe concentration may improve detection sensitivity. This adjustment can be effective and cost-free when users have an excess of probes. Second, the size of the probe-set can be increased. In our experiments, we used a set of 20 split-initiator probe pairs, each targeting a total of 52 bases within an mRNA type (i.e., 26 bases for each split-initiator probe of a pair). Using our protocol, a probe-set size of 20 consistently yielded reliable signals for a variety of target genes. However, increasing the probe-set size to 30 or more may further enhance detection, especially for transcripts with low levels of gene expression. Third, confocaling under higher magnification can improve the signal-to-noise ratio. Finally, adjusting the amplification conditions may also enhance sensitivity. In our experiments, we typically used an amplification time of 16 h. Although longer amplification time beyond overnight is not expected to strengthen HCR signals, increasing the hairpin concentration may help (H. Choi, personal communication).

4.3 Cost

Another major advantage of in situ HCR over other mRNA-FISH protocols [5, 6] is in the cost. Currently, other published mRNA-FISH protocols on whole-mount *Drosophila* nervous systems [5, 6] utilize probe-sets supplied by Stellaris™, which start at US$675 and provide enough material for 200–400 experiments [6]. At the time of writing this chapter, a set of 20 customized HCR probe pairs starts at US$252 for academic and nonprofit users, which allows ~250 in situ. With the amplifier mix currently starting at US$97.50, the cost of probing the mRNA expression of a single gene is notably more affordable using HCR than using Stellaris-based approaches. The savings in cost are more substantial if users wish to detect the expression of multiple genes, as a single amplifier mix can be used with multiple HCR probe-sets. Furthermore, if one wishes to visualize transcripts using a different fluorophore, it can be easily achieved by adopting a different amplifier mix instead of purchasing a new probe-set.

Acknowledgements

We thank Dr. Lisha Shao for initially bringing the HCR method to our attention and sharing useful tips during our protocol development, Dr. Harry Choi from Molecular Instruments™ for valuable inputs, and Dr. Justin Walsh for scRNA-Seq analyses.

References

1. Dirks RM, Pierce NA (2004) Triggered amplification by hybridization chain reaction
2. Choi HMT, Chang JY, Trinh LA et al (2010) Programmable in situ amplification for multiplexed imaging of mRNA expression. Nat Biotechnol 28:1208–1212. https://doi.org/10.1038/nbt.1692
3. Choi HMT, Beck VA, Pierce NA (2014) Next-generation in situ hybridization chain reaction: higher gain, Lower Cost, Greater Durability https://doi.org/10.1021/nn405717p
4. Choi HMT, Schwarzkopf M, Fornace ME et al (2018) Third-generation in situ hybridization chain reaction: multiplexed, quantitative, sensitive, versatile, robust. https://doi.org/10.1242/dev.165753
5. Long X, Colonell J, Wong AM et al (2017) Quantitative mrna imaging throughout the entire drosophila brain. Nat Publ Gr 14:703. https://doi.org/10.1038/nmeth.4309
6. Yang L, Titlow J, Ennis D et al (2017) Single molecule fluorescence in situ hybridisation for quantitating post-transcriptional regulation in drosophila brains. Methods 126:166–176. https://doi.org/10.1016/j.ymeth.2017.06.025
7. Shao L, Chung P, Wong A et al (2019) A neural circuit encoding the experience of copulation in female drosophila. Neuron 102:1025–1036.e6. https://doi.org/10.1016/j.neuron.2019.04.009
8. Lacin H, Chen HM, Long X et al (2019) Neurotransmitter identity is acquired in a lineage-restricted manner in the drosophila CNS. elife 8. https://doi.org/10.7554/eLife.43701
9. Allen AM, Neville MC, Birtles S et al (2020) A single-cell transcriptomic atlas of the adult drosophila ventral nerve cord. elife 9. https://doi.org/10.7554/eLife.54074
10. Tamura K, Subramanian S, Kumar S (2004) Temporal patterns of fruit Fly (drosophila) evolution revealed by mutation clocks. Mol Biol Evol 21:36–44. https://doi.org/10.1093/molbev/msg236
11. Li H (2020) Single-cell RNA sequencing in drosophila: technologies and applications. Wiley Interdiscip Rev Dev Biol
12. Truman JW, Schuppe H, Shepherd D, Williams DW (2004) Developmental architecture of adult-specific lineages in the ventral CNS of drosophila. Development 131:5167–5184. https://doi.org/10.1242/dev.01371
13. Shepherd D, Harris R, Williams DW, Truman JW (2016) Postembryonic lineages of the drosophila ventral nervous system: Neuroglian expression reveals the adult hemilineage associated fiber tracts in the adult thoracic neuromeres. J Comp Neurol 524:2677–2695. https://doi.org/10.1002/cne.23988
14. Harris RM, Pfeiffer BD, Rubin GM, Truman JW (2015) Neuron hemilineages provide the functional ground plan for the drosophila ventral nervous system. elife 4. https://doi.org/10.7554/eLife.04493
15. Shepherd D, Sahota V, Court R et al (2019) Developmental organization of central neurons in the adult drosophila ventral nervous system. J Comp Neurol 527:2573–2598. https://doi.org/10.1002/cne.24690
16. Davie K, Janssens J, Koldere D et al (2018) A single-cell transcriptome atlas of the aging drosophila brain. Cell 174:982–998. https://doi.org/10.1016/j.cell.2018.05.057

Chapter 2

Mapping Brain-Wide Mesoscale Connectome from Single Animals with BRICseq

Longwen Huang

Abstract

The connectivity patterns of neural circuits determine how information is transmitted and processed in the nervous system and play central roles in circuit function and animal behaviors. Therefore, understanding the logic that relates distinct modalities of the nervous system, including connectome, neural activity, gene expression, and animal behaviors, is one of the core questions in neuroscience. Here I describe methods of BRICseq, which allows for high-throughput mapping of brain-wide mesoscale connectome from single animals. In combination with other methods to map brain-wide gene expression or neural activity, BRICseq allows comparative and integrative studies of the nervous system across animal models of various manipulations, diseases, and species.

Key words Connectome, Mesoscale, Barcoding, Neuroanatomy, High-throughput sequencing

1 Introduction

Structure determines function. This is a general principle in biology. In neuroscience, brain function is determined by connections among brain areas, and disrupted connectivity may result in a variety of neuropsychiatric disorders, including autism and schizophrenia [1, 2]. Thus, connectome is a key that bridges gene expression to neural activity, circuit function, animal behaviors, and neural disorders. In particular, high-quality mesoscale connectivity maps that describe brain-wide inter-regional connectivity are essential to modern neuroscience [3–5].

A number of conventional methods allow for mapping of mesoscale connectome. Among them, the typical approach is to use fluorescent tracers (e.g., a virus encoding fluorescent protein) to trace from a given source area and determine its projection targets via fluorescent imaging. Then, this procedure will be repeated to target different source areas of interest in multiple separate animals, and the results will be pooled to determine brain-wide mesoscale connectome [3, 4]. Because hundreds to

Daisuke Yamamoto (ed.), *Behavioral Neurogenetics*, Neuromethods, vol. 181, https://doi.org/10.1007/978-1-0716-2321-3_2,

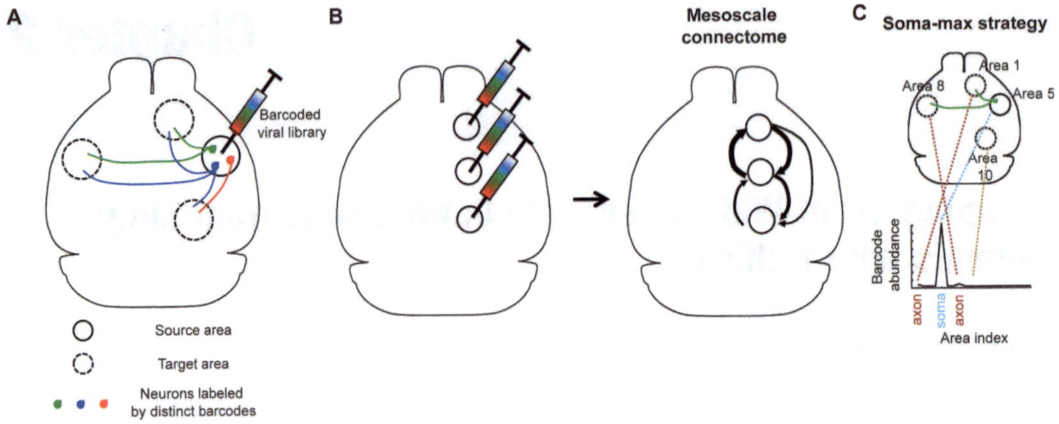

Fig. 2.1 Principles of BRICseq. (**a**) In the barcoded viral library, each viral particle in principle harbors a unique RNA barcode. When the virus is injected into a brain area of interest, individual cells are uniquely labeled by distinguishable barcodes. The RNA barcodes can also be trafficked to axon terminal in infected cells, allowing axonal tracing by sequencing. (**b**) In BRICseq, a high-diversity barcoded viral library is injected into multiple brain regions in a single animal, allowing multiplexed mapping of inter-regional connectivity. (**c**) In BRICseq, the "soma-max" strategy can be used to determine the soma location for each barcode

thousands of single animals are required to obtain such a map, this approach is usually low-throughput, expensive, and labor-intensive.

To overcome this bottleneck, we have recently reported BRICseq (brain-wide individual animal connectome sequencing) [5], which implements DNA barcoding [6, 7] to multiplex projection tracing and converts this neuroanatomy problem into a sequencing problem (Fig. 2.1). In BRICseq, a Sindbis viral library is prepared to harbor random and distinguishable RNA sequences, or "barcodes." This library of high diversity of barcodes is next used to uniquely label individual neurons (Fig. 2.1a). In order to trace axons, the virus is also engineered so that these RNA barcodes can be trafficked into axon terminals. To map brain-wide connectome, this viral library is injected to multiple brain areas for multiplex tracing experiments in single animals (Fig. 2.1b). Importantly, because most of the RNA barcodes reside at the soma of the infected cells, and only a small fraction of them are trafficked into axon terminals, the soma location of individual barcode can be determined based on a "soma-max" strategy (Fig. 2.1c). Uniquely labeling single neurons, widespread injection of the barcode library, and implementation of the "soma-max" strategy therefore allow for high-throughput mapping of brain-wide mesoscale connectome from single animals. The application of BRICseq for mapping connectome in various animal models will usher in a new era of comparative connectomics.

This chapter will include general protocols and notes for performing BRICseq.

2 Materials

2.1 Reagents

- Sindbis virus with RNA barcodes [5, 6, 8].
- Toluidine blue.
- RNA extraction reagents or kits (here I use ThermoFisher RNA-queous micro kit).
- RNaseZap.
- Reverse transcriptase (here I use ThermoFisher SuperScript IV).
- Barcoded reverse transcription primer: (5')CTT GGC ACC CGA GAA TTC CAX XXX XXX XXX XXZ ZZZ ZZZ ZTG TAC AGC TAG CGG TGG TCG(3′). Here X and Z stand for random nucleotides, where X_{12} is a unique molecular identifier [9], and Z_8 is a cubelet-specific identifier. The length of X and Z can be customized according to experiment needs.
- dNTP.
- Spike in RNA: (5')GUC AUG AUC AUA AUA CGA CUC ACU AUA GGG GAC GAG CUG UAC AAG UAA ACG CGU AAU GAU ACG GCG ACC ACC GAG AUC UAC ACU CUU UCC CUA CAC GAC GCU CUU CCG AUC UNN NNN NNN NNN NNN NNN NNN NNN NAU CAG UCA UCG GAG CGG CCG CUA CCU AAU UGC CGU CGU GAG GUA CGA CCA CCG CUA GCU GUA CA(3′). Here N stands for random nucleotides.
- RNasin.
- Size selection beads (here I use Beckman Coulter SPRIselect beads).
- *E. coli* DNA ligase.
- DNA polymerase I.
- RNase H.
- T4 DNA polymerase.
- Exonuclease I.
- High-fidelity DNA polymerase (here I use AccuPrime Pfx).
- Nested PCR primer: first, forward, (5′)GGA CGA GCT G(3′); first, reverse, (5′) CAA GCA GAA GAC GGC ATA CGA GAT CGT GAT GTG ACT GGA GTT CCT TGG CAC CCG AGA ATT CCA(3′); second, forward, (5′)AAT GAT ACG GCG ACC ACC GA(3′); second, reverse, (5′) CAA GCA GAA GAC GGC ATA CGA(3′).
- Gel purification kit.

2.2 Equipment

- Centrifuge.
- Thermocycler.
- Real-time PCR system.
- Bioanalyzer.
- Tissue lyser (here I use Qiagen TissueLyser II).
- Cryostat.
- Small animal stereotaxic instrument.
- Nanoject III.
- Laser dissection microscope.

2.3 Software

- Matlab.
- Bowtie [10].

3 Methods

3.1 Viral Injection and Tissue Dissection

Animal surgery

1. Animal procedures must follow institutional and national ethics standards.
2. Sterilize surgical tools.
3. Induce anesthesia with isoflurane (4%, mixed with oxygen, 0.5 L/min).
4. After the animal has been anesthetized, shave and clean its head, and apply ophthalmic ointment onto the eyes to keep the moisture. Mount the animal in a stereotaxic setup using proper ear bars, with a heating pad set at the appropriate temperature under its abdomen, and administer meloxicam (2 mg/kg), dexamethasone (1 mg/kg), and Baytril (10 mg/kg) subcutaneously.
5. Use 7.5% Betadine and 70% ethanol alternatively three times to clean the head skin. Use forceps and scissors to make an incision along the midline of the head skin to expose the skull. Adjust head positioning, and make sure that the head is correctly positioned in the rostral-caudal and left-right directions.
6. To expose brain areas for injection, etch the edge of the target area with a dental drill, and remove the etched skull with forceps. Fill the Nanoject pipette with the barcoded Sindbis virus. Make multiple pipette penetrations to cover the entire exposed cortical hemisphere (penetration sites are spaced by 0.5 mm, in the AP and ML axes), and use Nanoject to inject the virus (~20 μL, 2×10^{10} GC/mL) at three

to four depths per penetration site (every 200–250 μm). Note that the regions to expose and target depend on brain areas of interest. Here I aim to inject the entire right cortical hemisphere.

7. After injection, apply Kwik-Cast to cover the exposed brain area for protection, and close the skin with sutures.
8. Administer one dose of meloxicam (2 mg/kg), dexamethasone (1 mg/kg), and Baytril (10 mg/kg) subcutaneously.
9. Put the animal into a clean, pre-warmed cage above a heating pad for recovery.
10. Monitor animal condition, and administer one dose of meloxicam (2 mg/kg), dexamethasone (1 mg/kg), and Baytril (10 mg/kg) subcutaneously every 12 h post-surgery.

Cryostat sectioning

11. Forty-four hours after viral injection, clean field and tools with 75% ethanol and RNaseZap. Then dissect and fresh freeze the brain at −80 °C.
12. Clean the cryostat and brushes with 100% ethanol. Use cryostat to make 300 μm coronal slices (or other appropriate thickness). After collecting each slice, change the cutting location of the blade, and clean the blade holder with 100% ethanol to avoid cross-sample contamination.
13. Mount each slice in a laser microdissection-compatible slide. Fix the slice in 75% ethanol at 4 °C for 3 min. Fixation helps to deactivate RNase.
14. Briefly dip the slice in MilliQ H_2O, and transfer the slice to 0.5% toluidine blue solution to stain neurons for 30s at room temperature.
15. Wash the slice in MilliQ H_2O for three times (in three different chambers or tubes) at room temperature, 15 s each time.
16. Fix the slice again in 75% ethanol twice (in two different chambers or tubes) at room temperature, 2 min each time.
17. After fixation, leave the slide in a vacuum desiccator for at least 30 min.
18. Put another clean slide on top of the mounting slide to tightly hold the slice between slide membranes, and tape the two slides (optional). Then store the sandwiched slice in the vacuum desiccator at room temperature until laser microdissection (LMD). For long-term storage (>1 week), the sandwiched slices should be stored at −80 °C in a desiccated container.

Laser microdissection

19. Clean field and slide holders before laser microdissection with 75% ethanol and RNaseZap. Use laser to dissect and collect brain areas of interest (here I refer to these dissected areas as cubelets).
20. Transfer dissected cubelets into collection tubes with pre-loaded homogenizing beads, and add 100 μL lysis solution (ThermoFisher RNAqueous micro kit, or other appropriate reagents) into each collection tube.
21. Keep collected samples in dry ice. After LMD is completed, store the entire samples at −80 °C.

3.2 BRICseq Library Preparation and Sequencing

Note in this section, all the H_2O refers to purified MilliQ H_2O.

RNA extraction

1. Thaw collected samples at room temperature. If there are too many samples to be processed, the RNA extraction steps (1–3) can be repeated for multiple times. Here I extract RNA from 20 to 30 samples at a time.
2. Homogenize the collected tissue using TissueLyser II for 6 min at 20 Hz.
3. Follow the ThermoFisher RNAqueous Micro Total RNA Isolation Kit protocol to extract RNA from each sample. Elute RNA in 20 μL elution solution. The DNase treatment step can be omitted because genomic DNA does not influence following steps. Other appropriate reagents/kits and protocols can also be used to extract RNA from samples.

Reverse transcription

4. Add 1 μL dNTP (10 mM), 1 μL barcoded reverse transcription primer, 1 μL spike in RNA (10^{-9} μg/μL), and 10 μL extracted RNA to an RNase-free tube (or a plate well) for each extracted RNA sample. Mix them well, keep the mix at 70 °C for 10 min, and immediately move it onto ice to stay for 5 min.
5. Add 4 μL SuperScript IV buffer, 1 μL DTT (dithiothreitol, 0.1 M), 1 μL RNasin and 1 μL SuperScript IV into each mix. Mix them well, and incubate the mix at 55 °C for 10 min and then 80 °C for 10 min. Other appropriate reverse transcriptase can also be used, with proper reaction temperature and duration. Note that after step 5, the reverse transcribed DNA molecules have already been labeled by cubelet-specific identifiers from the barcoded reverse transcription primers. So cross-contamination between cubelets

is now of less concern because each DNA molecule can be assigned to a specific cubelet based on the cubelet-specific identifier.

Product cleanup and concentration.

6. Pool products of every 12 reverse transcription reactions (240 μL in total), and add 240 μL H_2O to every pool to decrease magnesium concentration.
7. Add 864 μL (1.8×) SPRIselect beads into each product (240 μL), and follow SPRI-based size selection (left side size selection) protocol to elute product in 96 μL H_2O.
8. Add 173 μL (1.8×) SPRIselect beads into each eluted product (96 μL). Follow SPRI-based size selection (left side size selection) protocol to elute product in 17 μL H_2O. Steps 6–8 aim to clean up and concentrate the reverse transcribed product. Note that the number of reverse transcription reactions per pool and the elution solution volume can be adjusted. For each round of cleanup (step 7 and step 8), I recommend that the ratio of SPRIselect bead volume to elution solution volume is not greater than 10:1, and if the product can be eluted in 17 μL (meet the abovementioned ratio requirement) in step 7, step 8 can be omitted.

Second strand synthesis

9. Add 2.4 μL first strand buffer, 0.6 μL DTT (0.1 M), 5.6 μL second strand buffer, 0.75 μL dNTP (10 mM), 0.25 μL *E. coli* DNA ligase, 1 μL DNA polymerase I, and 0.25 μL RNase H into each eluted product (17 μL). Incubate the mix at 16 °C for 2 h. Add 1 μL T4 DNA polymerase, and incubate the mix at 16 °C for another 10 min.

Product cleanup and concentration.

10. Add 1.8× SPRIselect beads to each second strand synthesis product, and follow SPRI-based size selection (left side size selection) protocol to elute product in 16 μL H_2O. Note samples can be pooled together before cleanup.
11. Add 2 μL exonuclease I buffer and 2 μL exonuclease I into each cleaned-up sample (16 μL). Incubate the mix at 37 °C for 1 h, and then heat inactivate the enzyme at 80 °C for 20 min.

Nested PCR

12. Add 5 μL AccuPrime buffer, 0.5 μL nested PCR first forward primer (100 μM), 0.5 μL nested PCR first reverse primer (100 μM), 0.4 μL AccuPrime Pfx enzyme, and 23.6 μL H_2O into each exonuclease I-treated product (20 μL). Perform first PCR as follows:
    (i) Denature: 95 °C, 2 min.
    (ii) Denature: 95 °C, 15 s.
    (iii) Anneal: 68 °C, 30s.
    (iv) Extend: 68 °C, 2 min.
    Repeat ii) to iv) for 5 times in total.
    (v) Extend: 68 °C, 5 min.
13. Add 10 μL exonuclease I to each reaction, incubate the mix at 37 °C for 30 min, and heat inactivate exonuclease I at 80 °C for 20 min.
14. Pool all of the exonuclease I-treated 1st PCR product.
15. Add the following components for 2nd PCR:
    360 μL exonuclease I-treated 1st PCR product
    1200 μL AccuPrime Pfx buffer (10×)
    120 μL nested PCR second forward primer (100 μM)
    120 μL nested PCR second reverse primer (100 μM)
    96 μL AccuPrime Pfx enzyme
    10,104 μL H_2O
    Note that I use high volume for 2nd PCR reaction in order to reduce template switching errors [5, 11].
16. Perform 2nd PCR as follows:
    (i) Denature: 95 °C, 2 min.
    (ii) Denature: 95 °C, 15 s.
    (iii) Anneal: 68 °C, 30s.
    (iv) Extend: 68 °C, 2 min.
    Repeat (ii) to (iv) for five to ten times in total.
    (v) Extend: 68 °C, 5 min.

Note that the number of cycles in 2nd PCR should be adjusted according to initial DNA concentration. The product is not enough for sequencing when the cycle number is too low, and nonspecific products as well as template switching errors can be introduced when too many cycles are performed. I expect to get ~0.2 pmol DNA product after nested PCR and product purification (step 19). I recommend using a small volume (5 μL) of 1st PCR product to run 2nd PCR to optimize the cycle number. Note because only 5 μL 1st PCR product is used for the testing experiment, the optimized cycle number in the testing experiments should be converted to the cycle number in the actual experiments based on the input ratio (5 μL:360 μL).

17. Concentrate nested PCR product using SPRIselect beads (1.8× beads, left side size selection protocol) to a final volume of 25 μL. I recommend that the ratio of SPRIselect bead volume to elution solution volume is not greater than 10:1, so multiple rounds of bead cleanup should be performed.
18. Run the concentrated PCR product using 2% agarose gel, cut out the 230 bp band (when unique molecular identifiers are 12mers and cubelet-specific identifiers are 8mers), and recover DNA using a gel purification kit. Elute the product in 12 μL H_2O.
19. Check quality and concentration of the product using Bioanalyzer. The expected amount of ~230 bp DNA product is ~0.2 pmol. The sequencing library is ready.

Sequencing

20. Sequence the library on Illumina Nextseq (high output flow cell, paired ends 36 bp). Use standard Illumina read 1 primer for read 1 and Illumina small RNA read 2 primer for read 2.

3.3 Reconstruction of Mesoscale Connectome

1. The raw Illumina data consisted of 32-nt BC (barcode) sequences in paired end 1 and 12-nt UMI (unique molecular identifier) and 8-nt CSI (cubelet-specific identifier) sequences in paired end 2. Merge them to get the full BC-UMI-CSI sequences and de-multiplex them according to CSI. Now each BC-UMI sequence is assigned to a given cubelet. Sequences with ambiguous bases (N) are excluded.
2. For each CSI (cubelet), there can be multiple reads of the same BC-UMI sequence. For digital counting [5, 9], I collapse all the identical reads to get unique BC-UMI sequences and then count the number of UMIs for each BC in each cubelet.
3. To account for substitution errors, three mismatches are allowed for a given BC sequence. I use *bowtie* to uncover groups of BC sequences with <3 mismatches and collapse all the BC sequences within the same groups.
4. After digital counting and BC collapsing, generate a barcode matrix, where each row represents a unique BC sequence and each column represents a cubelet (CSI). The values in the matrix represent UMI counts of a given BC in a given cubelet. An additional noise threshold may be applied to this matrix, and values not greater than the threshold are considered as noises. In a previous study, the noise threshold was 1 [5].
5. For each row, find the maximum element. The cubelet with the highest element is the putative soma location. In practice, I also

use an additional soma threshold, and a cubelet is considered as soma when (i) the UMI count in the given cubelet is the highest and (ii) the UMI count in the given cubelet is greater than the soma threshold. In a previous study, the soma threshold was 250 [5].

6. After soma calling, each barcode can be assigned to a neuron in a given source cubelet. So each row of the matrix represents the projection patterns from a single neuron in a source cubelet to all the other target cubelets.
7. Find all the neurons whose somata reside in the same source cubelets, and average the projection patterns over all these neurons to determine bulk projection patterns from source cubelets to target cubelets. This will generate an inter-regional mesoscale connectivity matrix.

4 Notes

1. In BRICseq, RNA samples should be handled carefully to avoid RNA degradation and contamination between samples (before reverse transcription). Tools, bench surface, and laboratory gloves should be cleaned with 75% ethanol and RNaseZap to inactivate RNase. Tools or gloves that may have touched an RNA sample should be thoroughly cleaned or changed before processing another sample. The experimenter should minimize the time that the samples are exposed in the air and keep the samples in low-temperature and low-humidity environment for long-term storage.
2. Here I implement BRICseq to map corticocortical connectivity. In practice, the details of viral injection, cryostat sectioning, and laser microdissection can be modified to meet experimental needs and map connectivity between brain areas of interest. Particularly, because the spatial resolution of the connectivity map is completely determined by laser microdissection, experimenters should carefully choose dissection strategy (such as the number, size, and location of cubelets) according to experiment needs.
3. Two major factors that may result in false-positive projection errors are template switching (hybrid molecules that are generated during PCR) and reused barcodes (the same barcodes that infect more than one neurons) [5]. To minimize these artifacts, I use 12 mL volume for second PCR (to reduce the template switching rate), and the diversity of barcodes in the viral library greatly outnumbers the infected cells (to reduce the reused barcode rate). Moreover, additional thresholds and criteria can be used to further reduce errors, such as noise

threshold (UMI counts not greater than this threshold are considered as noises), soma threshold (cubelets with UMI counts not greater than this threshold are not considered as soma cubelets), and axon threshold (if the UMI count of a barcode in a target cubelet is higher than this threshold, this barcode is excluded for analysis). These thresholds can be determined based on the UMI count distribution in the collected dataset. For instance, in a previous study, the noise threshold, soma threshold, and axon threshold were 1, 250, and 20, respectively [5].

References

1. Geschwind DH, Levitt P (2007) Autism spectrum disorders: developmental disconnection syndromes. Curr Opin Neurobiol 17:103–111
2. Kubicki M, McCarley R, Westin CF et al (2007) A review of diffusion tensor imaging studies in schizophrenia. J Psychiatr Res 41: 15–30
3. Zingg B, Hintiryan H, Gou L et al (2014) Neural networks of the mouse neocortex. Cell 156:1096–1111
4. Oh SW, Harris JA, Ng L et al (2014) A mesoscale connectome of the mouse brain. Nature 508:207–214
5. Huang L, Kebschull JM, Fürth D et al (2020) BRICseq bridges brain-wide interregional connectivity to neural activity and gene expression in single animals. Cell 182:177–188
6. Kebschull JM, Garcia da Silva P, Reid AP et al (2016) High-throughput mapping of single-neuron projections by sequencing of barcoded RNA. Neuron 91:975–987
7. Kebschull JM, Zador AM (2018) Cellular barcoding: lineage tracing, screening and beyond. Nat Methods 15:871–879
8. Kebschull JM, Garcia da Silva P, Zador AM (2016) A new defective helper RNA to produce recombinant Sindbis virus that infects neurons but does not propagate. Front Neuroanat 10: 56
9. Shiroguchi K, Jia TZ, Sims PA, Xie XS (2012) Digital RNA sequencing minimizes sequence-dependent bias and amplification noise with optimized single-molecule barcodes. Proc Natl Acad Sci U S A 109:1347–1352
10. Langmead B, Trapnell C, Pop M, Salzberg SL (2009) Ultrafast and memory-efficient alignment of short DNA sequences to the human genome. Genome Biol 10:R25
11. Kebschull JM, Zador AM (2015) Sources of PCR-induced distortions in high-throughput sequencing data sets. Nucleic Acids Res 43: e143

Chapter 3

Visualization of Mating-Dependent Activation of Neurons and Oogenesis in *Drosophila melanogaster*

Yuto Yoshinari and Ryusuke Niwa

Abstract

Females of the fruit fly *Drosophila melanogaster* exhibit drastic physiological and behavioral changes after mating, including an increase in the number of germline stem cells in the ovaries. The mating-induced increase in the number of female germline stem cells is regulated by a small subset of neurons that directly project to the ovary. After mating, the neurons release the neurotransmitter octopamine, which is received by the octopamine receptor expressed in the ovary, followed by the activation of germline stem cell niche signaling. In this chapter, we describe two indispensable methods for analyzing octopamine-dependent germline stem cell number increase in female *D. melanogaster*. One method is based on an in vivo analytical tool to monitor the ovary-projecting octopaminergic neuronal activity. The other method is for an ex vivo assay system to assess germline stem cell numbers with or without octopamine administration. In this chapter, we describe the basic protocols of these methodologies. These methods can be expanded to study the mechanisms of female germline stem cell number increase regulated by not only octopamine but also other neurotransmitters and hormones.

Key words *Drosophila melanogaster*, Mating, Germline stem cell, Octopamine, Calcium, TRIC, Neurotransmitter, Hormone

1 Introduction

The nervous system of animals consists of a very large number of neurons, which are closely connected to each other to control complex behavioral patterns. To investigate how the neuronal circuits contribute to the regulation of animal behaviors, the fruit fly, *Drosophila melanogaster*, has served as a good model organism, because it allows the use of many elegant and sophisticated genetic tools, and its nervous system consists of a relatively small number of cells [1–3].

The nervous system of *D. melanogaster* shows dimorphism between males and females and a group of nerves control sex-specific behaviors such as mating behavior [4]. Female *D. melanogaster* exhibits physiological and behavioral changes

Daisuke Yamamoto (ed.), *Behavioral Neurogenetics*, Neuromethods, vol. 181, https://doi.org/10.1007/978-1-0716-2321-3_3,

after mating, including increased food intake, metabolic changes, a significant increase in the number of eggs laid, and refusal to re-mate with males [5, 6]. A male seminal fluid peptide called the sex peptide plays an important role in these changes [7, 8]. The peptide is received by sex peptide receptor (SPR) expressed in a small subset of SPR-positive sensory neurons (SPSNs), whose cell bodies are located in the uterine lumen and their afferent axons are sent to the abdominal ganglion [9, 10]. To understand how physiological changes after mating are caused in females, recent studies have taken advantage of the GAL4/upstream activation sequence (UAS) and LexA/LexAop systems to induce tissue-specific gene expression and suppression [11]. For example, our group has recently reported that an increase in the number of germline stem cells (GSCs), which produce eggs, is induced after mating in *D. melanogaster* [12–15]. In this study, we also found that the ovary-projecting neurons that are directly connected to SPSNs play an important role in the increase of GSC number [13]. The neurons projecting to the ovary express the *tyrosine decarboxylase 2* (*Tdc2*) gene, which encodes the enzyme responsible for octopamine (OA) biosynthesis [16]. Therefore, a transgenic line in which *GAL4* is driven under the control of the *Tdc2* enhancer sequence (*Tdc2-GAL4*) can be used to perform genetic manipulations in the small subset of neurons that project to the ovary [16]. In combination with *Tdc2-GAL4*, we utilized the transcriptional reporter of intracellular Ca^{2+} (TRIC) line [17], which induces the reporter gene expression in a Ca^{2+} influx-dependent manner (Fig. 3.1a). This system enables us to monitor the activity of the *Tdc2*$^{+}$ neurons and help find whether mating induces the activation of the ovary-projecting neurons.

One way is to examine whether the target tissues anatomically dissected from the body can respond to an artificial administration of the neurotransmitter of interest. Indeed, such an ex vivo culture system has been used as a powerful tool for studying the function of neurotransmitters and hormones [18–21]. To verify the function of the ovary-projecting *Tdc2*$^{+}$ neurons of *D. melanogaster*, we applied an ex vivo ovary culture system, where we could investigate the direct effect of OA on GSC number increase in the ovaries. We have successfully revealed that OA treatment of culture media resulted in the increase in female GSC number [13]. Our method can be expanded to other neurotransmitters and hormones, as we have reported that the administration of neuropeptide F (NPF) synthetic peptides also evoked an increase in female GSC number [20].

In this article, we describe TRIC, a method used to evaluate the activation of the nervous system before and after mating (Method 3-1) and the method of ex vivo tissue culture system using the ovary (Method 3-2).

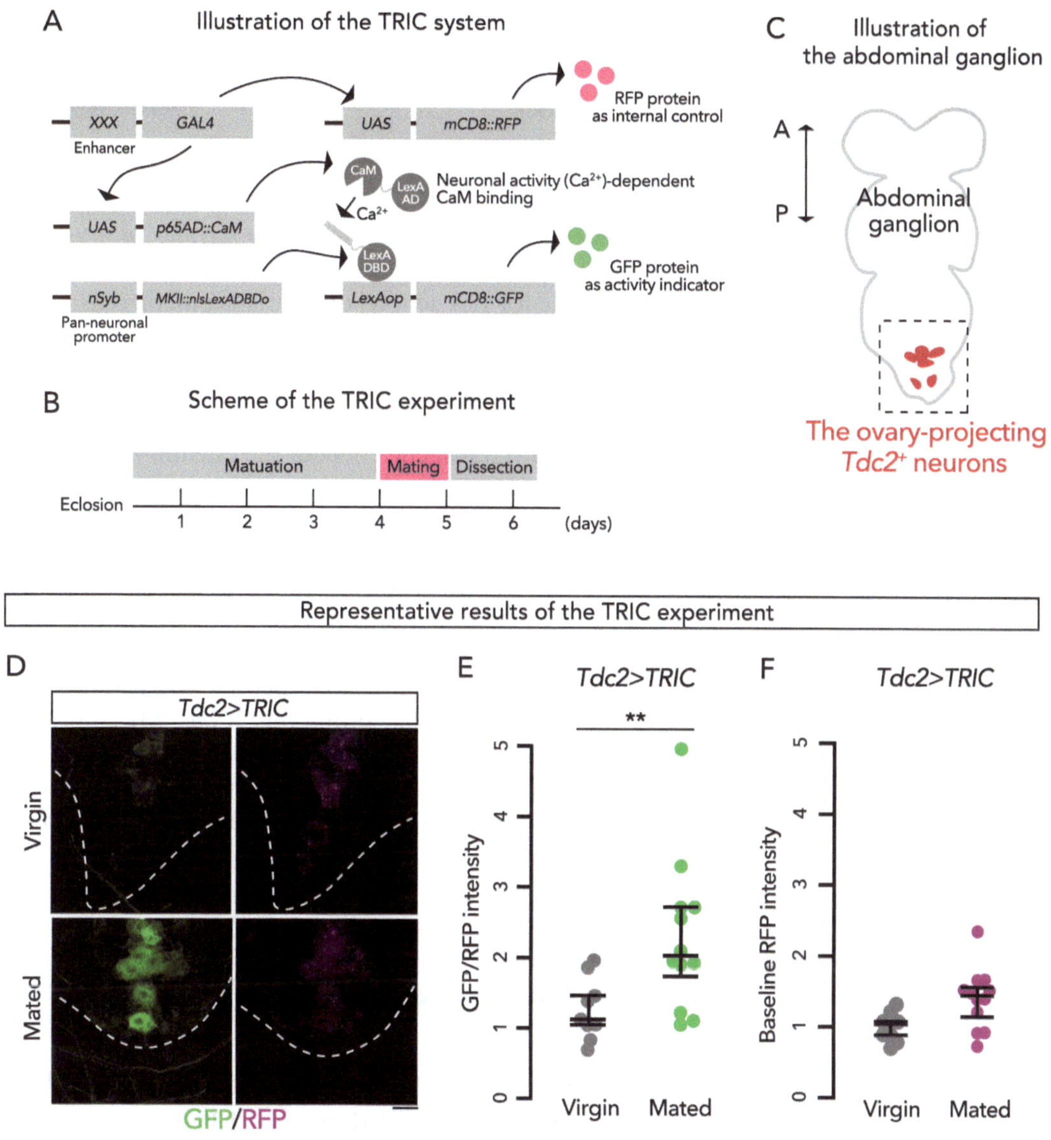

Fig. 3.1 Visualization of neuronal activity by TRIC method. (**a**) An illustration of the TRIC system. AD::CaM is driven by GAL4/UAS, whereas *LexA::DBD* is expressed under *nSyb*, the pan-neuronal promoter sequence. Ca^{2+} mediates the binding of AD::CaM and its targeting peptide, thereby bringing a transcriptional AD of AD:: CaM to the DBD of the transcription factor LexA (LexA::DBD). The Ca^{2+}-mediated activation of LexA promotes the transcription of effector that is under the control of *LexAop*. (**b**) Scheme of TRIC experiment. Virgin female flies were collected within 8 h (25 °C) or 16 h (17 °C) after eclosion. After 4–5 days of maturation, eight to ten female flies were reared with or without eight to ten males for 24 h, followed by dissection. (**c**) A cartoon of the ventral nerve cord of *D. melanogaster*. The ovary-projecting *Tdc2*$^+$ neurons are represented in red. Arrows indicate the anterior (A)-posterior (P) axis. (**d**) Representative images of the posterior tip of ventral nerve cord corresponding to an area marked by a dashed box in (**c**). In *Tdc2*>*TRIC* flies, TRIC transgenes are expressed only in the small subset of neurons expressing *Tdc2*. GFP signal (green) is increased in the neurons of mated females, whereas baseline RFP signal (magenta) is not affected. Scale bar, 10 μM. (**e**) Quantified data of GFP/RFP signal intensity. Note, mating significantly increases (230%) GFP/RFP ratio of *Tdc2*$^+$ neurons. (**f**) Quantified data of RFP signal intensity. Note, the RFP signal is not significantly affected by mating contrasting to the GFP/RFP intensity. Statistics; Wilcoxon rank sum test. $^{**}p < 0.01$

2 Materials

2.1 Animal Preparations

D. melanogaster stock

1. Control strain, w^{1118} (Bloomington Stock Center (BDSC) #5905).
2. A genetic deletion mutant of *octopamine receptor in mushroom bodies* (*Oamb*$^{\Delta}$) [13]. We previously demonstrated that Oamb is indispensable for OA actions on ovarian somatic cells.
3. *Tdc2-GAL4* (BDSC #9313).
4. TRIC (*w*, UAS-mCD8::RFP, LexAop2-mCD8::GFP; nSyb-MKII::nlsLexADBDo; UAS-p65AD::CaM*), (BDSC #61679). The *p65AD::CaM* transgene encodes the calcium-sensing protein calmodulin (CaM) fused with the p65 transcriptional activation domain (AD). *p65AD::CaM* is under the control of the *UAS* promoter, which can be activated by exogenous transcription factor GAL4. In contrast, *nSyb-MKII::nlsLexADBDo* expresses a DNA-binding domain (DBDo) of the other exogenous transcription factor LexA, fused with the nuclear localization signal (nls) and the CaM-target peptide MKII, under the control of neuron-specific promoter *nSyb*. MKII::nlsLexADBDo is translocated from the cytoplasm to the nucleus when calcium signaling is elevated in the cells, resulting in transcriptional activation of the membrane-tethered green fluorescence protein (mCD8::GFP) that is under the control of the *LexAop2* promoter, which is bound by LexA (Fig. 3.1a).

2.2 Other Reagents and Equipment

1. Paraformaldehyde.
2. Phosphate-buffered saline (PBS).
3. PBT (0.1%Triton X-100 in PBS).
4. Blocking solution (2% bovine serum albumin (BSA) in PBT).
5. Chicken anti-GFP antibody (Abcam; #ab13970).
6. Rabbit anti-red fluorescence protein (RFP) antibody (Medical and Biological Laboratories, PM005).
7. Mouse anti-Hu-li tai shao (Hts) monoclonal antibody 1B1 (Developmental Study Hybridoma Bank (DSHB)).
8. Rat anti-*D*E-Cad monoclonal antibody DCAD2 (DSHB).
9. Goat anti-chicken IgG Alexa Fluor® 488 (Thermo Fisher Scientific #A11039).
10. Goat anti-rabbit IgG Alexa Fluor® 555 (Thermo Fisher Scientific #A32732).
11. Goat anti-mouse IgG Alexa Fluor® 488 (Thermo Fisher Scientific #A11001).

12. Goat anti-rat IgG Alexa Fluor® 546 (Thermo Fisher Scientific #A11081).
13. FluorSave™ Reagent (Merck-Millipore #345789).
14. 35 mm dish (IWAKI #3910-035).
15. Schneider's *Drosophila* medium (Thermo Fisher Scientific #21720024). For cell culture experiments, the medium was supplemented with fetal bovine serum and penicillin-streptomycin solution.
16. Fetal bovine serum (FBS) (Biowest #S1820-500).
17. Penicillin-streptomycin solution, ×100 (Fujifilm Wako #168-23191).
18. Octopamine hydrochloride (Sigma-Aldrich #O-0250).
19. Parafilm® (Amcor #P7793-1EA).
20. LSM 700 laser confocal microscopy (Carl Zeiss).
21. ImageJ (Fiji) [22].

3 Methods

3.1 TRIC System Allows the Visualization of Mating-Dependent Activation of Neurons

3.1.1 Preparation of Transgenic Flies

1. Cross *Tdc2-GAL4* line with TRIC line and rear their progeny at 25 °C to become adults.
2. Collect the virgin females carrying the genotype *Tdc2* > *TRIC*, which corresponds to *w*, UAS-mCD8::RFP, LexAop2-mCD8::GFP/+ (X), Tdc2-GAL4/nSyb-MKII::nlsLexADBDo (II), UAS-p65AD::CaM/+ (III)*, and rear them 4–5 days after eclosion at 25 °C (see Note 1-1).
3. Put 8–10 *Tdc2>TRIC* virgin females with 8–10 w^{1118} male flies in a single vial, and allow them to mate for 24 h (see Note 1-2) (Fig. 3.1b).
4. In parallel, 8–10 *Tdc2>TRIC* virgin females are also kept and raised without males in a single vial for 24 h as controls.

3.1.2 Dissection and Immunostaining of Abdominal Ganglion of D. Melanogaster

1. Anesthetize female flies with CO_2 gas, and transfer them to a dish filled with PBS.
2. Peel away the thorax cuticle to expose the abdominal ganglia using forceps (see Note 1-3).
3. Collect the abdominal ganglion from the fly body, and fix it in the tube containing 4% paraformaldehyde in PBS for 30 min.
4. Replace the fixative solution with PBT and quickly rinse the samples thrice.
5. Wash the samples twice with 0.1% PBT for 5 min each at room temperature (RT, around 25 °C).

6. Replace PBT with blocking solution. Incubate the samples for 1 h at RT.
7. Replace the blocking solution with primary antibody solution (chicken anti-GFP antibody (1:4000) and rabbit anti-RFP antibody (1:2000) in blocking solution).
8. Incubate the samples on a rocker overnight at 4 °C.
9. Wash the samples thrice with PBT for 5 min each at RT.
10. Replace PBT with the secondary antibody solution (anti-chicken IgG Alexa Fluor 488 (1:200) and anti-rabbit IgG Alexa Fluor 555 (1:200) in blocking solution).
11. Incubate the samples for 2 h on a rocker at RT in the dark.
12. Wash the samples thrice with PBT for 15 min each at RT in the dark.
13. Transfer the samples to a glass slide and mount with the mounting reagent FluorSave.

3.1.3 Acquisition of TRIC Signals with Microscope

1. Place the slide glass containing the stained abdominal ganglia under a laser confocal microscope (Fig. 3.1c).
2. Position the sample and bring it into focus.
3. Capture fluorescent images using a 40× objective with a confocal microscope equipped with the following filter sets: laser excitation 488 nm, emission 488 nm (GFP channel); laser excitation 561 nm, emission 568 nm (RFP channel) (*see* **Notes 1-4**).

3.1.4 Data Analysis of TRIC Signal

1. Open the analysis software (ImageJ) and import data file.
2. Superimpose the image by using Image/stack/Z-project in ImageJ.
3. Select multiple cell bodies as regions of interest (ROIs) using Analyze/Tools/ROI Manager in ImageJ.
4. Measure the pixel intensity of the GFP and RFP channels, each by clicking More/Multi Measure/OK in the ROI Manager window.
5. Repeat the experiment using multiple abdominal ganglia samples.
6. Calculate the relative GFP signal intensity by dividing with the RFP signal intensity that shows general activity of each neuron (*see* **Notes 1-5**).

3.1.5 Note

Note 1: Females are not mature immediately after eclosion. Therefore, females should be allowed 4–5 days to reach maturity.

Note 2: Using either small number of males or aged males results in a low mating frequency. Therefore, it is suitable to use equal number of young (3–7 days after eclosion) males as the number of females.

Note 3: We usually start dissecting the abdomens from the ventral side. We recommend that the legs and thoracic cuticles be carefully removed at the beginning of the dissection process. As the legs and thoracic cuticles are tightly connected with the abdominal ganglia, we peel the abdominal cuticle to expose the abdominal ganglia more easily from the samples without legs and thoracic cuticles. Do not pinch the center of the thoraxes to avoid histological damage to the abdominal ganglia.

Note 4: Z-stack imaging enables the measurement of fluorescent intensity with a superimposed image. The Z-axis imaging range is set to acquire maximum intensity Z-projections of several consecutive confocal stacks (1-μm-thick each for Fig. 3.1d) covering the entire soma of each neuron.

Note 5: In the TRIC method, a GAL4-driven RFP is used as an internal control. GFP immunostaining signals can be normalized by RFP immunostaining signals, even when there is a difference in the degree of immunostaining between samples (Fig. 3.1a).

As shown in Fig. 3.1, we dissected and fixed the abdominal ganglia of mated and virgin female *D. melanogaster* and monitored TRIC signals of the ovary-projecting neurons at the tip of the abdominal ganglia (Fig. 3.1c, d). We found that the GFP/RFP intensity ratio was significantly higher after mating than before mating, indicating that the ovary-projecting neurons in the abdominal ganglion were activated by mating (Fig. 3.1e). The intensity of the RFP did not significantly change before and after mating (Fig. 3.1f), indicating that this TRIC-based system can be used robustly.

3.2 Ex Vivo Culture Experiment Reveals Direct Effect of Neurotransmitter and Hormones to the Peripheral Tissues

3.2.1 Preparation of Flies

1. Collect virgin females carrying the appropriate genotype and rear them 4–5 days after eclosion at 25 °C (Fig. 3.2a).

3.2.2 Dissection and Ex Vivo Culture of the Ovary of D. melanogaster

1. Dissect the ovaries from the females in Schneider's *Drosophila* medium (*see* **Notes 2-1** and **2-2**) (Fig. 3.2b).
2. Prepare a 35 mm dish filled with 3 mL of Schneider's *Drosophila* medium with 0, 1, 10, 100, or 1000 μM OA.
3. Put the dissected ovaries from five to eight flies into the prepared dishes (*see* **Notes 3**).

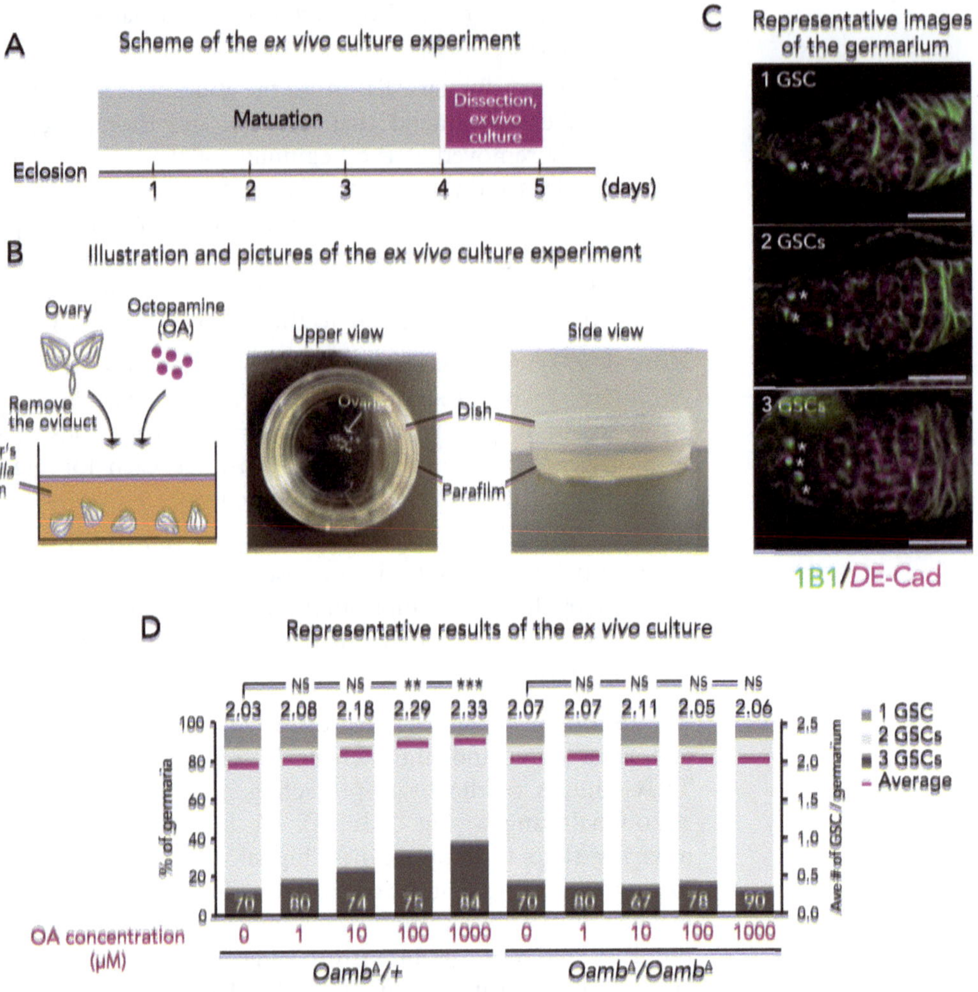

Fig. 3.2 Ex vivo culture system reveals the direct action of neurotransmitter on the ovary. (**a**) Scheme of ex vivo culture experiment. Virgin female flies were collected within 8 h (25 °C) or 16 h (17 °C) after eclosion. After 4–5 days of maturation, dissected ovaries were cultured. (**b**) Illustration and pictures of equipment of ex vivo culture experiment. (Left) A schematic representation of ex vivo culture system. The ovaries separated from the oviduct were cultured in dishes filled with Schneider's *Drosophila* medium. To examine the function of OA, we supplemented octopamine into the medium. (Middle) Upper view of the dish containing the ovary and medium. (Right) Side view of the dish containing the ovary and medium. The side of the dish was wrapped with parafilm. (**c**) Representative image of the germarium, a chamber-like structure containing GSCs (asterisk). The GSCs attaching to the tip of the germarium are distinguished by their morphology from other cell types. The boundary of GSC niche cells at the tip of the germarium was stained by *D*E-cadherin antibody (magenta). The spectrosome (green) is in contact with the boundary, representing a GSC-specific configuration. (**d**) Representative results of ex vivo culture experiments showing frequencies of germaria containing 1, 2, and 3 GSCs (left vertical axis) and the average number of GSCs per germarium (right vertical axis) in the ovaries dissected from virgin female flies. The ovaries were cultured with several concentrations of OA and exhibited concentration-dependent GSC number increase in control (*Oamb*$^{\Delta}$/+). In homozygous *Oamb* null mutant animals (*Oamb*$^{\Delta}$/*Oamb*$^{\Delta}$), OA did not induce the OA-dependent GSC number increase. Statistics; Wilcoxon rank sum test with Holm's correction. $^{**}p < 0.01$, $^{***}p < 0.001$; *NS*, nonsignificant ($p > 0.05$)

4. Wrap the dishes with parafilm and culture them at 25 °C for 16 h (*see* **Note 4**).

3.2.3 Immunostaining of the Ovary and Germline Stem Cells

1. Transfer the ovaries from the culture medium to a tube containing 4% paraformaldehyde in PBS and fix them for 30 min.
2. Replace with 4% paraformaldehyde in PBT, and quickly rinse the samples thrice.
3. Wash the samples twice with PBT for 5 min each at RT.
4. Replace PBT with blocking solution. Incubate the samples for 1 h at RT.
5. Replace the blocking solution with primary antibody solution (mouse anti-Hts antibody (1:50) and rat anti-*D*E-Cad antibody (1:50) in blocking solution).
6. Incubate the samples overnight at 4 °C.
7. Wash the samples thrice with PBT for 5 min each at RT.
8. Replace the PBT with secondary antibody solution (anti-mouse IgG Alexa Fluor 488 (1:200) and anti-rat IgG Alexa Fluor 555 (1:200) in blocking solution).
9. Incubate the samples for 2 h on a rocker at RT in the dark.
10. Wash the samples thrice with PBT for 15 min each at RT. Store the samples in the dark until further analysis.
11. Transfer the samples to a glass slide and mount with the mounting reagent FluorSave (*see* **Notes 2-5**).

3.2.4 Imaging and Counting of Germline Stem Cells

1. Place the glass slide containing the ex vivo cultured ovaries under a laser confocal microscope.
2. Position the tip of the ovary and bring it into focus.
3. Capture fluorescent images using a 40× objective with the laser confocal microscope equipped with the following filter sets: laser excitation 488 nm, emission 517 nm (Hts); laser excitation 561 nm, emission 568 nm (*D*E-Cad).
4. Count the number of GSCs, which can be identified based on the morphology and positioning of their anteriorly anchored spherical spectrosome (*see* **Notes 2-6**) (Fig. 3.2c).

3.2.5 Notes

Note 1: To avoid indirect effects from other tissues, such as the oviduct and fat body, try to dissect the ovaries without any associated tissues as much as possible (Fig. 3.2b).

Note 2: Using Schneider's *Drosophila* medium is important for maintaining healthy ovaries. We do not recommend the use of PBS for this dissection.

Note 3: Culturing more than ten ovaries might lead to bacterial growth and poor culture conditions.

Note 4: Wrapping the side of the dish with parafilm prevents the evaporation of the medium.

Note 5: Separate the strings of ovarioles from each other using forceps. Do not injure the tips of ovarioles that contain the germarium, a chamber-like structure containing the GSCs.

Note 6: GSCs can be distinguished by their morphology that attaches to the tip of the germarium [23]. For instance, the boundary of the GSC niche cells is visualized by the anti-*D*E-cadherin antibody (magenta in Fig. 3.2c). Spectrosome, an intracellular spectrin-rich structure unique to the GSC and its daughter cells, is visualized by the anti-Hts antibody (green in Fig. 3.2c). Only the spectrosome of GSC attaches to the GSC niche boundary, allowing unequivocal identification of the GSC [24]. As an alternative method to distinguish GSCs, anti-phosphorylated Mad antibody (Abcam #ab52903, 1:1000 dilution) can be used because pMad protein is specifically localized in the nucleus of the GSC [25].

As shown in Fig. 3.2, we set the concentration of OA in the medium from 0 to 1000 μM and found that 100 and 1000 μM OA treatment significantly increased the number of GSCs in control animals ($Oamb^{A}$/+) (Fig. 3.2d). This method allowed us to examine the effects of OA on GSCs directly in the ovary and to rule out the effects of OA on tissues other than of the ovary. Furthermore, this culture system can be used not only for ovaries but also for various other tissues such as the brain, adipose tissue, and testes [18, 20, 26, 27]. Moreover, by performing tissue-specific gene knockdown and reporter gene expression in the tissue, the effects of the substances added to the tissue can be genetically analyzed in detail. For example, null mutant animals of the octopamine receptor ($Oamb^{A}$/$Oamb^{A}$) did not exhibit an OA-dependent increase in GSC number (Fig. 3.2d), suggesting that the Oamb in the ovary is required for OA-dependent GSCs. In conclusion, ex vivo culture systems are very compatible with *Drosophila* genetics and can be of great help to study the effects of specific neurotransmitters on specific tissues.

Acknowledgments

We thank Takayuki Kuraishi, the Bloomington Stock Center, and the Developmental Studies Hybridoma Bank for providing stocks and reagents. Y.Y. was a recipient of a fellowship from the Japan Society for the Promotion of Science. This work was supported by grants from AMED-CREST, AMED (21gm1110001h0005) to R. N., and KAKENHI 18 J20572 to Y.Y. We would like to thank Editage (www.editage.com) for English language editing.

References

1. Raji JI, Potter CJ (2021) The number of neurons in *Drosophila* and mosquito brains. PLoS One 16:e0250381. https://doi.org/10.1371/journal.pone.0250381
2. Brand AH, Perrimon N (1993) Targeted gene expression as a means of altering cell fates and generating dominant phenotypes. Development 118:401–415. https://doi.org/10.1242/dev.118.2.401
3. Dietzl G, Chen D, Schnorrer F et al (2007) A genome-wide transgenic RNAi library for conditional gene inactivation in *Drosophila*. Nature 448:151–156. https://doi.org/10.1038/nature05954
4. Yamamoto D (2008) Brain sex differences and function of the fruitless gene in *Drosophila*. J Neurogenet 22:309–332
5. Rezával C, Pavlou HJ, Dornan AJ et al (2012) Neural circuitry underlying *Drosophila* female postmating behavioral responses. Curr Biol 22: 1155–1165. https://doi.org/10.1016/j.cub.2012.04.062
6. White MA, Chen DS, Wolfner MF (2021) She's got nerve: roles of octopamine in insect female reproduction. J Neurogenet:1–22. https://doi.org/10.1080/01677063.2020.1868457
7. Liu H, Kubli E (2003) Sex-peptide is the molecular basis of the sperm effect in *Drosophila melanogaster*. Proc Natl Acad Sci U S A 100:9929–9933. https://doi.org/10.1073/pnas.1631700100
8. Kubli E (2003) Sex-peptides: seminal peptides of the *Drosophila* male. Cell Mol Life Sci 60: 1689–1704. https://doi.org/10.1007/s00018-003-3052
9. Häsemeyer M, Yapici N, Heberlein U, Dickson BJ (2009) Sensory neurons in the *Drosophila* genital tract regulate female reproductive behavior. Neuron 61:511–518. https://doi.org/10.1016/J.NEURON.2009.01.009
10. Yapici N, Kim Y-J, Ribeiro C, Dickson BJ (2008) A receptor that mediates the post-mating switch in *Drosophila* reproductive behaviour. Nature 451:33–37. https://doi.org/10.1038/nature06483
11. Del Valle RA, Didiano D, Desplan C (2012) Power tools for gene expression and clonal analysis in *Drosophila*. Nat Methods 9:47–55
12. Ameku T, Niwa R (2016) Mating-induced increase in germline stem cells via the neuroendocrine system in female *Drosophila*. PLoS Genet 12:e1006123. https://doi.org/10.1371/journal.pgen.1006123
13. Yoshinari Y, Ameku T, Kondo S et al (2020) Neuronal octopamine signaling regulates mating-induced germline stem cell increase in female *Drosophila melanogaster*. elife 9. https://doi.org/10.7554/eLife.57101
14. Ameku T, Yoshinari Y, Fukuda R, Niwa R (2017) Ovarian ecdysteroid biosynthesis and female germline stem cells. Fly (Austin) 11. https://doi.org/10.1080/19336934.2017.1291472
15. Yoshinari Y, Kurogi Y, Ameku T, Niwa R (2019) Endocrine regulation of female germline stem cells in the fruit fly *Drosophila melanogaster*. Curr Opin Insect Sci 31. https://doi.org/10.1016/j.cois.2018.07.001
16. Cole SH, Carney GE, McClung CA et al (2005) Two functional but noncomplementing *Drosophila* tyrosine decarboxylase genes: distinct roles for neural tyramine and octopamine in female fertility. J Biol Chem 280: 14948–14955. https://doi.org/10.1074/jbc.M414197200
17. Gao XJ, Riabinina O, Li J et al (2015) A transcriptional reporter of intracellular Ca 2+ in *Drosophila*. Nat Neurosci 18:917–925. https://doi.org/10.1038/nn.4016
18. Géminard C, Rulifson EJ, Léopold P (2009) Remote control of insulin secretion by fat cells in *Drosophila*. Cell Metab 10:199–207. https://doi.org/10.1016/j.cmet.2009.08.002
19. Britton JS, Edgar BA (1998) Environmental control of the cell cycle in *Drosophila*: nutrition activates mitotic and endoreplicative cells by distinct mechanisms. Development 125: 2149–2158. https://doi.org/10.1242/dev.125.11.2149
20. Ameku T, Yoshinari Y, Texada MJ et al (2018) Midgut-derived neuropeptide F controls germline stem cell proliferation in a mating-dependent manner. PLOS Biol 16:e2005004. https://doi.org/10.1371/journal.pbio.2005004
21. Sano H, Nakamura A, Texada MJ et al (2015) The nutrient-responsive hormone CCHamide-2 controls growth by regulating insulin-like peptides in the brain of *Drosophila melanogaster*. PLoS Genet 11:e1005209. https://doi.org/10.1371/journal.pgen.1005209
22. Schindelin J, Arganda-Carreras I, Frise E et al (2012) Fiji: an open-source platform for biological-image analysis. Nat Methods 9: 676–682
23. de Cuevas M, Spradling AC (1998) Morphogenesis of the *Drosophila* fusome and its

implications for oocyte specification. Development 125:2781–2789. https://doi.org/10.1242/dev.125.15.2781

24. Ables ET, Drummond-Barbosa D (2010) The steroid hormone ecdysone functions with intrinsic chromatin remodeling factors to control female germline stem cells in *Drosophila*. Cell Stem Cell 7:581–592. https://doi.org/10.1016/j.stem.2010.10.001

25. Kai T, Spradling A (2004) Differentiating germ cells can revert into functional stem cells in *Drosophila melanogaster* ovaries. Nature 428:564–569. https://doi.org/10.1038/nature02436

26. Gärthner SMK, Rathke C, Renkawitz-Pohl R, Awe S (2014) Ex vivo culture of *Drosophila* pupal testis and single male germ-line cysts: dissection, imaging, and pharmacological treatment. J Vis Exp:51868. https://doi.org/10.3791/51868

27. Kim J, Bilder D, Neufeld TP (2018) Mechanical stress regulates insulin sensitivity through integrin-dependent control of insulin receptor localization. Genes Dev. https://doi.org/10.1101/gad.305870.117

Chapter 4

Measuring Sleep in Drosophila

Takaaki Miyazaki, Julie A. Williams, and Hirofumi Toda

Abstract

In this chapter, we introduce the method by which we measure and quantify sleep in *Drosophila melanogaster. Drosophila* is a powerful genetic model system that has been instrumental in contributing to our understanding of numerous biological processes including development, immunity, the circadian clock, and, more recently, sleep at the molecular and mechanistic level. The past two decades of sleep research has provided enormous advances in our understanding of circuit and molecular mechanisms that underlie sleep. Indeed, this simpler genetic model has allowed us to unravel the complexity of sleep and improve understanding of its biological function and impact on many aspects of physiology across the life span. Here we describe a straightforward, high-throughput method that is most commonly used for measuring sleep in *Drosophila*.

Key words *Drosophila*, Sleep, Genetics, Screening

1 Introduction

Sleep remains one of the biggest mysteries in biology. While all animals with nervous systems that have been investigated so far show behavioral quiescence that qualifies as sleep, the fundamental function of sleep remains poorly understood [9, 30, 56]. Considering the fact that animals increase arousal threshold in response to sensory stimuli and reduce locomotor activity while asleep, sleep itself renders them at risk of danger such as an attack by predators. Despite this risk, sleep must have a fundamental role, as this behavior is conserved across a wide range of vertebrate and invertebrate species. It is evident that sleep loss leads to many fundamental malfunctions in memory formation, metabolic processes, immune response, and prolonged sleep loss ultimately leads to death [48, 57]. However, how sleep restores and maintains robust health remains unclear.

Molecular mechanisms of sleep are also not well understood. From a phenomenological point of view, sleep is regulated by two main systems, the internal circadian clock and the homeostatic

Daisuke Yamamoto (ed.), *Behavioral Neurogenetics*, Neuromethods, vol. 181, https://doi.org/10.1007/978-1-0716-2321-3_4,

system. The circadian clock determines the time of day that sleep occurs, whereas the homeostatic system determines how much sleep occurs depending on need.

Drosophila melanogaster has been instrumental in answering many biological questions including those relating to behavior. Screening for disrupted timing in eclosion, for example, identified two core clock genes, *period* [34] and *timeless* [55]. The development of a commercially available locomotor activity assay in 1987 (TriKinetics, Inc., Waltham MA, and see below), which utilized IR beam breaks to quantify daily locomotor activity patterns in flies, greatly facilitated behavioral screening, and led to the identification of core circadian clock gene components.

By the year 2000, a sleep like state had been characterized in flies, which necessitated updates to this system. These updates involved an increase in sampling rate (from a minimum of 30 min to 1 s), as well the introduction of newer instrument models capable of measuring more refined activity patterns [14]. Despite these advances, the molecular mechanisms of the sleep homeostatic system remain elusive.

Since the 1950s, electroencephalography (EEG) has been a common approach to quantifying various sleep parameters in freely moving animals, particularly mammalian species including humans and mice [64]. Prior to this, animals (typically larger species such as cats or dogs) were anesthetized or restrained to facilitate recording. However, this approach is not practical in smaller freely moving animals such as insects. Therefore, behavioral criteria were used to characterize sleep in *Drosophila*, which included behavioral quiescence at the nadir of the circadian cycle, species specific posture, increased arousal threshold in response to sensory stimuli, and homeostatic response to deprivation, where flies show a compensatory rebound sleep in response to sleep loss [25]. It was further shown that the stimulus threshold needed to wake flies up reached a maximum when flies were inactive for 5 consecutive minutes [25]. Thus, 5 min or longer of inactivity is considered to be a sleep like state and is now universally used to define sleep [4, 38, 42]. This assay system for sleep in *Drosophila* opened up a venue to investigate the genetic factors regulating sleep since *Drosophila* is a fantastic model system with a battery of genetic resources such as RNAi lines [23], mutagenesis [47, 66], and silencing/activating tools for specific subsets of neurons [33, 45].

The *Drosophila* Activity Monitoring (DAM) System (TriKinetics, Waltham, MA) remains the most commonly used assay system to quantify this behavior. Several video-based assays have also been developed [11, 16, 18, 22, 37], but are not practical for conducting large-scale genetic screens. Combined with a battery of genetic resources of *Drosophila*, the DAM system has successfully allowed us to identify novel genetic components regulating sleep in *Drosophila* [2, 8, 21, 26, 28, 32, 35, 39, 47, 49, 58–60, 62, 66, 67]. Here, step-by-step, we describe how to prepare and analyze the data obtained from the DAM system.

1.1 Methods to Quantify Sleep Amount in *Drosophila*

Because locomotor activity of animals is one of the most prominent representations of daily activity or wakefulness, this is often used as the primary index to evaluate sleep [20, 27]. The DAM system measures activity of flies in a 6.5 cm glass tube (5 mm diameter) bisected by an infrared (IR) light beam, which by definition is outside the spectra for visible light and does not affect behavior. Each time an individual fly crosses this beam, it is digitally recorded. Thus, hundreds if not thousands of flies can be simultaneously monitored for many days. Video-tracking methods monitor behavior in planar, circular arenas, which can detect two-dimensional movements, whereas beam breaks in the DAM system detect one-dimensional walking in the narrow tubes [18]. Video monitoring also offers much higher resolution and can detect more subtle movements other than overt motor activity (such as proboscis extensions [3, 24]). Both methods are capable of extracting several parameters such as total time asleep, the number of sleep episodes or bouts, duration of sleep bouts, and sleep latency. These methods are also used to evaluate sleep quality by measuring responses to arousal stimuli such as tapping, light, or smell [62]. Deeper sleep is associated with lowered responsiveness or arousability to these sensory stimuli. In this section, we describe the method for measuring *Drosophila* sleep using the DAM system. The advantage of the DAM system is that it is less labor-intensive, is high-throughput, and is relatively inexpensive. We can thus apply this method for large-scale genetic screens to search molecular components that regulate sleep. Indeed, loss-and gain-of-function genetic screens have successfully identified novel genes that are important/instructive for sleep [2, 8, 21, 26, 28, 32, 35, 39, 47, 49, 58–60, 62, 66, 67]. Moreover, this assay has been used for pharmacological screening to evaluate the therapeutic potential of various compounds for treatment of sleep disturbance associated with human disease [63].

2 Materials

2.1 Overview of *Drosophila* Activity Monitoring (DAM) System

The TriKinetics DAM system has been described in detail elsewhere [6, 14, 50, 54] and so we briefly summarize here. The DAM system consists of activity monitors, cable connectors, and a power supply and interface unit (PSIU) with a PC (or Mac) dedicated to data recording. The activity monitors are housed in an incubator to control the environmental condition (see the next section). A separate PC is used for data analysis. The activity monitor has up to 32 channels, each of which is equipped with an IR light source and a photosensor to detect movements of individual flies in a glass tube. The monitor detects IR beam breaks by an individual fly within each of 32 channels. The PSIU digitizes the count information from the monitors and transmits these data to a PC for recording at every time increment as set by the user (usually once per minute). Telephone cables are used to connect the activity monitors to the interface unit, while the interface unit connects to the data collection PC through a USB cable.

2.2 A Laboratory and Incubators for Sleep Measurements

Environmental control is necessary for stable monitoring of sleep behavior. To achieve this, the assays are conducted in an incubator to control for light cycles, humidity, and temperature.

Light Typical experiments are conducted in 12 h:12 h light-dark cycles. However, in some cases, we need to run multiple sleep measurements in different light/dark conditions and sometimes need to access an incubator during the flies' dark phase. Light, particularly blue light, is capable of degrading clock genes in flies when exposed at nighttime [61] and disturbs both activity rhythms and sleep. To avoid any perturbations to animals in the dark or nighttime phase, the laboratory room accommodating the incubators must be kept dark whenever the door of the "dark" incubator is open. Any light sources such as control panels of incubators, indicators of laboratory devices, computer screens, etc. must be covered with opaque materials to prevent exposure of flies to the light. The dark room can be equipped with a darkroom "safelight," a dim red light with a wavelength > 600 nm, compatible for use in developing photographic film.

Humidity Humidity inside the incubator should be maintained between 50% and 60%. *Drosophila* desiccate in low humidity, which alters behavior [52] and shortens life span [43]. Usually, a beaker or tray containing water is placed in the incubator to achieve this condition. Be careful not to over-humidify as condensation buildup inside the incubator can lead to mold growth and potentially leak into the DAMS electronics causing the system to short out. Occasional cleaning of the incubator with mild detergent (no bleach) is recommended.

Temperature The temperature of the incubator is normally set to 25 °C. However, some sleep measurements are carried out under changing temperatures for mimicking day-night oscillations of temperature, thermal control of gene expression for manipulation of neuronal function [41], or analyses of effects of temperature on sleep. Higher temperatures reduce sleep in flies, particularly at night [46]. A thermometer and hygrometer should be placed in each incubator to monitor and ensure stable environmental conditions.

2.3 Animals

We must control for genetic background, which affects sleep [68] and other behaviors in flies. Using an isogenic fly stock serves to minimize effects of genetic variation. *Iso*^{31} is an isogenic strain derived from w^{1118} that has been used by some labs for sleep assays due to its reported robust activity rhythms and similarity to wild-type flies [51]. However, we have found this line to be insensitive to stress, such that it does not show a robust increase in sleep in response to infection or heat stress (Toda and Williams, unpublished), likely due to its high level of daytime sleep. Canton Special (CS), w^{1118}, or other alternate wild-type strain should be considered depending on the nature of the experiment. In any case, transgenic flies should be isogenized to the strain of choice, where possible, by backcrossing at least four to five generations to minimize uncontrolled genetic variation. It is sometimes useful to employ opto-/thermo-/chemogenetics, because such techniques enable to changing the animals' status between experimental and control groups of the same genotype.

2.4 Apparatus, Accessories, and PCs for DAM System

A glass locomotor tube consists of food at one end, covered with a plastic cap (or paraffin wax), while the other end is plugged with a small piece of wool, cotton, or acrylic yarn. The glass tube is 6 mm outside diameter and 65-mm-long, available from TriKinetics Inc. (Waltham, MA USA). We can also prepare glass tubes by cutting Pyrex glass tubes into the designated length. The agar food, which is a minimal medium that contains 5% sucrose, sustains a fly for 1–2 weeks. For experiments of longer duration, it is possible to transfer flies to freshly prepared tubes to prevent mold or accumulation of waste. Depending on the nature of the experiment; some investigators use normal protein-rich food. Flies show less sleep on this type of food [4, 13]. Moreover, larvae from mated females in this assay will survive and potentially interfere with behavior. The plastic cap or seal prevents the agar food from drying out, while the yarn keeps the animals inside of the tube with ventilation. The plastic cap is also available from TriKinetics Inc. (CAP5-Black). Agar and sugar used routinely for fly food are suitable for this protocol.

There are several versions of commercially available activity monitors. The DAM2 model is used in the protocols described in this chapter. This simplest version, which has a single IR beam per tube, is sufficient to measure flies' sleep and ideal for high-throughput screening. We can put multiple DAM2 monitors, each of which accommodates 32 locomotor tubes, in an incubator to compare sleep patterns of many flies. A multibeam monitor (model MB5) is also available and provides higher resolution, as it contains 17 beams per tube per channel. The Insomniac3 software is capable of analyzing data from these monitors as long as they are processed the same way as for the single beam.

A 60-cm-long telephone cable supplied with each monitor is used to connect the activity monitor to the interface unit. A single PSIU9 accommodates up to 120 monitors. Phone splitters and 6P4C phone cables can be purchased separately (*see* **Note 1**). We need a USB cable to connect the PSIU9 to the data collection PC (*see* **Note 2**). The PSIU9 also connects to an external power supply (AC 100–240 V, 50/60 Hz) through an AC adaptor supplied with the interface unit. A similar set up is used for the MB5 model, but it currently requires its own PSIU, cables, and a separate version of data collection software.

For data collection, we need to install DAMSystem3 software and a USB driver for the interface unit into a PC running Windows 7/8/10 OS. An entry-level machine with a USB port suffices, either 32- or 64-bit OS. We can download the DAMSystem3 software and the USB driver from the TriKinetics Inc. website (https://trikinetics.com/) (DAMSystem311 along with the USB Driver1808, current as of April, 2021). Follow "QuickStart Guide" for the DAMSystem3 software, also available from the TriKinetics website. Do not forget to set the system preferences to inhibit power shutdown or windows updates (where applicable) of the computer.

To process the raw data, we use DAMFileScan software to generate time-stamped raw output files and use these files for subsequent quantitative analysis with the Insomniac3 program, described below. The former is available from TriKinetics Inc. (DAMFileScan113, current as of April, 2021), while the latter can be obtained upon request (Insomniac3.10.zip, current as of April, 2021, only available for Windows).

3 Methods

3.1 Preparation of Locomotor Tubes

1. Bundle the desired number of locomotor tubes with a rubber band in a container. It is easier to put them loosely so that they can be taken out from the container easily.
2. Microwave water including 5% sucrose in 2% agar to melt the agar. This minimal food medium can also be autoclaved in a 500 mL bottle and stored long term at 4 degrees. When needed for preparing tubes, add a small amount of water, enough to moisten the surface, and microwave at medium/low power until the solution is melted or boils.
3. Once the food is liquefied, allow to cool to ~60 °C if a drug (e.g., mifepristone, commonly used for chemically induced gene expression) is to be added. Pour the mixture into a beaker or weigh boat to no more than 1 cm high, and place the bundles of the glass tubes into the container.

4. Allow the food to solidify, and then remove and unbind the glass tubes. Remove the excess agar around the tubes and plug the end with tube caps. If using paraffin wax, melt the wax in a metal container (such as a recycled 5 oz. aluminum can) or in a glass beaker placed in a larger container of water, and gently heat until the wax is melted. Dip the end of the food tube into the wax and then into ice water to immediately solidify, and repeat.

   The locomotor tubes prepared as above can be stored at 4 °C for long-term storage in a sealed plastic bag.

3.2 Preparation of Flies

Flies to be used for behavioral experiments or screening should be age-matched, since sleep changes across the life span [10, 44]. Unless age itself is an experimental variable, flies are typically aged 3–7 days after eclosion. Their sleep patterns are usually recorded for about 7–10 days. In addition, the animals are entrained to the daily 12 h:12 h light-dark cycle.

To load flies in to the DAM2 monitors, follow the steps below.

1. If the locomotor tubes have been stored at 4 °C, warm them at room temperature to allow condensation (if any) to dry.
2. Anesthetize flies with carbon dioxide or on ice. Load a single animal into each locomotor tube using a suction tube, paint brush, or blunt forceps. Plug the open end of each tube with a small piece of yarn using forceps.
3. Set each activity tube containing a fly in the channel opening of the DAM2 monitor. Center the tube roughly across the IR sensors in the DAM2 monitor, and ensure that food does not obstruct the IR beam. Fix the locomotion tubes in place with rubber bands.
4. Check if the DAMSystem3 is running on the data collection PC and that the interface unit is working normally (see the "QuickStart Guide" for the DAMSystem3). Before connecting the DAM2 monitors, the status boxes of the DAMSystem3 application that correspond to the monitor ID numbers should be colored red.
5. Place DAM2 monitors in the incubator, and connect them with the phone cables at the sockets of DAM2 monitors. After DAM2 monitors are connected, make sure the corresponding status boxes of the DAMSystem3 application on the PC turn to yellow, indicating the program is recognizing DAM2 monitors. Once recordings are successful, yellow turns into green. Collect DAM file data for the desired time period (e.g., 3–7 days) for sleep analysis.

3.3 Analysis of Sleep Using "Insomniac3" Program

As discussed above, the TriKinetics data collection software contains the number of activity counts recorded per unit time (1 min) for each fly across all monitors for several days. A challenge in the field has been to automate quantification of the absence of activity counts to assess sleep and various parameters (bout length, duration, latency, etc.). It has been necessary to develop software to process this activity data using the definition of sleep as activity counts of zero for a minimum of 5 consecutive minutes. Several software programs have been developed [1, 7, 17, 19, 31, 53, 65, 69, 70] and may be accessed through Trikinetics.com. Here we describe a simple program developed by a volunteer software/firmware professional.

An earlier version of Insomniac3 was described elsewhere [36]. This software was originally written to facilitate survival analyses of flies in the DAMS assay that had been subjected to bacterial infection as well as life span metrics in general. It was then merged with a program "Insomniac," originally written for MatLab by Dr. Lesley Ashmore (Univ. Pittsburgh), into Microsoft Visual Basic 6 for the purpose of analyzing both sleep and survival. There is now a third module which analyzes responses to arousal stimuli. We describe here the steps needed to prepare the data for using this software, as well as some basic steps for how to use the program itself. It is a stand-alone program that can run on any PC and suitable for analyzing data from single or multibeam monitors. The program is freely available upon request and will be provided with a user manual, documentation, and tutorials with an accompanying sample experiment [15].

An advantage to using this program is that all output files are numeric and can be exported into any program of choice for statistical analyses or generation of publishable figures.

To process DAMS raw data files, follow the steps below:

1. Copy the DAM files that were recorded in the data collection PC, and transfer data to another PC that has the DAMFile Scan program installed. The file names of the collected raw data are "MonitorN.txt," where "N" is the ID number of DAM2 monitor. Save these raw monitor files to a folder with a unique experiment name. The default name in the DAMFileScan program is "MyRun."
2. Copy one monitor file from any raw monitor files in the same folder, and rename it to "MonitorLC.txt." This is necessary to generate a legacy "Daylight" file which contains the time stamps for every data point collected in each 1-min bin.
3. Run the DAMFileScan program (detailed instructions for using this program are also found on the TriKinetics website) by double-clicking DAMFileScan113.exe (as of April, 2021).

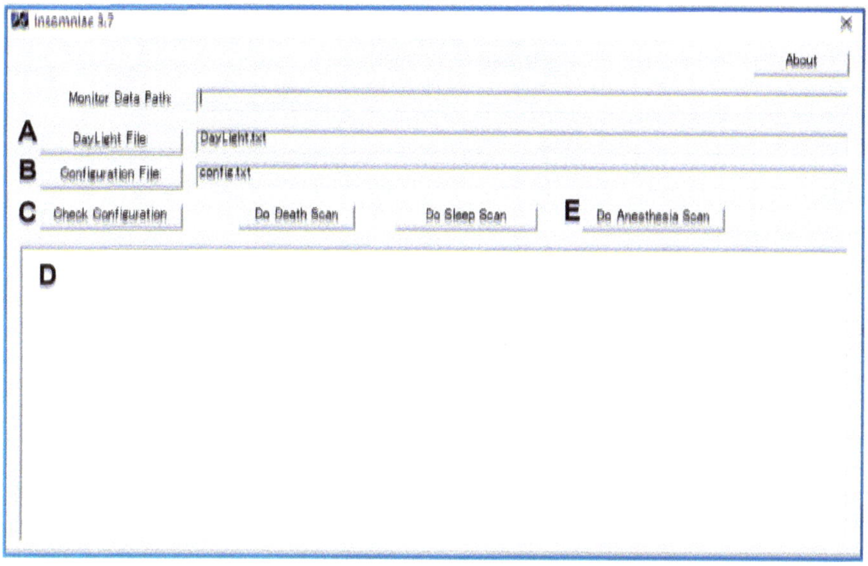

Fig. 4.1 Insomniac3 program window as it appears on a PC desktop. (*A*) Press the "DayLight File" button, and navigate to the desired data folder that contains the DayLight file to analyze. Select the file and the "Monitor Data Path" will automatically populate. (*B*) Press the "Configuration File" button, and navigate to the same folder to select your "config" file. (*C*) Run the program: Press the "Check Configuration" button, and the dialog window (*D*) will automatically fill. Next, press the "Do Death Scan button," and then "Do Sleep Scan" button. (*E*) For arousal threshold assays or sleep latency, select the appropriate "config" file, and then press the program buttons as follows: (1) "Check Configuration"; (2) "Do Death Scan"; and (3) "Do Anesthesia Scan" (skip the "Do Sleep Scan" button). The program completes after a few seconds. Close the program window, navigate to your data folder, and view the output files

4. (a) Click "Select Input Data Folder," and specify the folder containing "MonitorN.txt" and "MonitorLC.txt." Click "Scan" button. When it is finished scanning, you will see a "scan complete" message. You can now generate output data files.

   (b) Use the scroll bars to select "First bin to save" and "Last bin to save." Select the first lights-on period (or subjective lights-on if working in constant dark conditions) to save your file. (*see* **Note 3**)

5. Select the checkbox of "Make Channel Files" and input "Run Name." Use your unique folder name where your raw monitors are located, and click "Save." You will see a warning message, "Run name files already exist." Click "save" a second time, and the program will generate text channel files that correspond to each fly. If you wish to save your processed files to a separate folder, use a different "Run Name." This same protocol can be used for the MB5 monitors. However, these monitors are capable of generating different file types. For our purposes, we generate one set each of "channel files" and "move files." "Moves" in the MB5 are detected only when

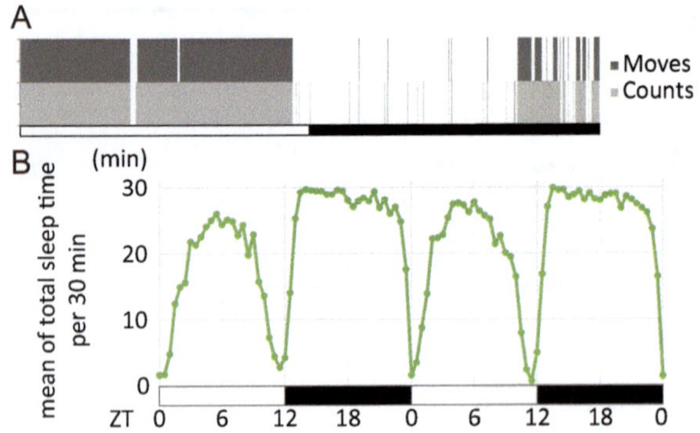

Fig. 4.2 Activity profile (**a**) or sleep graph (**b**) generated using the Insomniac3 program. (**a**) Activity counts and movements of a Canton-S female fly recorded each minute by a multibeam (MB5) monitor. Data are from output files located in the "sleep" folder. Waking minutes in which activity counts or moves were recorded were given a score of "1" and are plotted as a stacked column chart. Note that "counts" are more frequent than "moves," thus consolidated sleep in this fly occurred during the early part of the night, with few activity counts during movement "rest". (**b**) Mean daily sleep per 30 min is plotted against time in *iso*[31] female flies ($n = 24$). ZT (zeitgeber time) is shown on the x-axis. Horizontal bar on the x-axes in both (**a**) and (**b**) correspond to the day: night cycle

the fly moves from one beam to the next, whereas "channel counts" contain all activity counts, similar to the counts in the single-beam monitor, where in-place counting from the same beam will be included. Thus overt movement can be distinguished from "quiet" activity. This will be discussed in further detail below.

6. The data folder will now contain a "DayLight file" (MyRunCt-DayLight.txt) and "Channel files" (MyRunCtM[xxx]C[yy].txt, [xxx] is the ID numbers of DAM2 monitors (1–120) and [yy] is the ID numbers of channels.). "Move files" (*Mv) will also be contained in this folder if the MB5 model monitor was used. Then, Quit the DAMFileScan program. Your files are now ready to analyze using Insomniac3.

7. Insomniac3 uses a configuration file, generated by the user, that instructs the software on how to name the experimental groups, how they are assigned to monitors, how to set desired time increments (e.g., 30 min, 12 h, etc.), and how to qualify a fly as "dead," key time points, such as time of infection or other treatments, and so on. This configuration file is a simple Notepad *.txt file with the word "config" in the file name and should be saved in the same folder as your DayLight and channel files. To make a "Config file" for Insomniac3 program, work from a template (included in the software package), and make changes to the file that are custom to your experiment. It

is highly recommended to follow instructions in the tutorial, "How to make a config file" (*see* **Note 4**).

Information provided in the config file should be as follows:

(a) The name of your experiment (use your unique experiment name as designated in step 1), along with an indicator, such as bin size. This tells the program how to name your output files.

(b) The names of your experimental groups – genotypes, drug condition, etc.

(c) How each of the groups were assigned to the monitors.

(d) Indicate important time points using the "TP" command. For example, if you forgot to indicate your desired "First bin to save" in the DAMFileScan program, you can correct that here by using the "TP ALT_START" command and indicate the start time for your data analysis. The manual that accompanies the software contains a complete list of commands that can be used.

(e) (Optional) Indicate an event, such as an infection or sleep deprivation. The "event" command will refer to a time point "TP" defined by the user and will calculate survival duration from this point forward. If this command is not used, the program will, by default, report survival from the start of the data recording.

(f) Select the desired analysis output files. There are a total of 15 files that will appear in the "descriptive" folder (see below). These files will show values in analysis intervals (any size, ranging from 1 min to daily bins or longer). Analysis files include total activity, total sleep time, sleep bout duration, number of sleep bouts, waking activity levels, etc.

8. Open the Insomniac3 program. This will appear as a dialog window on your desktop (Fig. 4.1).
9. Tell the program where to find your data files: Click the button that says "DayLight file" and navigate to your DayLight File and select. Next, click the "Config file" button, and the software will automatically navigate to the same folder. Select your configuration file (*see* **Note 5**). Now you are ready to run the program!
10. Click "check configuration" button. The program will "review" your config file and stop if there are any errors. If it freezes or doesn't run, scroll down inside the dialog box, which will show the last line that it read in your config file. It is on this line that you made a mistake. Correct the mistake and try again.

11. Next, click the "Do Death Scan" button. Next, click the "Do Sleep Scan" button (see Fig. 4.1). The program will finish the analysis within a few seconds. Close the program and navigate back to your data folder. There will be four output folders containing your analyzed files. Report files will either be *.txt or *.csv which can be opened into a spreadsheet program. Activity plot data and sleep data are available in each folder (Fig. 4.2) (*see* **Note 5**). The four output folders include:

Descriptive: This folder contains sleep analysis and other behavioral output files (selected in 7(f)). Each report file contains individual fly data and data averaged across each group minus the dead flies according to the assignments and algorithm specified by the user in the config file. At this time, the "sleep latency" file is faulty and should be ignored. This parameter should be turned off in the config file and can instead be analyzed using the "Anesthesia" module (see below).

Survival: This folder contains three files generated from each config file. One file (with "survival" in the filename) contains the number of hours survived by each fly from the time that you specified in the config file. The file named "survivalhisto" contains the percent flies for each group surviving each hour, and the "KaplanMeierInput" contains the number of hours survived for each fly in each group along with censure values, useful for exporting to a statistical analysis program.

Death: This folder contains one file for each monitor in the experiment and a "class death" file across all monitors. Each monitor file shows the activity counts of each fly across every hour in the experiment. It is sometimes useful to scroll through quickly to verify the time of the death for each fly. The "class death" file contains average activity counts for each group of flies for each hour in the assay, but only includes *living* flies at each time point. It also reports the number of flies ("n") remaining at each time point. This file is useful for seeing the activity patterns across the full range of living flies for the duration of the experiment.

Sleep: This folder also contains monitor files, and shows how every fly within that monitor was scored for sleep (designated "0") for every minute recorded during the experiment. Waking minutes are scored as "1." These files are useful for cross-checking the data to ensure the program is functioning accurately. We have also found them useful for evaluating the number of channel counts that occur while a fly is not moving (obtained from the MB5 or any multibeam monitor only). These data may reflect sleep quality of the fly (Fig. 4.2a).

12. Anesthesia Module

This module is useful for measuring arousal threshold responses and generates "kick" files in the "Descriptive folder" described above. Separate config files should be generated as described in detail in the tutorial. The precise time (date and time of day) and strength of the arousing stimulus (such as a light or vibratory pulse) are specified in the config file, and the analysis interval is set to "1 min." For measuring sleep latency to daily lights off, specify the exact time of lights off and each date if more than one night of sleep latency measurement is desired.

The "kick" report output file contains how each fly was scored in response to the stimulus, where "BUSY" indicates that the fly was already awake at the time of the stimulus, "SLEEP" indicates that the fly did not wake up, "WAKE1" indicates that the fly awakened within the first poststimulus period (usually the first minute), and "WAKE2" indicates the fly did not awaken until the second poststimulus period. Scrolling down below these scores are sleep latency results for each fly at each time point specified in the config file. Latency times are in minutes. Scores of "−0.01" are nonsense scores and indicate that the fly was already asleep at this time point and is not counted toward the average sleep latency in that group. Below the individual fly scores are data that summarize the percent flies awakened by the stimuli at each time point for each group, as well as average sleep latency values.

All output files generated by Insomniac3 can be further processed in an application of choice, such as Microsoft Excel, for additional quantitation, generation of charts (Fig. 4.2b), or statistical analyses.

4 Conclusion

The DAM system has provided us with the means to quantify the time-course of locomotor activity of many individual flies simultaneously from days to weeks. In combination with a plethora of genetic resources available in *Drosophila*, this method is an especially powerful high-throughput technique useful for "hunting" genes and neuronal components that have effects on amount, timing, and depth of sleep in *Drosophila*. To find animals which exhibit different sleep patterns, we can start from a large set of flies mutagenized by chemicals or transposon elements or a spontaneous event [2, 8, 32, 35, 39, 58, 59, 66]. We can also express protein-coding genes or noncoding RNAs (including RNAs artificially designed for the gene knockdown as well as naturally occurring regulatory RNAs) in a subset of cells using binary Gal4/UAS expression systems available in *Drosophila* [5, 21, 26, 28, 49, 60, 62, 67].

Although the simple beam-break method is useful for identifying key components of *Drosophila* behavior, a limitation of this system is that it lacks the high resolution that can be achieved with video tracking. However, video-recording of flies requires more disk space for data storage, has a more limited capacity for recording individual flies, and takes more time to analyze. Therefore, it is often easier for us to handle and analyze data generated from DAM system than video-tracking for measuring sleep in *Drosophila*. Moreover, the multibeam models indeed add higher resolution and to some extent address this single-beam limitation.

The DAM system can also be readily combined with opto- or thermo-genetics, which have been used to dissect neuronal mechanisms of many types of behaviors in Drosophila [29, 33, 40, 45]. These techniques allow us to activate/silence a specific subset of neurons, in which light-/temperature-sensitive neuronal stimulators/blockers are ectopically expressed. For these purposes, illumination with a specific wavelength and changes of temperature, respectively, is applied on a daily basis. These methods have been applied to sleep research using *Drosophila* and have been successful in identifying sleep mechanisms at the level of molecules or neurons [2, 5, 12, 29, 40].

Numerous software programs have been developed to analyze data collected on the DAMS. Insomniac3 has many advantages, particularly the detailed numeric output allows for huge flexibility in analyzing multiple aspects of sleep and locomotor activity. One current limitation is adjusting to a crude user interface. User errors are easily introduced but fortunately do not interfere with program output, performance, or accuracy. The multiple file outputs facilitate cross-checking with raw data files. Several aspects of this program also remain in development, particularly for accommodating large datasets generated by multibeam monitors. Currently, it is our preferred analysis software due to its ease of use; rapid analysis (with output files generated within seconds); the wide range of analysis output, including survival; and flexibility.

In conclusion, the DAM system to measure sleep in *Drosophila*, which employs the same beam-cutting principle used in the first report of the clock gene in *Drosophila* half a century ago, is still a powerful tool to call up untapped flies which carry important but unaddressed components for sleep.

5 Notes

Note 1: The cables connecting the splitters to the interface unit or other splitters have to be short as possible because they transmit analog signals, which are sensitive to noise and easily attenuate depending on distance. Do not use a 3-m-long extension cable and a coupler, which are supplied with the interface unit. If you need to extend cable length between the monitors and the data collection PC, you must extend the USB cable between the interface unit and the PC, instead of the phone cable between the monitors and the interface unit (see the next note).

Note 2: You can extend the USB cable between the interface unit and the data collection PC, because it transmits digitized data, which are noise-resistant.

Note 3: Although monitors are connected immediately after loading flies, in most cases it is necessary to allow an adjustment period. For setting the start time in "Filescan," we typically begin the data analysis period from the first lights-on, the next day after the monitors are plugged in. Alternatively, the Inomniac3 program has a feature where you can set a start date and time alternative to that in Filescan (see step 7(d)), as long as it is within the recording window.

Note 4: The "config" file template contains several command lines that are still in development, but do not interfere with the performance of the program itself.

Note 5: multiple "config" files can be generated and saved within the same data folder. This is necessary if you want to try different analysis parameters, such as different analysis time increments (30 min, 12 h, etc.). The program will independently generate output files of a root name that you specify, so be sure to indicate unique output file names in each config file. Output files will be saved within the same output folders. For multibeam monitors "*Mv" data (for channel "moves" can be stored in the same folder as the "*.Ct" (channel counts) data.)

Acknowledgments

This research was supported by WPI, JSPS KAKENHI Grant Numbers JP20K21441 and 20H03291, Kanae foundation, Inamori foundation to H.T. and NIH #R01-GM123783 to J.W.

References

1. Abhilash L, Sheeba V (2019) RhythmicAlly: your R and shiny–based open-source ally for the analysis of biological rhythms. J Biol Rhythm 34:551–561
2. Afonso DJS, Liu D, Machado DR, Pan H, Jepson JEC, Rogulja D, Koh K (2015) TARANIS functions with cyclin a and Cdk1 in a novel arousal center to control sleep in drosophila. Curr Biol 25:1717–1726. https://doi.org/10.1016/j.cub.2015.05.037
3. van Alphen B, Semenza ER, Yap M, van Swinderen B, Allada R (2021) A deep sleep stage in Drosophila with a functional role in waste clearance. Sci Adv 7:2999. https://doi.org/10.1126/sciadv.abc2999
4. Catterson JH, Knowles-Barley S, James K, Heck MMS, Harmar AJ, Hartley PS (2010)

Dietary modulation of Drosophila sleep-wake behaviour. PLoS One 5:e12062. https://doi.org/10.1371/journal.pone.0012062
5. Cavanaugh DJ, Geratowski JD, Wooltorton JRA, Spaethling JM, Hector CE, Zheng X, Johnson EC, Eberwine JH, Sehgal A (2014) Identification of a circadian output circuit for rest: activity rhythms in Drosophila. Cell 157:689–701. https://doi.org/10.1016/j.cell.2014.02.024
6. Chiu JC, Low KH, Pike DH, Yildirim E, Edery I (2010) Assaying locomotor activity to study circadian rhythms and sleep parameters in Drosophila. J Vis Exp:2157. https://doi.org/10.3791/2157
7. Cichewicz K, Hirsh J (2018) ShinyR-DAM: a program analyzing drosophila activity, sleep and circadian rhythms. Commun Biol 1:25. https://doi.org/10.1038/s42003-018-0031-9
8. Cirelli C, Bushey D, Hill S, Huber R, Kreber R, Ganetzky B, Tononi G (2005) Reduced sleep in Drosophila shaker mutants. Nature 434:1087–1092. https://doi.org/10.1038/nature03486
9. Concetti C, Burdakov D (2021) Orexin/hypocretin and MCH neurons: cognitive and motor roles beyond arousal. Front Neurosci 15:639313
10. Curran JA, Buhl E, Tsaneva-Atanasova K, Hodge JJL (2019) Age-dependent changes in clock neuron structural plasticity and excitability are associated with a decrease in circadian output behavior and sleep. Neurobiol Aging 77:158–168. https://doi.org/10.1016/j.neurobiolaging.2019.01.025
11. Donelson N, Kim EZ, Slawson JB, Vecsey CG, Huber R, Griffith LC (2012) High-resolution positional tracking for long-term analysis of drosophila sleep and locomotion using the "tracker" program. PLoS One 7:e37250. https://doi.org/10.1371/journal.pone.0037250
12. Dubowy C, Moravcevic K, Yue Z, Wan JY, van Dongen HPA, Sehgal A (2016) Genetic dissociation of daily sleep and sleep following thermogenetic sleep deprivation in drosophila. Sleep 39:1083–1095. https://doi.org/10.5665/sleep.5760
13. Duhart JM, Baccini V, Zhang Y, Machado DR, Koh K (2020) Modulation of sleep-courtship balance by nutritional status in drosophila. eLife 9:1–23. https://doi.org/10.7554/eLife.60853
14. Garbe DS, Bollinger WL, Vigderman A, Masek P, Gertowski J, Sehgal A, Keene AC (2015) Context-specific comparison of sleep acquisition systems in Drosophila. Biol Open 4:1558–1568. https://doi.org/10.1242/bio.013011
15. Gardner B, Strus E, Meng QC, Coradetti T, Naidoo NN, Kelz MB, Williams JA (2016) Sleep homeostasis and general anesthesia: are fruit flies well rested after emergence from propofol? Anesthesiology 124:404–416. https://doi.org/10.1097/ALN.0000000000000939
16. Geissmann Q, Beckwith EJ, Gilestro GF (2019) Most sleep does not serve a vital function: evidence from Drosophila melanogaster. Sci Adv 5:9253. https://doi.org/10.1126/sciadv.aau9253
17. Geissmann Q, Rodriguez LG, Beckwith EJ, Gilestro GF (2019) Rethomics: an R framework to analyse high-throughput behavioural data. PLoS One 14:e0209331. https://doi.org/10.1371/journal.pone.0209331
18. Gilestro GF (2012) Video tracking and analysis of sleep in drosophila melanogaster. Nat Protoc 7:995–1007. https://doi.org/10.1038/nprot.2012.041
19. Gilestro GF, Cirelli C (2009) PySolo: a complete suite for sleep analysis in drosophila. Bioinformatics 25:1466–1467. https://doi.org/10.1093/bioinformatics/btp237
20. Gizowski C, Bourque CW (2020) Sodium regulates clock time and output via an excitatory GABAergic pathway. Nature 583:421–424. https://doi.org/10.1038/s41586-020-2471-x
21. Goodwin PR, Meng A, Moore J, Hobin M, Fulga TA, van Vactor D, Griffith LC (2018) MicroRNAs regulate sleep and sleep homeostasis in Drosophila. Cell Rep 23:3776–3786. https://doi.org/10.1016/j.celrep.2018.05.078
22. Guo F, Yu J, Jung HJ, Abruzzi KC, Luo W, Griffith LC, Rosbash M (2016) Circadian neuron feedback controls the Drosophila sleep-activity profile. Nature 536:292–297. https://doi.org/10.1038/nature19097
23. Heigwer F, Port F, Boutros M (2018) Rna interference (RNAi) screening in Drosophila. Genetics 208:853–874. https://doi.org/10.1534/genetics.117.300077
24. Hendricks JC, Finn SM, Panckeri KA, Chavkin J, Williams JA, Sehgal A, Pack AI (2000) Rest in Drosophila is a sleep-like state. Neuron 25:129–138. https://doi.org/10.1016/S0896-6273(00)80877-6
25. Huber R, Hill SL, Holladay C, Biesiadecki M, Tononi G, Cirelli C (2004) Sleep homeostasis in Drosophila melanogaster. Sleep 27:628–639. https://doi.org/10.1093/sleep/27.4.628

26. Itoh TQ, Matsumoto A (2012) Genome-wide RNA interference screening for the clock-related gene of ATP-binding cassette transporters in Drosophila melanogaster (Diptera: Drosophilidae). Appl Entomol Zool 47:79–86. https://doi.org/10.1007/s13355-012-0091-0
27. Jaggard JB, Lloyd E, Lopatto A, Duboue ER, Keene AC (2019) Automated measurements of sleep and locomotor activity in Mexican cavefish. J Vis Exp:e59198. https://doi.org/10.3791/59198
28. Jin X, Tian Y, Zhang ZC, Gu P, Liu C, Han J (2021) A subset of DN1p neurons integrates thermosensory inputs to promote wakefulness via CNMa signaling. Curr Biol. https://doi.org/10.1016/j.cub.2021.02.048
29. Jones JR, Tackenberg MC, McMahon DG (2021) Optogenetic methods for the study of circadian rhythms. Methods Mol Biol 2130: 325–336. https://doi.org/10.1007/978-1-0716-0381-9_24
30. Keene AC, Duboue ER (2018) The origins and evolution of sleep. J Exp Biol 221
31. Klarsfeld A, Leloup JC, Rouyer F (2003) Circadian rhythms of locomotor activity in Drosophila. In: Behavioural Processes. Elsevier, Amsterdam, pp 161–175
32. Koh K, Joiner WJ, Wu MN, Yue Z, Smith CJ, Sehgal A (2008) Identification of SLEEPLESS, a sleep-promoting factor. Science 321: 372–376. https://doi.org/10.1126/science.1155942
33. Kohsaka H, Nose A (2021) Optogenetics in Drosophila. In: Advances in experimental medicine and biology. Springer, Cham, pp 309–320
34. Konopka RJ, Benzer S (1971) Clock mutants of Drosophila melanogaster. Proc Natl Acad Sci U S A 68:2112–2116. https://doi.org/10.1073/pnas.68.9.2112
35. Kume K, Kume S, Park SK, Hirsh J, Jackson FR (2005) Dopamine is a regulator of arousal in the fruit fly. J Neurosci 25:7377–7384. https://doi.org/10.1523/JNEUROSCI.2048-05.2005
36. Kuo TH, Pike DH, Beizaeipour Z, Williams JA (2010) Sleep triggered by an immune response in Drosophila is regulated by the circadian clock and requires the NFκB relish. BMC Neurosci 11:1–12. https://doi.org/10.1186/1471-2202-11-17
37. Qiao B, Li C, Allen VW, Shirasu-Hiza M, Syed S (2018) Automated analysis of long-term grooming behavior in Drosophila using a k-nearest neighbors classifier. eLife 7: e34497. https://doi.org/10.7554/eLife.34497
38. Lin FJ, Pierce MM, Sehgal A, Wu T, Skipper DC, Chabba R (2010) Effect of taurine and caffeine on sleep-wake activity in Drosophila melanogaster. Nat Sci Sleep 2:221–231. https://doi.org/10.2147/NSS.S13034
39. Liu S, Lamaze A, Liu Q, Tabuchi M, Yang Y, Fowler M, Bharadwaj R, Zhang J, Bedont J, Blackshaw S, Lloyd TE, Montell C, Sehgal A, Koh K, Wu MN (2014) WIDE AWAKE mediates the circadian timing of sleep onset. Neuron 82:151–166. https://doi.org/10.1016/j.neuron.2014.01.040
40. Men J, Li A, Jerwick J, Li Z, Tanzi RE, Zhou C (2020) Non-invasive red-light optogenetic control of Drosophila cardiac function. Commun Biol 3:1–10. https://doi.org/10.1038/s42003-020-1065-3
41. Montell C (2021) Drosophila sensory receptors – a set of molecular Swiss army knives. Genetics 217:1–34. https://doi.org/10.1093/GENETICS/IYAA011
42. Naidoo N, Casiano V, Cater J, Zimmerman J, Pack AI (2007) A role for the molecular chaperone protein BiP/GRP78 in Drosophila sleep homeostasis. Sleep 30:557–565. https://doi.org/10.1093/sleep/30.5.557
43. Nevo E, Rashkovetsky E, Pavlicek T, Korol A (1998) A complex adaptive syndrome in Drosophila caused by microclimatic contrasts. Heredity 80:9–16. https://doi.org/10.1046/j.1365-2540.1998.00274.x
44. de Nobrega AK, Lyons LC (2020) Aging and the clock: perspective from flies to humans. Eur J Neurosci 51:454–481
45. Owald D, Lin S, Waddell S (2015) Light, heat, action: neural control of fruit fly behaviour. Philos Trans R Soc B Biol Sci 370:20140211
46. Parisky KM, Agosto Rivera JL, Donelson NC, Kotecha S, Griffith LC (2016) Reorganization of sleep by temperature in Drosophila requires light, the homeostat, and the circadian clock. Curr Biol 26:882–892. https://doi.org/10.1016/j.cub.2016.02.011
47. Pfeiffenberger C, Allada R (2012) Cul3 and the BTB adaptor insomniac are key regulators of sleep homeostasis and a dopamine arousal pathway in drosophila. PLoS Genet 8: e1003003. https://doi.org/10.1371/journal.pgen.1003003
48. Rechtschaffen A, Gilliland MA, Bergmann BM, Winter JB (1983) Physiological correlates of prolonged sleep deprivation in rats. Science 221:182–184. https://doi.org/10.1126/science.6857280
49. Rogulja D, Young MW (2012) Control of sleep by cyclin a and its regulator. Science 335:1617–1621. https://doi.org/10.1126/science.1212476

50. Rosato E, Kyriacou CP (2006) Analysis of locomotor activity rhythms in Drosophila. Nat Protoc 1:559–568. https://doi.org/10.1038/nprot.2006.79
51. Ryder E, Blows F, Ashburner M, Bautista-Llacer R, Coulson D, Drummond J, Webster J, Gubb D, Gunton N, Johnson G, O'Kane CJ, Huen D, Sharma P, Asztalos Z, Baisch H, Schulze J, Kube M, Kittlaus K, Reuter G, Maroy P, Szidonya J, Rasmuson-Lestander Å, Ekström K, Dickson B, Hugentobler C, Stocker H, Hafen E, Lepesant JA, Pflugfelder G, Heisenberg M, Mechler B, Serras F, Corominas M, Schneuwly S, Preat T, Roote J, Russell S (2004) The DrosDel collection: a set of P-element insertions for generating custom chromosomal aberrations in Drosophila melanogaster. Genetics 167: 797–813. https://doi.org/10.1534/genetics.104.026658
52. Sayeed O, Benzer S (1996) Behavioral genetics of thermosensation and hygrosensation in Drosophila. Proc Natl Acad Sci U S A 93: 6079–6084. https://doi.org/10.1073/pnas.93.12.6079
53. Schmid B, Helfrich-Förster C, Yoshii T (2011) A new ImageJ plug-in "actogramJ" for chronobiological analyses. J Biol Rhythm 26: 464–467. https://doi.org/10.1177/0748730411414264
54. Seggio JA (2011) Utilizing Drosophila Activity Monitors (DAMs) in an undergraduate teaching and research setting. Dros Inf Serv 91: 170–173
55. Sehgal A, Price JL, Man B, Young MW (1994) Loss of circadian behavioral rhythms and per RNA oscillations in the Drosophila mutant timeless. Science 263:1603–1606. https://doi.org/10.1126/science.8128246
56. Shafer OT, Keene AC (2021) The regulation of Drosophila sleep. Curr Biol 31:R38–R49
57. Shaw PJ, Tortoni G, Greenspan RJ, Robinson DF (2002) Stress response genes protect against lethal effects of sleep deprivation in Drosophila. Nature 417:287–291. https://doi.org/10.1038/417287a
58. Shi M, Yue Z, Kuryatov A, Lindstrom JM, Sehgal A (2014) Identification of Redeye, a new sleep regulating protein whose expression is modulated by sleep amount. eLife 2014: 1473. https://doi.org/10.7554/eLife.01473
59. Stavropoulos N, Young MW (2011) Insomniac and cullin-3 regulate sleep and wakefulness in drosophila. Neuron 72:964–976. https://doi.org/10.1016/j.neuron.2011.12.003
60. Stoleru D, Peng Y, Agosto J, Rosbash M (2004) Coupled oscillators control morning and evening locomotor behaviour of Drosophila. Nature 431:862–868. https://doi.org/10.1038/nature02926
61. Suri V, Qian Z, Hall JC, Rosbash M (1998) Evidence that the TIM light response is relevant to light-induced phase shifts in Drosophila melanogaster. Neuron 21:225–234. https://doi.org/10.1016/S0896-6273(00)80529-2
62. Toda H, Williams JA, Gulledge M, Sehgal A (2019) A sleep-inducing gene, nemuri, links sleep and immune function in Drosophila. Science 363:509–515. https://doi.org/10.1126/science.aat1650
63. Wang Y-Y, Ma W-W, Peng I-F (2020) Screening of sleep assisting drug candidates with a Drosophila model. PLoS One 15:e0236318. https://doi.org/10.1371/journal.pone.0236318
64. Weber F, Dan Y (2016) Circuit-based interrogation of sleep control. Nature 538:51–59
65. Williams JA, Su HS, Bernards A, Field J, Sehgal A (2001) A circadian output in Drosophila mediated by neurofibromatosis-1 and Ras/-MAPK. Science 293:2251–2256. https://doi.org/10.1126/science.1063097
66. Wu MN, Koh K, Yue Z, Joiner WJ, Sehgal A (2008) A genetic screen for sleep and circadian mutants reveals mechanisms underlying regulation of sleep in Drosophila. Sleep 31: 465–472. https://doi.org/10.1093/sleep/31.4.465
67. Wu Y, Cao G, Nitabach MN (2008) Electrical silencing of PDF neurons advances the phase of non-PDF clock neurons in Drosophila. J Biol Rhythm 23:117–128. https://doi.org/10.1177/0748730407312984
68. Zimmerman JE, Chan MT, Jackson N, Maislin G, Pack AI (2012) Genetic background has a major impact on differences in sleep resulting from environmental influences in Drosophila. Sleep 35:545–557. https://doi.org/10.5665/sleep.1744
69. Research Tools | Vecsey Lab. https://academics.skidmore.edu/blogs/cvecsey/?page_id=57. Accessed 12 May 2021
70. ClockLab | Actimetrics. https://actimetrics.com/products/clocklab/. Accessed 12 May 2021

Chapter 5

Gut Microbes and *Drosophila* Behavior

Hibiki Kadoguchi, Aki Hori, and Takayuki Kuraishi

Abstract

Gut microbiota plays an essential role in gut–brain interactions. From insect to mammals, studies using germ-free animals have demonstrated that indigenous host microbes influence emotional activities, stress, and sexual behaviors. In addition, microbes appear to govern human mental disorders such as autism and depression. Bacterial infections of the gut also influence host behaviors. For example, food poisoning causes the cessation of food intake behavior. Interestingly, intestinal parasites facilitate their transmission by manipulating host behaviors. However, the underlying mechanisms of how gut microbes modulate brain functions remain to be elucidated. *Drosophila melanogaster* is an effective animal model in neurobiology, particularly the genetic dissection of instinctive behaviors. Here, we provide detailed methods of working with gut bacteria—endogenous as well as infectious—for *Drosophila* neurobiologists who are interested in microbiology.

Key words *Drosophila*, Gut microbes, Gut–brain interactions, Germ-free fly, Oral infection

1 Introduction

1.1 Gut–Brain Interactions and Indigenous Microbes

Essential animal behaviors include eating foods to sustain bodily activities and finding partners to reproduce. During feeding, valuable nutrients, as well as noxious materials, are taken up from the mouth and perceived in the intestine. The intestine is equipped with mechanisms recognizing the external signals; redirecting the information to other parts of the host's body, including the brain; and modulating the host's behaviors to adjust to the environment. The network of the intestine and brain is called "the gut–brain interactions," or the gut/brain axis; its importance has been known since ancient times. In the nineteenth century, clinicians thought that intestinal disorders evoked autotoxins that triggered the damage of the nervous system, thereby causing psychiatric diseases [1, 2]. Likewise, experience tells that anxiety or stress could trigger stomachache or diarrhea, an example of signaling from the brain to the gut. Conversely, from the gut to the brain, harmful stimuli in the gut, such as eating rotten meat, will cause food intake cessation.

Daisuke Yamamoto (ed.), *Behavioral Neurogenetics*, Neuromethods, vol. 181, https://doi.org/10.1007/978-1-0716-2321-3_5,

Mechanistically, gut–brain interactions can be mediated through bidirectional neuronal and hormonal circuits between the two organs. The gut has its nervous system, the enteric nervous system, which can function independently from other nervous systems. The enteric nervous system function is also modulated by the central, autonomic sympathetic, and parasympathetic nervous systems [3]. For example, Tsurugizawa et al. [4] showed that the intragastric administration of L-glutamate activated a specific region in the forebrain; such activation was abolished by removing the vagus nerve. Similarly, the gut and the brain communicate with each other via several hormones. From the brain, mainly from the hypothalamus and hypophysis in mammals, several hormones are released, such as corticotropin-releasing hormone involved in irritable bowel syndrome [5, 6].

From the gut, many intestinal hormones are released to control the body systemically. Incretins stimulate the release of insulin from the pancreas and slow the rate of nutrient absorption into the bloodstream by reducing gastric emptying, likely reducing food intake directly [7]. Intriguingly, some of the intestinal hormones are peptide hormones that are categorized as neuropeptides, e.g., somatostatin and tachykinins, being expressed and functioning in the nervous system; this phenomenon is phylogenetically conserved from insects to mammals [8, 9].

The concept of gut–brain interactions has been revisited at the initiation of the research on intestinal microbiota. Approximately 1000 bacterial species have been found to reside in the human intestine, and 10^{14} individual bacterial cells of at least 160 different species inhabit each individual's intestine [10–12]. Recently, there is much literature suggesting that intestinal microbiota plays a significant role not only in the nutritional and immune regulation of the host [13–15] but also in many aspects of host behavior, such as feeding, anxiety, and autism [16–18]. Sudo et al. [19] showed that the level of corticosterone, a stress hormone, in response to restraint stress is higher in germ-free mice. Furthermore, germ-free mice exhibited a reduced level of a brain-derived neurotrophic factor in the cortex and hippocampus compared with normal mice. The increased hypothalamic–pituitary–adrenal stress response of germ-free mice is reversed by gut microbial reconstitution with *Bifidobacterium infantis*, a commensal microbe, suggesting that gut microbiota can affect the gut/brain axis in mice. Now, an interaction between the intestinal microbiota and host behaviors is considered "the microbiome–gut–brain axis."

The systemic effect of gut microbiota is exerted directly or indirectly. Gut microbes produce bioactive metabolites such as short-chain fatty acids and polyamines [20, 21]. These molecules enter the bloodstream and affect remote organs. Conversely, gut microbes and enteric bacterial infections also impact animal behaviors through the gut nervous and endocrine systems [22]. For example, the anxiety- and depression-related behaviors of mice are suppressed by the feeding of lactic acid bacteria, in a manner dependent on the vagus nerve [23]. Enterochromaffin cells, a type of enteroendocrine cells in the gut, produce serotonin that acts as a neurotransmitter for local enteric neurons and exert systemic effects [24, 25]. Sugisawa et al. [26] demonstrated that single-stranded RNA from gut microbiota triggers serotonin production by the enterochromaffin cells via the Ca^{2+} channel Piezo1, revealing a mechanism of the microbiome–enteroendocrine–nerve interactions. It could be said that the molecular mechanisms underlying the effects of gut microbiota on host behaviors have been started to be demystified.

1.2 Gut-Associated and Gut-Invaded Microbes and Host Behaviors in Drosophila melanogaster

Drosophila melanogaster, a fruit fly, is a genetically tractable model animal used to study behaviors and their underlying neuronal mechanisms. Besides, a fruit fly is a valuable model to investigate the effect of gut microbiota on gut–brain interactions. Indeed, flies possess an enteric nervous system, enteroendocrine cells, and indigenous gut microbes that can control fly behaviors [27]. Several neurons, most of which are motor neurons controlling the movement of visceral intestinal muscles, are innervated in the gut of larvae and adults [28]. There are also sensory innervations in the gut, at least at the posterior part of the hindgut. Enteric neurons have been demonstrated to control the barrier function of the gut that is important for the defense against bacteria invasion [29]. In addition, gut-innervating neurons are involved in food intake during reproduction [30]. In addition to using neurons, the fly intestine can communicate with other organs through the endocrine system. The fly gut secretes several peptide hormones produced by the enteroendocrine cells [31, 32]. These hormones can have long-range systemic effects. Moreover, the *Drosophila* gut contains indigenous microbes of relatively low diversity, at approximately 1–30 species, and in a small number of cells, at approximately 10^3–10^5 per fly [33, 34]. These three players have certain homology in terms of possible connections with animal behaviors. Thus, flies can be used for investigating the fundamental mechanisms underlying gut microbiota's control of the gut/brain axis.

It is necessary to generate germ-free animals to study the roles of intestinal microbiota [35, 36]. One can compare any given phenotype in the conventionally reared animals with that in the germ-free animals, and if a difference is observed, the responsible bacteria can be identified by reintroducing them into germ-free animals to make gnotobiotic animals. Germ-free flies are believed to be easily generated by treating embryos with bleach [37, 38]. Alternatively, gut microbes can be significantly reduced using antibiotics. Also, the gut-associated microbes in the fly play a crucial role under a poor nutrient condition; for example, the development of the germ-free larvae on a low-yeast diet is significantly delayed [39, 40]. Last, the gut microbes are an important food source as well as provide amino acids, cholesterol, and vitamins under certain conditions [27].

In addition to being nutritional modifiers, the gut microbiota is involved in fly behaviors. Sharon et al. [41] demonstrated that adult flies exhibited homogametic mating preference. Flies prefer to select partners eating the same foods; this behavior is associated with the endogenous microbial communities. The odors of microbes may be responsible for this phenomenon. Although there are studies supporting this finding [42, 43], there are also contradictory observations [44]. The odors of the gut microbes also govern other host behaviors, such as an appetite for foods and egg-laying locations. Volatile compounds from a specific microbe and their concerted production by multiple microbes regulate feeding and egg-laying behaviors [45–49]. The taste of amino acids from commensal microbes may also influence food choice behaviors [50]. In addition to odor-related traits, gut microbes are responsible for locomotor behaviors. For example, Schretter et al. [51] reported that germ-free flies walked faster and had an increase in daily activity, which is recovered by mono-association with a commensal microbe *Lactobacillus brevis*. However, there are also conflicting reports [52, 53]. Last, gut-associated microbes are also involved in social behavior in a fly autism model [54].

The enteric infection of pathogenic bacteria also affects host behaviors. In rodents, anxiety-like behavior is induced by the oral infection with *Campylobacter jejuni*, one of the bacteria causing food poisoning [55]. Other enteric infections also induce behavioral change mediated by peripheral sensory neurons [56]. Parasite infections have similar effects. Toxoplasma infection of rodents specifically blocks the aversion behavior for the smell of cat urine without affecting olfaction and the general anxiety-like behaviors [57, 58]. In flies, oral infection is related to eating behaviors, particularly the cessation of feeding. *Pseudomonas entomophila*, an entomopathogen utilized as a model bacterium that infects the fly gut [59–63], is known to be involved in this phenomenon. Oral infection with *P. entomophila* causes a cessation of food uptake in larvae [64]. However, the underlying neuronal mechanism requires clarification.

In summary, commensal and infectious microbes are essential factors for host behavioral characteristics. Here, we provide protocols on dealing with indigenous and infectious bacteria in *Drosophila melanogaster* for neurobiology researchers who are interested in handling microbes.

2 Materials, Methods, and Notes

2.1 Reduction of Indigenous Microbes Using Antibiotic Treatment

The first step of investigating the effect of gut-associated microbes on fly behavior is to remove the microbes from the host flies and compare them with those of the conventionally reared flies. Microbe-free flies, or "semi-germ-free flies," because they are not entirely germ-free, can be generated quickly by treating the flies with antibiotics. An antibiotic cocktail is added to the typical fly food to rear flies [65, 66].

Materials

1. 10 × antibiotics cocktail

• Ampicillin sodium salt (final. 100 μg/mL)	1.0 g
• Vancomycin hydrochloride (final. 50 μg/mL)	0.5 g
• Neomycin sulfate (final. 100 μg/mL)	1.0 g
• Metronidazole (final. 100 μg/mL)	1.0 g
• Tetracycline hydrochloride (final. 50 μg/mL)	0.5 g

Up to 1000 mL with Milli-Q water.

Methods

1. Add the 10 × antibiotic cocktails to fly food while fly food is being made and before it solidifies or to the fly food after it is melted in the microwave oven to achieve the final concentration of 1×.
2. Rear the flies with this food for three generations.
3. Use the flies as semi-germ-free flies.

2.2 Generation of Germ-Free Flies via Dechlorination

One can apply the hypochlorite treatment to embryos and raise the flies with autoclaved fly food to obtain flies that are more germ-free than those treated with antibiotics [67].

Materials

1. Apple agar plates
   - Solution 1

– 100% apple juice (Dole)	250 mL
– D-glucose (Nacalai Tesque)	25 g

(i) Warm solution one to approximately 50 °C and stir well.

- Solution 2

– Milli-Q water	250 mL
– Agar (Nacalai Tesque)	6.25 g

(ii) Autoclave solutions 1 and 2 at 121 °C for 20 min.

(iii) Mix the two solutions.

(iv) Pour the mixed solution onto Petri dishes.

2. Yeast pellet
   - Dry yeast (Oriental Yeast) 11 g
   - Milli-Q water 13 mL

(i) Vortex well and keep at 4 °C to ferment overnight.

3. The yeast pellet is placed onto the apple agar plate before use.
4. Normal fly food is autoclaved for 121 °C for 20 min and used as sterile food.

Methods

1. Put many flies into an Embryo Collection Cage (Flystuff, USA) with the apple agar plate.
2. Let the flies lay eggs for 6 h at 25 °C or 12 h at 18 °C.
3. Collect the embryos with a brush to a Cell Strainer (BD, USA).
4. In a biosafety cabinet and with gloves, soak the collected embryos in 70% ethanol for 1 min, 2.5% sodium hypochlorite (Wako) for 2 min, and 70% ethanol for 2 min.
5. Wash the treated embryos extensively with Otsuka distilled water (Otsuka).
6. Place the embryos into tubes with sterile food, and raise them.

Notes

1. A fresh hypochlorite solution should be used.
2. The developmental stage of the embryos needs to be carefully monitored; the embryos should not be too young, but keep the number of larvae to a minimum.

2.3 Endosymbiont Wolbachia

Wolbachia pipientis is a symbiont bacterium that resides inside the cells of *Drosophila*. It manipulates several, particularly sex-related, host traits [68, 69]. Besides, *Wolbachia* is reported to affect gut microbiota [70, 71]. Thus, it is recommended that the bacterium be removed before dechorination when performing behavioral analyses of the flies [72].

Methods

1. Prepare the 10× antibiotic cocktail as described earlier (or a 10× tetracycline hydrochloride solution).
2. Add the 10× antibiotic cocktail to the fly food it is being made before it solidifies or to the fly food after it is melted in a microwave oven to achieve the final concentration of 1×.
3. Rear the flies with this food for at least three generations.
4. Keep the flies without antibiotics for at least six generations before for use.
5. The absence of *Wolbachia* can be confirmed using DAPI staining of the flies' ovaries (Fig. 5.1).

2.4 Confirmation of the Absence of Germs with the Colony-Forming Unit (CFU) Assay

After germ-free flies or larvae have been generated and maintained for experiments, it is recommended to confirm the absence of germs in them. The presence of bacteria can be checked by culturing bacteria on agar plates or liquid medium or detecting bacteria-specific DNA using conventional, quantitative, or perhaps digital PCR [73, 74]. The culture method with agar plates of appropriate media may be the most sensitive way to check for gut microbes in flies.

Each bacterial species has its nutritional requirement and thus should be cultured in an optimized media. Also, the presence of

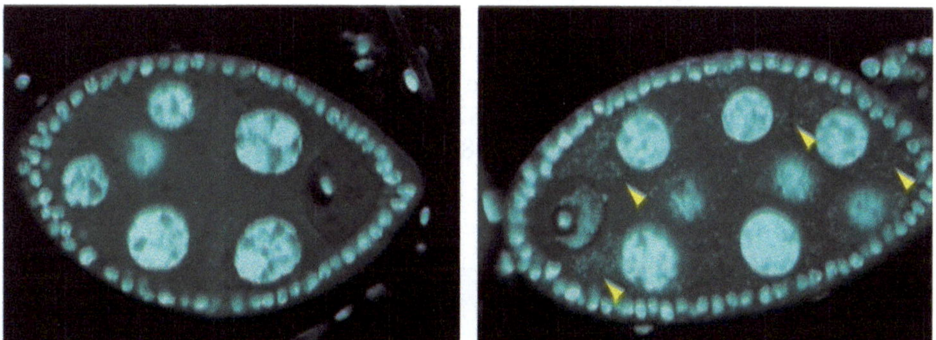

Fig. 5.1 DAPI staining of ovary that shows *Wolbachia* infection. Left, without *Wolbachia*; right, with *Wolbachia*. White arrowheads indicate many nuclei of *Wolbachia*

oxygen affects bacterial growth. Some *Lactobacillus* species found in the fly gut [75] prefer less oxygen even though they can grow under normal atmospheric conditions. Therefore, one may use microaerophilic conditions to culture them or confirm the absence of germs in the flies. Though it is considered that flies do not possess obligate anaerobic bacteria that can grow in the oxygen concentration of 0.5%, one might check for them if one can use an anaerobic chamber.

Materials

1. Potato dextrose broth for checking fungi and yeast.

• BD™ Difco™ Dehydrated Culture Media: Potato Dextrose Broth (BD 254920)	5 g
• Milli-Q water	200 mL

   (i) Dissolve the media, and autoclave at 121 °C for 15 min.

   (ii) Make agar plates by adding 1.5 g of agar to the broth, autoclaving, and pouring into Petri dishes.

2. Brain Heart Infusion broth for checking bacteria in general.

• BD™ Bacto™ Brain Heart Infusion (BD4128960)	4.8 g
• Milli-Q water	200 mL

   (i) Dissolve the media, and autoclave at 121 °C for 15 min.

   (ii) When making an agar plate, add 1.5 g of agar to the broth before autoclave and pour into Petri dishes.

3. MRS broth for *Lactobacillus* and *Acetobacter* species.

• Difco Lactobacilli MRS Broth (BD288130)	14 g
• Milli-Q water	200 mL

   (i) Dissolve the media, and autoclave at 121 °C for 15 min.

   (ii) When making an agar plate, use Difco Lactobacilli MRS Agar (BD288210) instead of the Difco Lactobacilli MRS Broth (BD288130), and pour the mixture into Petri dishes.

4. Cooked meat medium for anaerobic bacteria.

• Bacto™ Agar	14 g
• Milli-Q water	200 mL

   (i) Dissolve the Bacto Agar in Milli-Q water.

   (ii) Boil the solution for 20 min.

(iii) Dispense to tubes with calcium carbonate.
(iv) Boil again for 20 min while stirring with a pestle to remove the bubbles.
(v) Put the tubes into an anaerobic chamber.
(vi) Replace the air in the anaerobic chamber with N_2 and plug the tubes.
(vii) Take out the tubes from the chamber, and autoclave them at 121 °C for 15 min.

5. Thioglycolate medium for anaerobic bacteria.

• Thioglycolate medium	6 g
• Milli-Q water	200 mL

(i) Dissolve the thioglycolate medium in Milli-Q water.
(ii) Boil the solution for 20 min.
(iii) Dispense to tubes with calcium carbonate.
(iv) Boil again for 20 min while stirring with a pestle to remove the bubbles.
(v) Put them into an anaerobic chamber.
(vi) Replace the air in the anaerobic chamber with N_2 and plug the tubes.
(vii) Take out the tubes from the chamber, and autoclave them at 121 °C for 15 min.

Methods

1. Collect 30 larvae or adults in a safety cabinet.
2. Wash them with 70% ethanol and then with Milli-Q water.
3. Put them into 2-mL tubes (SARSTEDT 72.693) with 500 μL of PBS and 100 μL of 0.1-mm glass beads (Bertin Technologies).
4. Homogenize the larvae or adults with Precellys 24 (Bertin Technologies).
5. Make serial dilutions of the mixture with sterile PBS, and spread the solutions onto appropriate agar plates.
6. Incubate the plates at 29 °C for several days, and count the number of colonies.
7. When culturing the *Lactobacillus* species, use an anaerobic chamber or an Anaerobic Bacterial Culture Kit (Mitsubishi Gas Chemical) for microaerophilic conditions.

Notes: Caveat for Using Germ-Free Flies

Generating a germ-free animal has a long history [76], and there are well-established methods to create germ-free mammals, such as mice, rats, and pigs [77–79]. Because a safety cabinet is not suitable for keeping animals germ-free in the long term, an apparatus called an isolator (also called a vinyl isolator because it is generally made of vinyl) can be used to keep any animal germ-free [80]. Germ-free flies may be easily generated by applying the hypochlorite treatment to animals as described; however, the absence of germs may not be maintained during the entire life span of the flies and for the next generations. Notably, there have been multiple contradictory observations of the phenotypes of germ-free adults [27, 81], including those on the life span or behaviors of germ-free flies [53, 82]. Therefore, special care should be taken for flipping the flies to new vials. Also, the germ-free status of the flies should be checked and confirmed with an appropriate method.

2.5 DNA Extraction from Gut Microbes

If one discovers a behavioral phenotype of germ-free flies, one may want to identify the bacterium responsible for the phenotype. Thus, one needs to isolate gut microbes from fly stocks. Before isolating gut bacteria, one could obtain the catalog of gut bacteria of one's flies. Analyzing bacterial composition is achieved by reading total gut bacterial DNA with a next-generation sequencer, called metagenomic analysis [83].

One can examine the composition of gut microbes using mammalian model animals, such as mice, and extract bacterial DNA from the feces for metagenomic analysis [84]. However, the feces of *Drosophila* are so tiny that one needs to use the whole gut to extract bacterial DNA. Bear in mind that a large amount of host genomic DNA is included with the gut bacterial DNA in this method. Unless one performs metagenomic analysis (16S sequencing) regularly, the outsourcing of sequencing analysis is recommended because of budgetary concerns.

Materials

- Filtrated TE10 (10 mM Tris–HCl, 10 mM EDTA, pH 8.0), sterilize with a 0.22-μm filter
- Lysozyme (Sigma)
- 1000 units/mL Achromopeptidase (Wako)
- 20 units/mL Proteinase K (Nacalai)

Methods

1. Dissect the gut from approximately 20 larvae or adult flies.
2. Put the guts into 2-mL tubes (SARSTEDT 72.693) with 250 μL of filtrated TE10 and 10 zirconia beads (Bertin Technologies).
3. Homogenize with Precellys-24 at 6000 rpm for 1 min, one cycle.
4. Collect the homogenate, and add 250 μL of the filtrated TE10 and 50 μL of 0.1-mm glass beads (Bertin Technologies).
5. Homogenize again under the same condition as step 3.
6. Add 7.5 mg of lysozyme to achieve the final concentration of 15 mg/mL.
7. Incubate at 37 °C for 1 h.
8. Add 10 μL of 1000 units/mL Achromopeptidase to achieve the final concentration of 20 units/mL).
9. Incubate at 37 °C for 30 min.
10. Add 56 μL of 10% SDS (pH 7.2) to achieve the final concentration of 1% and mix well.
11. Add 30 μL of 20 mg/mL Proteinase K to achieve the final concentration of 1 mg/mL.
12. Incubate at 55 °C for 1 h.
13. Add 70 μL of 8 M potassium acetate.
14. Incubate on ice for 30 min.
15. Centrifuge at 15,000 rpm and 4 °C for 15 min.
16. Collect the supernatant, add 500 μL of phenol/chloroform/isoamyl alcohol (25:24:1), and mix well.
17. Centrifuge at 15,000 rpm and 4 °C for 1 min.
18. Collect the supernatant, add 500 μL of phenol/chloroform/isoamyl alcohol (25:24:1), and mix well.
19. Centrifuge at 15,000 rpm and 4 °C for 1 min.
20. Collect the supernatant, add 400 μL of 2-propanol, and mix well.
21. Centrifuge at 15,000 rpm and 4 °C for 5 min.
22. Discard the supernatant, and add 500 μL of 70% ethanol to wash the pellet.
23. Centrifuge at 15,000 rpm and 4 °C for 5 min.
24. Discard the supernatant, dry the pellet, and resuspend it with 20 μL of filtrated TE10.

2.6 Isolation of Gut Microbes and Identification of Bacterial Species Using PCR

Once the list of gut microbes in one's flies is obtained, specific bacteria can be isolated by culturing them from the gut homogenate based on the gut microbial community's information. Then one can generate flies mono-associated with any given gut microbe. Generally, *Lactobacillus* and *Acetobacter* species can be grown in MRS medium. Note that several specific media for *Acetobacter* species are available. After selecting colonies, a bacterial species can be identified through the PCR amplification of its 16S rRNA gene followed by Sanger sequencing. Here, we describe the procedure of our method.

Methods

1. Obtain the colonies by performing the CFU assay.
2. Select as many colonies as one wants.
3. For colony PCR, add 20 μL of Milli-Q water into each well of a sterilized 96-well plate.
4. Pick a colony with a toothpick, and dissolve it in Milli-Q water for colony PCR.
5. Simultaneously, inoculate a colony in an appropriate media, and make a glycerol stock from the culture.
6. To identify a bacterial species, amplify its 16S rRNA gene region using colony PCR as follows.
   (i) Mix the following reagents on ice.

• Template (dissolved colony in Milli-Q water)	1.5 μL
• 2× Buffer	10 μL
• dNTP	4 μL
• 27F-Primer (5′-GAGTTTGATCCTGGCTCAG-3′)	0.6 μL
• 1492-R Primer (5′-ACCTTGTTACGACTT-3′)	0.6 μL
• KOD Fx neo (Toyobo)	0.4 μL
• Milli-Q water	2.9 μL

   (ii) Run the PCR (94 °C/120 s, [98 °C/10 s, 57 °C/30 s, 68 °C/90 s] × 35 cycles, 68 °C/90 s, 4 °C)
7. Check for specific amplification products with the target size of 1.5–2.0 bp with 1.2% agarose gel electrophoresis.
8. Purify the PCR products using AM Pure XP (NIPPON Genetics) by adding 30 μL of AM Pure XP to 20 μL of the PCR product.
9. Mix by pipetting ten times, and incubate the mixture at 25 °C for 5 min.

10. Set the 96-well plate onto a 96 Super Ring Magnet Plate (ALPAQUA).
11. Incubate for 2 min at room temperature.
12. Remove the supernatant.
13. Add 200 μL of 70% ethanol, incubate for 30 s, and remove the ethanol solution.
14. Repeat step 13.
15. Remove the 96-well plates from the 96 Super Ring Magnet Plate.
16. Add 80 μL of Milli-Q water, and mix by pipetting ten times.
17. Set the 96-well plates onto the 96 Super Ring Magnet Plate again.
18. Incubate at 25 °C for 1 min.
19. Collect the supernatant, which contains the purified DNA.
20. Check the DNA concentration using a Nanodrop (Thermo Scientific) spectrophotometer; the concentration is likely 20–40 ng/μL.
21. Perform Sanger sequencing on the purified DNA with the 27F-Primer.
22. Analyze the sequences with the Molecular Evolutionary Genetics Analysis software to identify the bacterial species [85, 86].

2.7 Oral Infection

P. entomophila is a Gram-negative bacterium isolated from *Drosophila* in the wild and used for oral infection experiments [59]. Its infection causes severe gut damage, food uptake cessation [64], and high mortality. The molecular mechanisms underlying the change in the feeding behavior of the infected flies remain unclear.

P. entomophila frequently undergoes mutation and loses its toxicity, which is correlated with the secretion of proteases from the bacteria. Thus, when one cultures *P. entomophila*, one needs to select a colony that produces proteases. Also, those proteases induce self-lysis of *P. entomophila*; therefore, one should not keep a Luria broth (LB) plate of the bacteria for more than a week. Once one obtains the bacterial liquid culture after incubation, one should place it on ice.

Materials

1. 10% skim milk in water

• Skim milk	10 g
• Milli-Q water	100 mL

(i) Autoclave at 121 °C for 20 min

2. LB plate containing 1% (w/v) skim milk.

• Hi polypeptone	1 g
• Yeast extract	0.5 g
• NaCl	0.5 g

(i) Dissolve in 90 mL of Milli-Q water.
(ii) Add approximately 500 μL of 1 M NaOH to adjust the pH to 7.4; measure the pH using a pH paper.
(iii) Bring the final volume to 100 mL with Milli-Q water.
(iv) Add 1.5 g of agar.
(v) Autoclave the mixture at 121 °C for 20 min.
(vi) Add 12 mL of 10% skim milk to 100 mL of LB medium so that the skim milk's final concentration is approximately 1%.
(vii) Mix well, and pour into Petri dishes.

3. The culturing of *P. entomophila*.
(i) Streak a DMSO stock of the wild-type strain L48 of *P. entomophila* onto LB plates containing 1% skim milk.
(ii) Culture at 29 °C overnight.
(iii) Select a colony that shows a "halo" that appears transparent around the colony.
(iv) Inoculate the colony in 10 mL of LB liquid medium to establish a preculture.
(v) Incubate at 29 °C overnight.
(vi) Inoculate 10 mL of the suspension culture into 500 mL of LB liquid medium to establish the main culture.
(vii) Incubate at 29 °C overnight.
(viii) Pellet the bacteria by centrifuging at approximately 2500 × *g* at 4 °C for 15 min.
(ix) Remove the supernatant (but keeping some of the supernatant with the pellet), and estimate the volume of the pellet.
(x) Mix the pellet and culture supernatant at 1:1 ratio as a concentrated solution of bacteria for oral infection experiments.

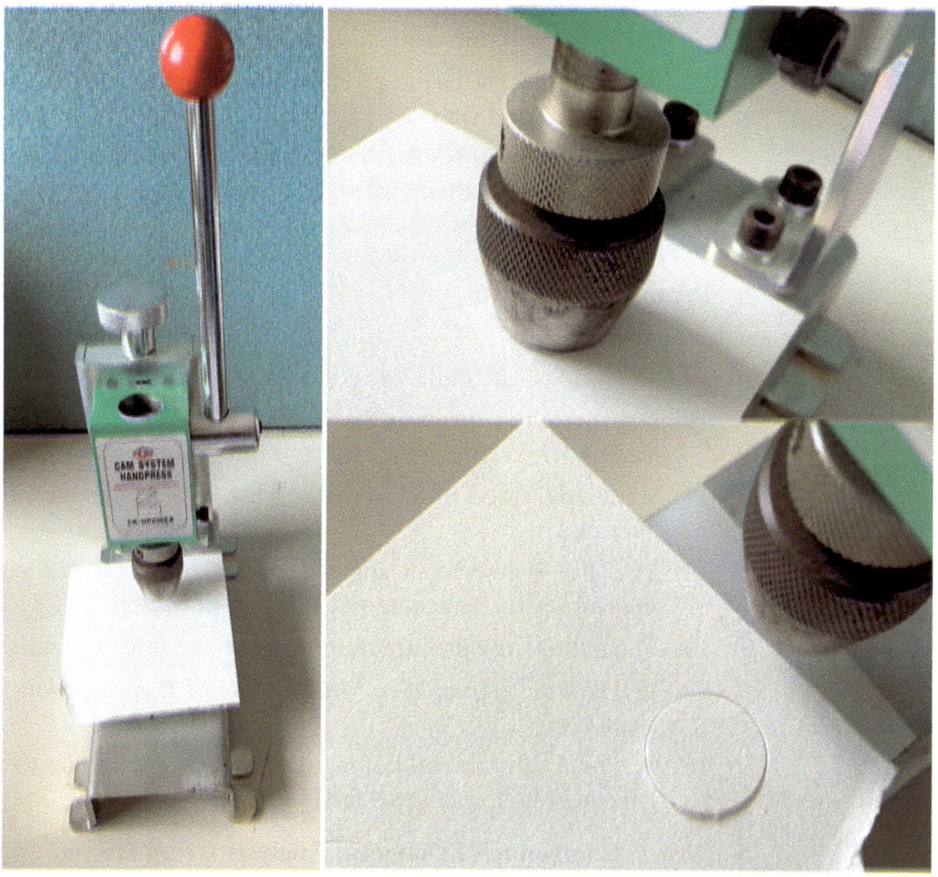

Fig. 5.2 An equipment to prepare filter disks for oral infection. Filter disks are cut with CAM SYSTEM HANDPRESS FK-HP200EX (FUKUI KIKO SHOKAI, http:/www.f-k.co.jp/, Japan)

Methods

1. Filter disks are created by cutting Whatman filter papers into disks with the same diameter as a fly vial's inner diameter (Fig. 5.2).
2. Crushed banana is produced by crushing a banana with a blender, aliquoting the mixture to 2-mL tubes, and storing at −20 °C.
3. Sponge plugs for the 2-mL tubes are produced by cutting any sponge into approximately 1-cm cubes.

Protocol for Larvae

4. The concentrated solution of bacteria is mixed with crushed banana in 2-mL tubes.
5. Larvae are taken out from the vials, washed with water, and put into the tubes with bacteria (the infection tubes).

6. The tubes are plugged with sponge plugs to prevent the larvae from crawling out.
7. After 30 min, the larvae are transferred from the infection tubes to normal food vials.
8. After incubating the larvae for a given period for continuous infection, wash them with water to stop the infection, and analyze.

Protocol for Adults

9. Several-day-old adults are collected into empty vials.
10. Dehydrate the adults at 29 °C for 2 h to ensure synchronous feeding when they are transferred into the infection vials.
11. Standard fly food vials but without live yeast are covered using a filter disk.
12. The disk is soaked in approximately 150 μL of the control solution or the concentrated solution of bacteria.
13. The dehydrated flies are transferred to the infection vials.
14. The flies are kept at 29 °C (in the case of *P. entomophila*) for a given period.
15. Optional: after several hours, transfer the infected flies to a new infection vial for them to feed on fresh bacteria.

Notes: Caution for Behavioral Analysis Using Larvae

Please be careful when dealing with larvae. Touching larvae induces a robust activation of their innate immune responses without an infection [73]. In other words, handling larvae with forceps induces the production of antimicrobial peptides and stress response molecules even in the absence of any infection. Thus, when one uses larvae for behavioral assays under infectious conditions or tests the effect of specific indigenous microbes, one should be careful with interpreting one's data on whether the phenotype is caused by the bacteria or the touching of the larvae.

References

1. Brower DR (1898) Auto-intoxication in its relations to the diseases of the nervous system. J Am Med Assoc 11:575–577
2. Bested AC, Logan AC, Selhub EM (2013) Intestinal microbiota, probiotics and mental health: from Metchnikoff to modern advances: part I – autointoxication revisited. Gut Pathog 5:5
3. Mayer EA (2011) Gut feelings: the emerging biology of gut–brain communication. Nat Rev Neurosci 12:453–466
4. Tsurugizawa T et al (2009) Mechanisms of neural response to gastrointestinal nutritive stimuli: the gut-brain axis. Gastroenterology 137: 262–273

5. Fukudo S, Nomura T, Hongo M (1998) Impact of corticotropin-releasing hormone on gastrointestinal motility and adrenocorticotropic hormone in normal controls and patients with irritable bowel syndrome. Gut 42:845
6. Kano M et al (2017) Altered brain and gut responses to corticotropin-releasing hormone (CRH) in patients with irritable bowel syndrome. Sci Rep 7:12425
7. Baggio LL, Drucker DJ (2007) Biology of incretins: GLP-1 and GIP. Gastroenterology 132:2131–2157
8. Nässel DR, Winther ÅME (2010) *Drosophila* neuropeptides in regulation of physiology and behavior. Prog Neurobiol 92:42–104
9. Elphick MR, Mirabeau O, Larhammar D (2018) Evolution of neuropeptide signalling systems. J Exp Biol 221:jeb151092
10. Whitman WB, Coleman DC, Wiebe WJ (1998) Prokaryotes: the unseen majority. Proc Natl Acad Sci 95:6578–6583
11. Ley RE, Peterson DA, Gordon JI (2006) Ecological and evolutionary forces shaping microbial diversity in the human intestine. Cell 124: 837–848
12. Qin J et al (2010) A human gut microbial gene catalogue established by metagenomic sequencing. Nature 464:59–65
13. Kau AL, Ahern PP, Griffin NW, Goodman AL, Gordon JI (2011) Human nutrition, the gut microbiome and the immune system. Nature 474:327–336
14. Valdes AM, Walter J, Segal E, Spector TD (2018) Role of the gut microbiota in nutrition and health. BMJ 361:k2179
15. Skelly AN, Sato Y, Kearney S, Honda K (2019) Mining the microbiota for microbial and metabolite-based immunotherapies. Nat Rev Immunol 19:305–323
16. Foster JA, Neufeld K-AM (2013) Gut–brain axis: how the microbiome influences anxiety and depression. Trends Neurosci 36: 305–312
17. Sampson TR, Mazmanian SK (2015) Control of brain development, function, and behavior by the microbiome. Cell Host Microbe 17: 565–576
18. Mayer EA, Tillisch K, Gupta A (2015) Gut/-brain axis and the microbiota. J Clin Invest 125:926–938
19. Sudo N et al (2004) Postnatal microbial colonization programs the hypothalamic–pituitary–adrenal system for stress response in mice. J Physiol 558:263–275
20. Tofalo R, Cocchi S, Suzzi G (2019) Polyamines and gut microbiota. Front Nutr 6:16
21. Dalile B, Oudenhove LV, Vervliet B, Verbeke K (2019) The role of short-chain fatty acids in microbiota–gut–brain communication. Nat Rev Gastroenterol 16:461–478
22. Kuwahara A et al (2020) Microbiota-gut-brain axis: enteroendocrine cells and the enteric nervous system form an interface between the microbiota and the central nervous system. Biomed Res 41:199–216
23. Bravo JA et al (2011) Ingestion of *Lactobacillus* strain regulates emotional behavior and central GABA receptor expression in a mouse via the vagus nerve. Proc Natl Acad Sci 108: 16050–16055
24. Racké K, Reimann A, Schwörer H, Kilbinger H (1995) Regulation of 5-HT release from enterochromaffin cells. Behav Brain Res 73: 83–87
25. Reigstad CS et al (2015) Gut microbes promote colonic serotonin production through an effect of short-chain fatty acids on enterochromaffin cells. FASEB J 29:1395–1403
26. Sugisawa E et al (2020) RNA sensing by gut piezo1 is essential for systemic serotonin synthesis. Cell 182:609–624.e21
27. Miguel-Aliaga I, Jasper H, Lemaitre B (2018) Anatomy and physiology of the digestive tract of *Drosophila melanogaster*. Genetics 210: 357–396
28. Kuraishi T, Kenmoku H, Kurata S (2015) From mouth to anus: functional and structural relevance of enteric neurons in the *Drosophila melanogaster* gut. Insect Biochem Mol 67: 21–26
29. Kenmoku H, Ishikawa H, Ote M, Kuraishi T, Kurata S (2016) A subset of neurons controls the permeability of the peritrophic matrix and midgut structure in *Drosophila* adults. J Exp Biol 219:2331–2339
30. Hadjieconomou D et al (2020) Enteric neurons increase maternal food intake during reproduction. Nature 587:1–5
31. Veenstra JA, Agricola H-J, Sellami A (2008) Regulatory peptides in fruit fly midgut. Cell Tissue Res 334:499–516
32. Guo X et al (2019) The cellular diversity and transcription factor code of *Drosophila* enteroendocrine cells. Cell Rep 29:4172–4185.e5

33. Kuraishi T, Hori A, Kurata S (2013) Host-microbe interactions in the gut of *Drosophila melanogaster*. Front Physiol 4:375
34. Broderick N, Broderick NA, Lemaitre B (2014) Gut-associated microbes of *Drosophila melanogaster*. Gut Microbes 3:307–321
35. Thompson GR, Trexler PC (1971) Gastrointestinal structure and function in germ-free or gnotobiotic animals. Gut 12:230
36. Uzbay T (2019) Germ-free animal experiments in the gut microbiota studies. Curr Opin Pharmacol 49:6–10
37. Neyen C, Bretscher AJ, Binggeli O, Lemaitre B (2014) Methods to study *Drosophila* immunity. Methods 68:116–128
38. Koyle ML et al (2016) Rearing the fruit fly *Drosophila melanogaster* under axenic and gnotobiotic conditions. J Vis Exp 30(113):54219
39. Shin SC et al (2011) Drosophila microbiome modulates host developmental and metabolic homeostasis via insulin signaling. Science 334: 670–674
40. Storelli G et al (2011) *Lactobacillus plantarum* promotes *drosophila* systemic growth by modulating hormonal signals through TOR-dependent nutrient sensing. Cell Metab 14:403–414
41. Sharon G et al (2010) Commensal bacteria play a role in mating preference of *Drosophila melanogaster*. Proc Natl Acad Sci 107: 20051–20056
42. Lizé A, McKay R, Lewis Z (2014) Kin recognition in *Drosophila*: the importance of ecology and gut microbiota. ISME J 8:469–477
43. Najarro MA, Sumethasorn M, Lamoureux A, Turner TL (2015) Choosing mates based on the diet of your ancestors: replication of non-genetic assortative mating in *Drosophila melanogaster*. PeerJ 3:e1173
44. Leftwich PT, Clarke NVE, Hutchings MI, Chapman T (2017) Gut microbiomes and reproductive isolation in *Drosophila*. Proc Natl Acad Sci 114:12767–12772
45. Venu I et al (2014) Social attraction mediated by fruit flies' microbiome. J Exp Biol 217: 1346–1352
46. Farine J-P, Habbachi W, Cortot J, Roche S, Ferveur J-F (2017) Maternally-transmitted microbiota affects odor emission and preference in *Drosophila* larva. Sci Rep 7:6062
47. Fischer CN et al (2017) Metabolite exchange between microbiome members produces compounds that influence *Drosophila* behavior. elife 6:213
48. Wong AC-N et al (2017) Gut microbiota modifies olfactory-guided microbial preferences and foraging decisions in *Drosophila*. Curr Biol 27:2397–2404.e4
49. Qiao H, Keesey IW, Hansson BS, Knaden M (2019) Gut microbiota affects development and olfactory behavior in *Drosophila melanogaster*. J Exp Biol 222:jeb192500
50. Leitão-Gonçalves R et al (2017) Commensal bacteria and essential amino acids control food choice behavior and reproduction. PLoS Biol 15:e2000862
51. Schretter CE et al (2018) A gut microbial factor modulates locomotor behaviour in *Drosophila*. Nature 563:402–406
52. Selkrig J et al (2018) The *Drosophila* microbiome has a limited influence on sleep, activity, and courtship behaviors. Sci Rep 8:10646
53. Silva V et al (2021) The impact of the gut microbiome on memory and sleep in *Drosophila*. J Exp Biol 224:jeb233619
54. Chen K et al (2019) *Drosophila* histone demethylase KDM5 regulates social behavior through immune control and gut microbiota maintenance. Cell Host Microbe 25(4): 537–552.e8
55. Goehler LE et al (2005) Activation in vagal afferents and central autonomic pathways: early responses to intestinal infection with *Campylobacter jejuni*. Brain Behav Immun 19: 334–344
56. Lyte M, Li W, Opitz N, Gaykema RPA, Goehler LE (2006) Induction of anxiety-like behavior in mice during the initial stages of infection with the agent of murine colonic hyperplasia *Citrobacter rodentium*. Physiol Behav 89: 350–357
57. Vyas A, Kim S-K, Giacomini N, Boothroyd JC, Sapolsky RM (2007) Behavioral changes induced by Toxoplasma infection of rodents are highly specific to aversion of cat odors. Proc Natl Acad Sci 104:6442–6447
58. Ezenwa VO, Gerardo NM, Inouye DW, Medina M, Xavier JB (2012) Animal behavior and the microbiome. Science 338:198–199
59. Vodovar N et al (2005) *Drosophila* host defense after oral infection by an entomopathogenic Pseudomonas species. Proc Natl Acad Sci USA 102:11414–11419
60. Opota O et al (2011) Monalysin, a novel ß-pore-forming toxin from the *Drosophila* pathogen *Pseudomonas entomophila*, contributes to host intestinal damage and lethality. PLoS Pathog 7:e1002259
61. Kuraishi T, Binggeli O, Opota O, Buchon N, Lemaitre B (2011) Genetic evidence for a protective role of the peritrophic matrix against intestinal bacterial infection in *Drosophila melanogaster*. Proc Natl Acad Sci 108: 15966–15971
62. Chakrabarti S, Liehl P, Buchon N, Lemaitre B (2012) Infection-induced host translational

blockage inhibits immune responses and epithelial renewal in the *Drosophila* gut. Cell Host Microbe 12:60–70
63. Nonaka S et al (2020) Molecular and functional analysis of pore-forming toxin Monalysin from Entomopathogenic bacterium *Pseudomonas entomophila*. Front Immunol 11:520
64. Liehl P, Blight M, Vodovar N, Boccard F, Lemaitre B (2006) Prevalence of local immune response against oral infection in a *Drosophila/Pseudomonas* infection model. PLoS Pathog 2: e56
65. Brummel T, Ching A, Seroude L, Simon AF, Benzer S (2004) *Drosophila* life span enhancement by exogenous bacteria. Proc Natl Acad Sci 101:12974–12979
66. Ryu J-H et al (2008) Innate immune homeostasis by the homeobox gene caudal and commensal-gut mutualism in *Drosophila*. Science 319:777–782
67. Bakula M (1969) The persistence of a microbial flora during post-embryogenesis of *Drosophila melanogaster*. J Invertebr Pathol 14:365–374
68. Saridaki A, Bourtzis K (2010) Wolbachia: more than just a bug in insects genitals. Curr Opin Microbiol 13:67–72
69. Ote M, Ueyama M, Yamamoto D (2016) Wolbachia protein TomO targets nanos mRNA and restores germ stem cells in *Drosophila* sex-lethal mutants. Curr Biol 26:2223–2232
70. Arbuthnott D, Levin TC, Promislow DEL (2016) The impacts of Wolbachia and the microbiome on mate choice in *Drosophila melanogaster*. J Evol Biol 29:461–468
71. Simhadri RK et al (2017) The gut commensal microbiome of *Drosophila melanogaster* is modified by the endosymbiont Wolbachia. MSphere 2:e00287-17
72. Chrostek E et al (2013) *Wolbachia* variants induce differential protection to viruses in *Drosophila melanogaster*: a phenotypic and phylogenomic analysis. PLoS Genet 9:e1003896
73. Kenmoku H, Hori A, Kuraishi T, Kurata S (2017) A novel mode of induction of the humoral innate immune response in *Drosophila* larvae. Dis Model Mech 10:271–281
74. Hori A, Kurata S, Kuraishi T (2018) Unexpected role of the IMD pathway in *Drosophila* gut defense against *Staphylococcus aureus*. Biochem Biophys Res Commun 495:1–6
75. Wong CNA et al (2011) Low-diversity bacterial community in the gut of the fruitfly *Drosophila melanogaster*. Environ Microbiol 13: 1889–1900
76. Gordon HA (1960) The germ-free animal. Am J Dig Dis 5:841–867
77. Gustafsson BE (1959) Lightweight stainless steel systems for rearing germfree animals. Ann N Y Acad Sci 78:17–28
78. Griffiths LA, Barrow A (1972) Metabolism of flavonoid compounds in germ-free rats. Biochem J 130:1161–1162
79. Meyer RC, Bohl EH, Kohler EM (1964) Procurement and maintenance of germ-free swine for microbiological investigations. Appl Microbiol 12:295–300
80. Trexler PC, Reynolds LI (1957) Flexible film apparatus for the rearing and use of germfree animals. Appl Microbiol 5:406–412
81. Douglas AE (2018) Contradictory results in microbiome science exemplified by recent *Drosophila* research. MBio 9:854
82. Lee H-Y, Lee S-H, Lee J-H, Lee W-J, Min K-J (2019) The role of commensal microbes in the lifespan of *Drosophila melanogaster*. Aging 11: 4611–4640
83. Kurokawa K et al (2007) Comparative metagenomics revealed commonly enriched gene sets in human gut microbiomes. DNA Res 14: 169–181
84. Morita H et al (2007) An improved DNA isolation method for metagenomic analysis of the microbial flora of the human intestine. Microbes Environ 22:214–222
85. Tamura K et al (2011) MEGA5: molecular evolutionary genetics analysis using maximum likelihood, evolutionary distance, and maximum parsimony methods. Mol Biol Evol 28: 2731–2739
86. Kumar S, Stecher G, Li M, Knyaz C, Tamura K (2018) MEGA X: molecular evolutionary genetics analysis across computing platforms. Mol Biol Evol 35:1547–1549

Chapter 6

Measurement of Thermoregulatory Behavior in *Drosophila melanogaster*

Takuto Suito, Kohjiro Nagao, Masahiro Kai, Naoto Juni, Takaaki Sokabe, and Masato Umeda

Abstract

Thermoregulation is a fundamental physiological function that maintains homeostasis in both endothermic and ectothermic animals. As a model organism, the fruit fly (*Drosophila melanogaster*) has markedly contributed to the study of behavioral genetics. Behavioral analyses using *Drosophila* have revealed the sensory neurons, sensor molecules, and other factors including energy metabolism that are closely linked to thermoregulation. In this chapter, we describe the equipment and procedures required for analyzing thermoregulatory behavior in *Drosophila*. We also provide guidelines for the selection of strategies to analyze behavioral thermoregulation in *Drosophila*. In addition, we provide representative results obtained with these protocols and show that changes in preferred temperature can be induced by altering the unsaturation of membrane phospholipids.

Key words Behavioral thermoregulation, *Drosophila melanogaster*, Temperature gradient assay, Thermosensation, Lipid unsaturation

1 Introduction

1.1 Study of Thermoregulatory Behavior in Drosophila

Thermoregulation is a key physiological function for maintaining homeostasis and survival. Endotherms, such as mammals and birds, can control body temperature by both a physiological response (e.g., shivering and non-shivering thermogenesis, sweating, and blood circulation) and a behavioral response (e.g., moving to favorable temperatures) [1]. In contrast, ectotherms, including insects and a wide variety of animal species, use primarily behavioral thermoregulation. Thus, behavioral thermoregulation is a fundamental thermoregulatory mechanism used by all animals.

The fruit fly, *Drosophila melanogaster*, is a widely used model organism because it is easy to rear and manipulate genetically. Because nearly 75 % of human disease-associated genes have *Drosophila* homologs [2], *Drosophila* genetics have contributed to our understanding of the molecular mechanisms of human disease and

Daisuke Yamamoto (ed.), *Behavioral Neurogenetics*, Neuromethods, vol. 181, https://doi.org/10.1007/978-1-0716-2321-3_6,

physiology. Behavioral genetics is a research field that uses genetic methods to elucidate the causative genes and underlying molecular mechanisms of certain behaviors. Genetic approaches to behavioral analysis using *Drosophila* have facilitated our understanding of the molecular mechanisms underlying circadian rhythm, courtship behavior, learning, memory, and phototaxis [3–5].

Drosophila have been frequently used as a model for behavioral thermoregulation, and factors associated with this mechanism can be readily studied at an animal level. The analysis of thermoregulatory behavior in *Drosophila* was initially reported by Sayeed et al. [6]. Thereafter, many thermoregulatory factors including thermosensory proteins and neuronal integration of thermal sensing have been identified. In this chapter, we introduce methods for studying thermoregulation using *Drosophila* larvae.

1.2 Thermosensory Proteins Involved in Thermoregulatory Behavior of Drosophila

Thermosensory proteins play dominant roles in thermoregulation. The most important thermosensory proteins are the family of transient receptor potential (TRP) channels, which are classified into seven subfamilies (i.e., TRPC, TRPM, TRPV, TRPA, TRPP, TRPML, and TRPN). Except for TRPN, six of these subfamilies occur in mammals and are involved in a variety of sensory processes, such as the perception of pungent chemicals, thermosensation, and mechanosensation [7, 8]. Several TRP channels have been reported to be involved in the thermoregulatory behavior of *Drosophila* (Fig. 6.1a). TRPA1 was identified as the receptor that controls thermoregulatory behavior in *Drosophila* [9, 10]. TRPA1 mutant flies exhibit a decreased ability to avoid warm temperatures [11, 12]. TRPA1 is also important for regulating heating-triggered rolling behavior in larvae [13]. Although the thermosensitivity of mammalian TRPA1 is controversial, *Drosophila* TRPA1, as well as its homologs in nonmammalian vertebrates, such as reptiles [14–16] and birds [17], can be directory activated by warm temperature. In *Drosophila*, five TRPA1-splicing variants have been identified. Of these, isoforms A and D are involved in warm temperature sensitivity [18]. Isoforms A and B are expressed in the brain, whereas the longer isoforms, C and D, are expressed in the peripheral neurons of larvae. Recently, isoform E was identified as a byproduct of alternative splicing, although its function is unknown [19]. In addition, as the receptor for nociceptive high temperature, two TRPA subfamily members, Pain [20, 21] and Pyx [22], were identified. Both are expressed in peripheral multidendritic neurons and directly activated by noxious hot temperatures.

Several TRP channels have been reported to be involved in cold sensation. TRPC subfamily members, TRP and TRPL, are required for cold avoidance [23]. A TRPV family member, Iav, is expressed in the chordotonal organs and is also involved in cold avoidance [24]. In adult flies, Brv (TRPP) is expressed in cold sensory neurons and is required for cold temperature avoidance [25]. Three TRP channels, Pkd2 (TRPP), NompC (TRPN), and TRPM, are

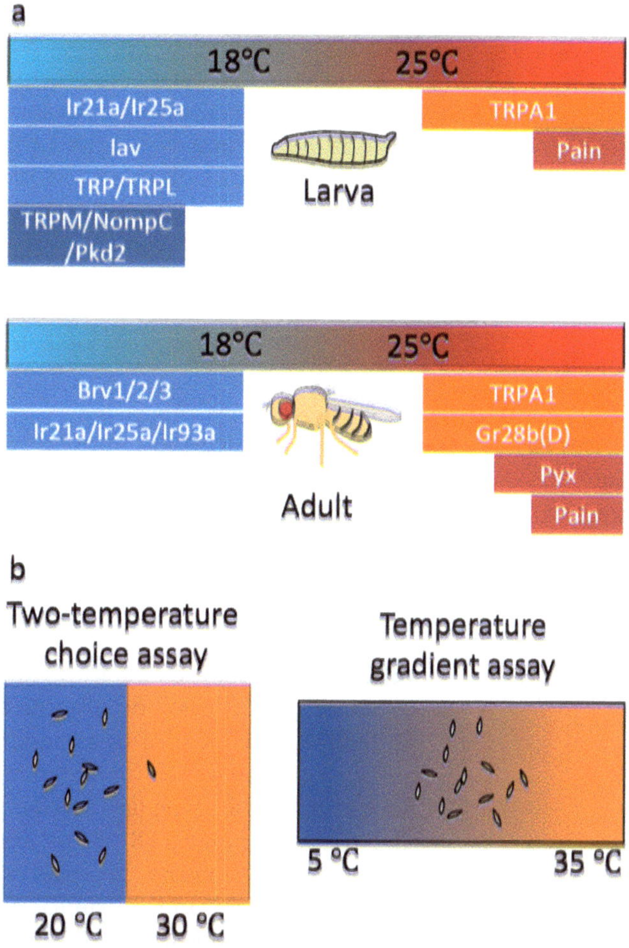

Fig. 6.1 Temperature sensors and assay methods for thermoregulatory behavior in *Drosophila*. (**a**) Sensory function-related molecules reported in the sensory neurons of *Drosophila* larvae (upper) and adults (lower). In larvae, TRPA1 was reported as the heat sensing protein. Ir21a/Ir25a, Iav, and TRP/TRPL represent the cold avoidance-associated proteins. In adults, TRPA1 and Gr28b (D) were identified as the sensory proteins that can be activated at temperatures above 25 °C. Brv and Ir21a/Ir25a/Ir93a were identified as cold avoidance-associated proteins. For sensing nociceptive high temperatures, Pain and Pyx were identified. TRPM/NompC/Pkd2 are associated with noxious cold sensing and body contraction behavior in larvae. (b) Two methods for analyzing thermoregulatory behavior of *Drosophila*. In the two-temperature choice assay (left panel), two different temperature zones (e.g., 20 °C and 30 °C) are prepared in one plate. Flies avoid an unfavorable temperature zone. In the temperature gradient assay (right panel), a continuous, linear temperature gradient (e.g., from 5 °C to 35 °C) is generated on the plate. Flies seek the temperature environment and settle down in a favorable temperature range

expressed in class III multidendritic sensory neurons and are required for the aversive response to noxious cold temperature [26].

Sensory receptor proteins other than TRP family members have been reported to be involved in thermoregulatory behavior in *Drosophila*. The gustatory receptor family protein, Gr28b(D), is expressed in warm-sensitive neurons in an antennal segment called arista in adults and is required for the rapid sensation of warm temperatures [27]. Two ionotropic receptors, Ir21a and Ir25a, are expressed in cold-sensitive dorsal organ neurons in larvae and are required for cold avoidance [28]. Although Ir21a and Ir25a are expressed in arista and also involved in cold sensation in adults, they appear to participate in the thermoregulatory behavior through a regulation of morphology of cold sensory neurons along with Ir93a [29].

1.3 Devices for Analyzing Thermoregulatory Behavior of Drosophila

As a simple model of thermoregulation, discriminating cooler and hotter temperatures in *Drosophila* determines their thermotactic behaviors (Fig. 6.1a). Thus, observation of the temperature preference of *Drosophila* in a temperature-controlled arena has been used for the analysis of thermoregulatory behavior. Two assay methods are primarily used: the two-temperature choice assay and the temperature gradient assay (Fig. 6.1b). In the two-temperature choice assay, two different temperatures are established in one plate to test which temperature is chosen by the flies. This assay is normally used for analyzing avoidance of an unfavorable temperature. The proportion of flies on each temperature zone is calculated for analysis. For example, wild-type flies avoid cold or hot temperatures outside their optimal range (18 °C to 25 °C), whereas flies harboring mutant thermosensory proteins undistinguishably distribute in both the ambient and uncomfortable temperature zones. Several genes that are associated with warm [12] and cold [23, 25] sensation have been identified using the two-temperature choice assay.

In contrast, the temperature gradient assay enables flies to choose their absolute preferred temperature [11, 22, 30]. In this assay, flies are placed on a plate forming a temperature gradient ranging from cold to hot (e.g., 5 °C to 35 °C). The change in thermoregulatory behavior is determined as a shift of preferred temperature on the gradient. For example, a clear shift of temperature preference depending on the development stage of larvae has been observed using a thermal gradient from 18 °C to 28 °C [31]. Early-stage third instar larvae (72 h after egg laying) preferred 24 °C, whereas late-stage third instar larvae (120 h after egg laying) preferred 18 °C [31].

In this chapter, we introduce the methodology of the temperature gradient assay using *Drosophila* larvae. The protocol for the two-temperature choice assay using larvae may be found in other reports [12, 23].

2 Materials

2.1 Fly Larvae

The fly strains of w^{1118} and UAS-desat1 RNAi (5887R-4) were obtained from NIG-FLY stock center (National Institute of Genetics, Japan). The strain of r4-GAL4 (#33832) was obtained from Bloomington Drosophila Stock Center (Indiana, USA). The w^{1118} strain was used as a control. Flies were kept in a 25 °C incubator under a 12/12-h light/dark cycle. The flies used in this protocol were cultured in medium containing the following ingredients: 80 g brewer's yeast powder, 100 g glucose, 40 g cornmeal, 7 g agar, 8.8 mL propionic acid, and 0.88 g butyl parahydroxybenzoate in 1 L of water.

2.2 Temperature Gradient Assay

2.2.1 Agarose-Covered Glass Plate

For the behavioral assay, larvae were placed on an agarose-covered glass plate (200 mm × 25 mm with 1.25 mm thickness) that was prepared by cutting a square glass plate. We used thin layer chromatography glass plates (TLC Silica gel 60, Merck) after removing the silica layer. Glass plate was covered with 2.2 % agarose gel (Iwai Chemicals).

2.2.2 Temperature Gradient Assay Instrument

The setup for the gradient assay is shown in Fig. 6.2a. For generating a linear temperature gradient (Fig. 6.2b), we used a Peltier device installed with a temperature controller (DIA Medical System Co.). Two Peltier devices were positioned below the two distal ends of an aluminum plate [250 mm (W) × 50 mm (D)]. For radiation of the Peltier device, a pump and reserve water were piped to the devices. Peltier devices were connected to a temperature controller so that the temperatures could be readily manipulated. The aluminum plate was surrounded by 9 mm walls and covered with an acrylic lid. For calculating the position of the larvae, a ruler was placed on the aluminum plate (see Fig. 6.3b).

2.2.3 Image Capture and Analysis

The distribution of larvae on the gradient plate (Fig. 6.2c) was captured using a digital camera (Nikon D90) with a wide-angle lens (35 mm, f/1.8, Nikkor). The camera was fixed directly above the assay plate using a tripod. Images were edited using Adobe Photoshop software, and the position of each larva was quantified using ImageJ (Fiji) software [32]. Finally, the calculation of larval position and the generation of histograms were performed using R [33] script.

2.3 Video Recording Setup for Tracking Drosophila Behavior on the Temperature Gradient

For tracking the movement of larvae on a temperature gradient, a video recording setup was used as an advanced method. Behavioral tracking of fly larvae was performed using the setup shown in Fig. 6.4. A temperature gradient assay instrument was placed in a box [400 mm (W) × 350 mm (D) × 650 mm (H)] made of 5-mm-thick black acrylic plates. A door was installed at the front

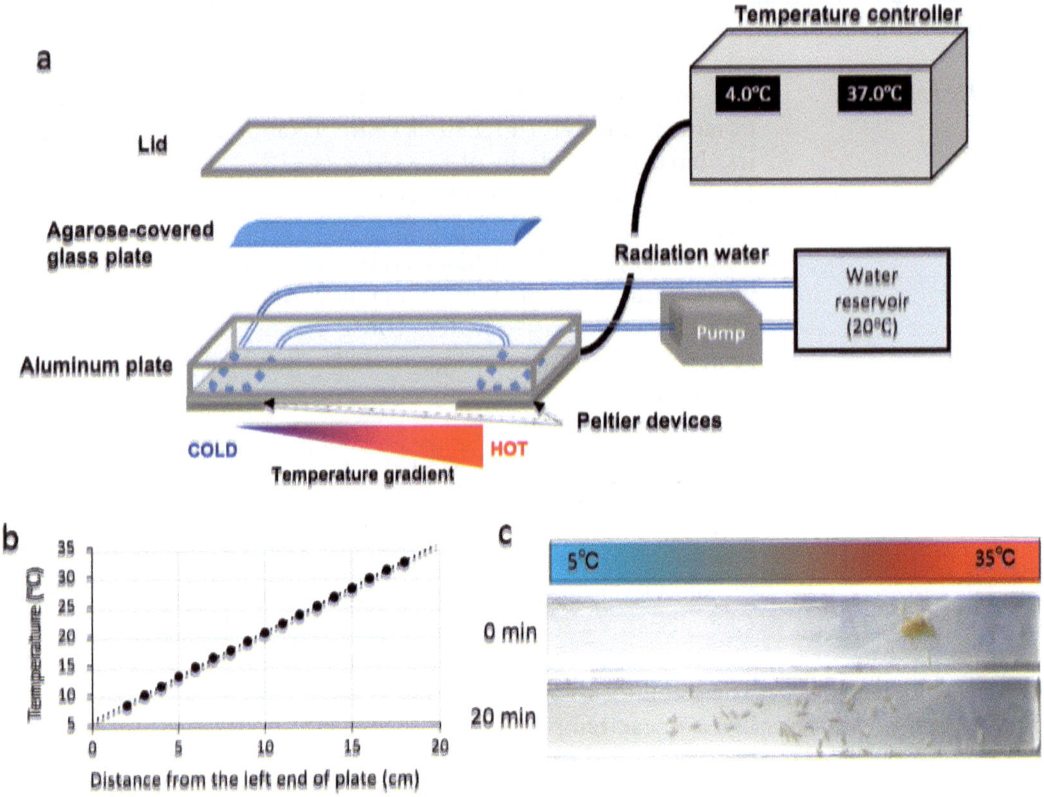

Fig. 6.2 Setup for the temperature gradient assay. (**a**) The temperature gradient is generated on an aluminum plate installed with Peltier devices at both ends. The temperature of the Peltier devices is regulated by a temperature controller. Radiation water (20 °C) is circulated by a pump from the water reservoir to stabilize the temperature of Peltier devices. An agarose-covered glass plate is placed on the aluminum plate. During the assay, a lid is used to prevent larvae from escaping and keep the humidity. (**b**) A temperature profile on the agarose gel plate. Dots represent the surface temperatures on the agarose gel measured by a thermocouple thermometer. A dashed line is the linear regression line of the plot. (**c**) Representative pictures of larvae on the agarose plate at 0 min (start of the assay) and 20 min (end of the assay)

of the box to provide access. LEDs (660 nm) were used for illuminating the inside of the box. A 5-megapixel CMOS digital camera (STC-MBS510U3V, OMRON Sentech) with a 12 mm prime focus lens (FL-CC1218-5MX, Ricoh) was installed to record the video. To avoid dew condensation on the lid of the temperature gradient assay plate during recording, an antifog film (MF-600, Fujifilm) was attached to the lid.

3 Methods

3.1 Preparation of Larvae

For behavioral analysis, wandering third instar *Drosophila* larvae (generally 5 days after egg laying) were used. Since the temperature preference of the larvae changes throughout developmental stages

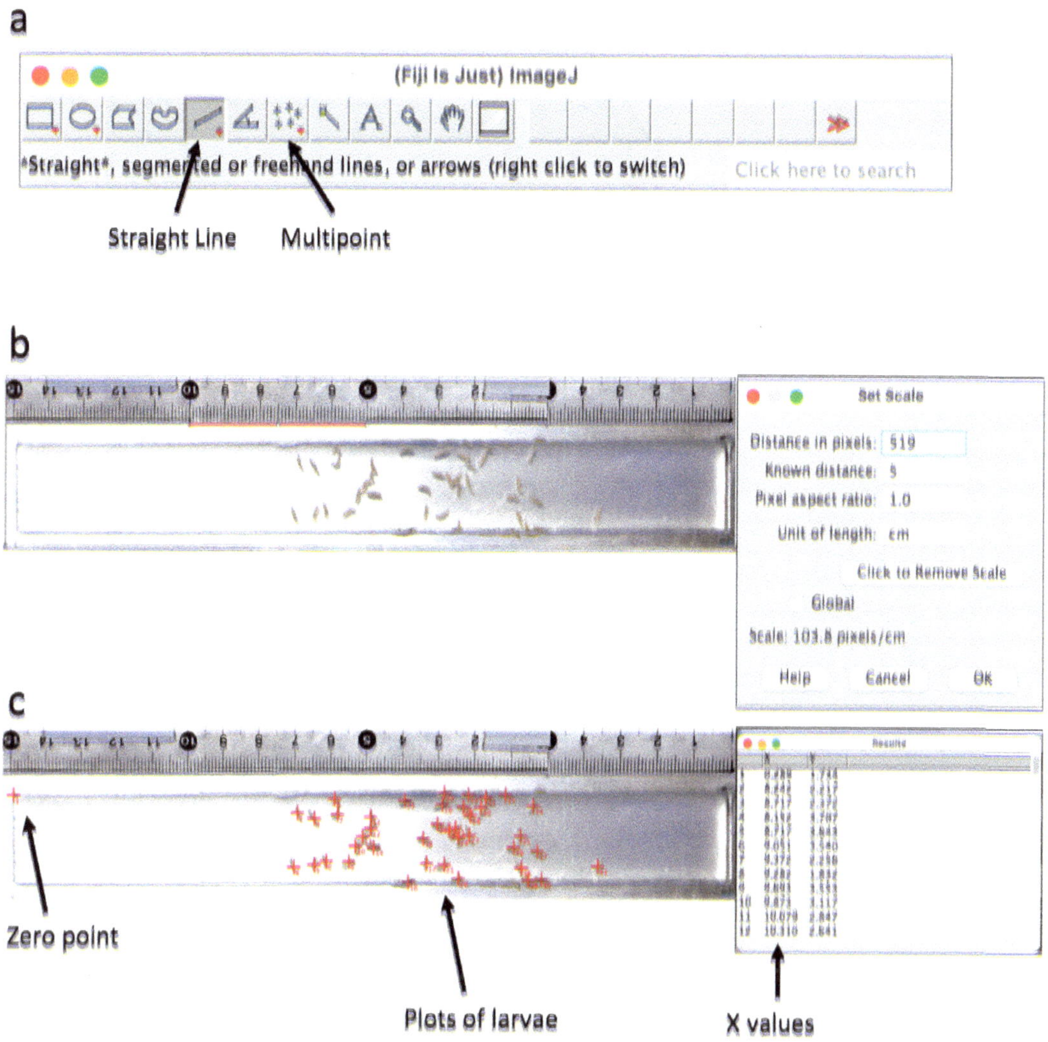

Fig. 6.3 Analysis of the distribution of larvae on the agarose plate. (**a**) The tool bar of ImageJ software (Fiji for macOS). "Straight line" in the line selection tool and "Multipoint" in the point tool are used for analysis. (**b**) A horizontal line (a red color) is drawn at 5 cm along the ruler in the picture. The "Set Scale" function is then used. (**c**) The position of the zero point and larvae are plotted using the "Multipoint" function. The "zero point" is at the left edge of the plate. After plotting the position of the larvae, the "Measure" function is used to display the position of the larvae in the "Results" window. The values in the X column are used to determine the temperature at which each larva is located

[31], larvae at the same developmental stage should be used for comparing the temperature preference among flies with different genotypes. Because egg-laying behavior and hatching rate differ among genotypes, several factors should be optimized (e.g., the number of mated females and the time for inoculation). We normally select the third instar larvae at the wandering stage that are moving on the wall of the culture vial and exclude larvae in the food and larvae at a prepupa stage.

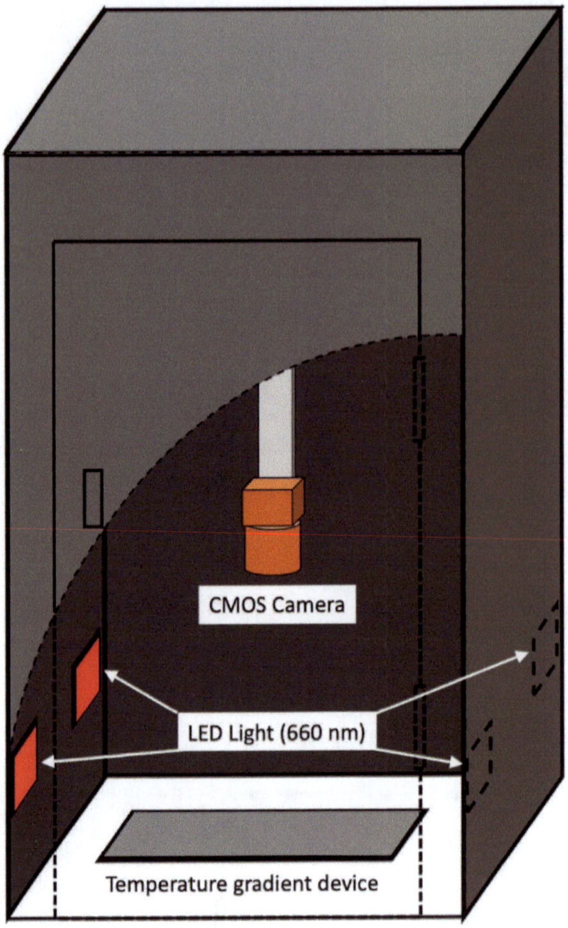

Fig. 6.4 Video recording setup. Box made of black acrylic plates is placed over the temperature gradient assay device. Inside the box, a CMOS camera is installed to record a movie of locomotion of larvae. A 660 nm LED illuminates the assay device from a low angle to prevent reflection

3.2 Temperature Gradient Assay

3.2.1 Preparation of Agarose-Covered Glass Plates

The day before the experiment, the agarose-covered glass plates are prepared. We normally prepare 10 to 20 plates at the same time. First, a 2.2 % agarose solution is microwaved until completely dissolved and melted. Then, 10 mL of melted agarose solution is gently poured onto a glass plate [200 mm (W) × 25 mm (D)] on a leveled table using a pipette. Because of surface tension, the agarose solution does not spill off the glass plate. It is important to make sure that the agarose solution covers the entire surface of the glass plate and that the surface of the agarose gel is smooth and level. After the agarose gel solidifies, it is wrapped with plastic wrap and stored at 4 °C. Before use, the agarose-covered glass plate is equilibrated to a room temperature. Prior to the assay, the plastic wrap from the agarose-covered glass plate is removed, and droplets of water from the surface of the agarose gel are wiped away using Kimwipes.

3.2.2 Preparation of the Peltier Device

For starting the temperature gradient assay, the Peltier device and a pump for radiation water are turned on. The temperature of the water reservoir should be kept at 20 °C or a stable temperature because water temperature affects the radiation efficiency of the Peltier device. One should wait 30 min until the temperature of the aluminum plate becomes stable. A linear thermal gradient may be generated on an aluminum plate by either warming or cooling at the end of the plate.

We normally use a 5 °C to 35 °C gradient on the surface of the 20 cm agarose-covered grass plate, which has a 1.5 °C/cm temperature slope (Fig. 6.2b). The temperature of the Peltier device and the surface temperature of the agarose plate are different. When the experiment is initiated with a new temperature gradient, the settings of the temperature controller and surface temperature of the agarose plate should be checked to ensure that the intended temperature gradient was generated on the agarose gel. The surface temperature of the agarose plate with 1 cm intervals may be measured using a thermocouple probe. Since evaporation from the agarose gel affects the surface temperature, temperature measurements should be performed only under closed-lid conditions. We prepared a lid with small holes and inserted a temperature probe through the hole to measure the surface temperature of the agarose gel. Using the plots and a linear regression line, we can calculate the slope of the gradient and the temperature at the left end of the plate (zero point) as shown in Fig. 6.2b.

3.2.3 Recoding and Analysis of the Distribution of Larvae

After putting the glass plate on the aluminum surface, one should wait for 5 min until the intended temperature gradient is generated on the surface of the agarose gel. During the 5 min waiting period, 30 to 40 wandering third instar larvae are collected from the walls of the vials using a paintbrush and transferred into a 35 mm petri dish. The larvae are washed in RO water, and the water is wiped off from the surface of the larvae using Kimwipes. Using a paintbrush, the larvae are transferred to the agarose plate at a region 5 cm from the right end of the plate where the surface temperature is approximately 28 °C (Fig. 6.2c). The lid is then closed to keep the humidity of the assay box constant. After 20 min, the lid is opened, and a snapshot of the larva distribution is captured on a digital camera (Fig. 6.2c). The agarose gel may be discarded and exchanged for a new agarose-covered glass plate for a subsequent experiment.

After the assay, the position of larvae is determined from the images using ImageJ software (Fig. 6.3). A 5 cm horizontal line is drawn over a ruler in the picture using the "Straight line" tool bar function (Fig. 6.3a). The "Analyze > Set Scale" function is then used to set the scale of the image to be analyzed (Fig. 6.3b). The "Distance in pixel" is the pixel length of the line drawn. The "Known distance" should be designated as "5." The unit of length can be set as "cm." Next, the "Multipoint" tool bar function is used to plot the position of the larvae and the zero point (Fig. 6.3c). The zero point is at the left end of the glass plate. The position of the larvae is manually selected as the center of the larval body. Larvae that are located outside of the agarose plate are not counted. Finally, the "Analyze > Measure" function is used to output the position of the larvae in cm units (Fig. 6.3c). The values in the X column in the "Results" window are used for the following procedure.

To calculate the temperature associated with the location of each larva, coefficients from linear regression line for the temperature gradient of the assay plate and the values of distance between the left end of the plate and the center of each larva are used. We use R script to calculate the temperature value. First, the X value in the "Results" window of ImageJ (Fig. 6.3c) for each larva is subtracted by the X value at the zero point to calculate the distance from the left end. Next, the temperature associated with the location of the larva is calculated using the following equation:

$$T = a\,x + t_o$$

where T is the temperature at the position of the larva (°C), a is the slope of the temperature gradient (°C/cm), x is the calculated distance between the zero point and the larva (cm), and t_o is the temperature (°C) at the zero point. Finally, a histogram is drawn to represent the distribution of the larvae.

3.2.4 Representative Results of the Temperature Gradient Assay

The distribution of control larvae in the temperature gradient displays a Gaussian distribution as shown in Fig. 6.5a. We have evaluated the effect of fatty acid unsaturation on thermoregulatory behavior by genetically manipulating the expression of fatty acid desaturases in *Drosophila*. A fatty acid desaturase generates a *cis* conformation of double bond, which results in a formation of a kinked structure in the hydrocarbon chain of fatty acids. This modulates the activity of membrane proteins by changing the physicochemical properties of the phospholipid membrane or through direct binding of membrane proteins and phospholipid molecules [34].

Transgenic third instar *Drosophila* larvae expressing *Caenorhabditis elegans* Δ12 fatty acid desaturase (FAT-2), which produces linoleic acid from oleic acid in neurons using the binary GAL4-UAS system [35], exhibited alterations in neuronal activity and temperature preference [36]. Compared with the temperature preference in the control flies, the average preferred temperature of flies expressing FAT-2 in the pan-neuron and TRPA1 neuron was significantly decreased by 3.5 °C (elav > w^{1118}: 22.1 °C and elav > FAT-2: 18.6 °C) and 4.9 °C (TRPA1 > w^{1118}: 22.2 °C and TRPA1 > FAT-2: 17.3 °C), respectively [36].

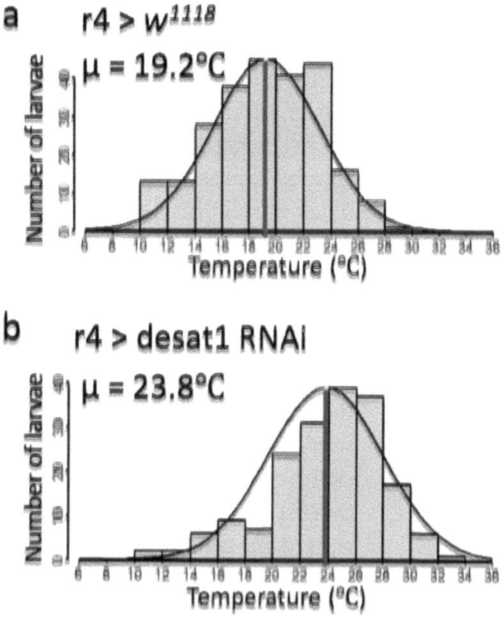

Fig. 6.5 Effect of phospholipid acyl chain unsaturation on thermoregulatory behavior. The histograms show the distribution of third instar larvae of (**a**) control (r4 > w^{1118}) and (**b**) *desat1* RNAi in fat body (r4 > desat1 RNAi) on temperature gradient. The larvae with *desat1* RNAi show a higher temperature preference. *r4-GAL4* is the driver for fat body. The distribution curve is presented as a black line, and the average temperature of the distribution is shown as a blue vertical line

Although *Drosophila* do not have the ability to produce polyunsaturated fatty acids, monounsaturated fatty acids, such as palmitoleic acid and oleic acid, can be produced by the Δ9 fatty acid desaturase. DESAT1 is the Δ9 fatty acid desaturase that regulates double-bond content in the fatty acid moiety of phospholipids and pheromone production in *Drosophila* [37]. The control flies (r4 > w^{1118}) showed an average preferred temperature of 19.2 °C (Fig. 6.5a). The RNAi-induced knockdown of the *desat1* gene in the lipogenic organ fat body (r4 > desat1 RNAi) resulted in an average preferred temperature of 23.8 °C (Fig. 6.5b), indicating that the preferred temperature was increased 4.6 °C by the suppression of DESAT1. These results suggest that the level of unsaturation of phospholipid acyl chains affects the thermoregulatory behavior of *Drosophila*.

3.3 Behavioral Tracking on a Temperature Gradient

In Sect. 3.2, the preferred temperature was measured from the picture taken at the end of a 20 min migration period as the index of thermoregulatory behavior. However, more information may be obtained by analyzing the locomotive behaviors during the period. The trajectory of larvae on a plate consists of forward movement and head sweeping behavior [38]. Moreover, several temperature-responding behaviors, such as body contraction [26] and rolling behavior [13, 20], were observed at nociceptive cold and hot temperatures. We developed an advanced device as shown in Fig. 6.4 to analyze the behavior of larvae on a temperature gradient. To exclude the effect of the light-dependent behavior in *Drosophila* larvae, the assay plate is shielded from environmental light by a black acrylic box. For illuminating the inside of the box, 660 nm LEDs were used since *Drosophila* do not have photoreceptors that can detect wavelengths above 600 nm [39]. The video is acquired at five frames per second at a 5-megapixel resolution. The trajectory of the larvae can be tracked by Ctrax [40] or other tracking software. In recent years, deep learning-based behavioral tracking methods [41, 42] have become available, which enable one to track the behavior of multiple animals in an assay area in detail. By combining the results of behavioral tracking with conventional picture-based distribution analyses, comprehensive information regarding thermoregulatory behavior can be obtained.

4 Conclusions

Research on thermoregulatory mechanisms using the *Drosophila* temperature preference assay has provided significant information with respect to the physiological functions of sensory neurons and proteins. In addition, although the underlying mechanism of behavioral thermoregulation has not been fully elucidated, neural circuits involving cold and warm sensation have recently been reported in *Drosophila* [43–46].

Experiments with *Drosophila* have provided evidences of the interaction between thermoregulation and various physiological factors. The dystroglycan mutant, *atsugari*, showed cryophilic behavior and aberrant mitochondrial oxidative energy metabolism [30]. In addition, Umezaki et al. reported that feeding state modulates temperature preference in *Drosophila* [47]. They also revealed that the insulin signaling cascade is involved in determining temperature preference [47]. These studies suggest that energy state contributes significantly to thermoregulation in *Drosophila*. In addition, we discovered that the level of fatty acid unsaturation affects the thermoregulatory behavior of *Drosophila* (Fig. 6.5) [36]. Changes in fatty acid unsaturation affect the molecular motion and the metabolism of fatty acid-containing lipids, such as phospholipid and triacylglycerol. Thus, it is likely that the regulation of fatty acid unsaturation is involved in various steps of thermoregulatory behavior in *Drosophila*. Tissue-specific alteration of fatty acid unsaturation may reveal the unknown regulatory mechanism of thermoregulation in *Drosophila*.

The unsolved mystery of thermoregulation is why organisms have a certain body temperature and/or temperature preference.

From simple behavioral assays using *Drosophila*, the mechanisms of body temperature regulation are being clarified by findings of many factors such as temperature sensors and metabolic states. Since the early twentieth century, *Drosophila* have been the foundation of many biological breakthroughs, and this small organism will continue to provide insights that are core of the mechanism of thermoregulation.

Acknowledgments

We thank Dr. Makoto Tominaga (National Institute for Physiological Science) for helpful discussion. We thank Toshiyuki Sazi (National Institute for Physiological Science) for creating the acrylic box for video recording. This work was supported by a Grant-in-aid for Scientific research 15H05930 (to M. U.), 15 K21744 (to M. U.), 20 K21388 (to M. U.), 21H02477 (to M. U.), 18 K05433 (to K.N), 21 K05391 (to K. N.), 21H02531 (to T. Sok.), 19 K23790 (for T. Sui.), and 21 K15192 (for T. Sui.) from the Japan Society for the Promotion of Science and the Ministry of Education, Culture, Sports, Science and Technology (MEXT).

References

1. Morrison SF, Nakamura K (2011) Central neural pathways for thermoregulation. Front Biosci (Landmark Ed) 16:74–104
2. Reiter LT, Potocki L, Chien S, Gribskov M, Bier E (2001) A systematic analysis of human disease-associated gene sequences in Drosophila melanogaster. Genome Res 11(6): 1114–1125
3. Benzer S (1967) BEHAVIORAL MUTANTS OF drosophila ISOLATED BY COUNTER-CURRENT DISTRIBUTION. Proc Natl Acad Sci U S A 58(3):1112–1119
4. Sokolowski MB (2001) Drosophila: genetics meets behaviour. Nat Rev Genet 2(11): 879–890
5. Link N, Bellen HJ (2020) Using drosophila to drive the diagnosis and understand the mechanisms of rare human diseases. Development 147(21). https://doi.org/10.1242/dev.191411
6. Sayeed O, Benzer S (1996) Behavioral genetics of thermosensation and hygrosensation in drosophila. Proc Natl Acad Sci U S A 93(12): 6079–6084
7. Tominaga M (2006) The role of TRP channels in Thermosensation. In: Liedtke WB, Heller S (eds) TRP Ion Channel function in sensory transduction and cellular Signaling cascades. Frontiers in Neuroscience. Taylor & Francis Group, LLC, Boca Raton, p 271
8. Venkatachalam K, Montell C (2007) TRP channels. Annu Rev Biochem 76:387–417
9. Rosenzweig M, Brennan KM, Tayler TD, Phelps PO, Patapoutian A, Garrity PA (2005) The drosophila ortholog of vertebrate TRPA1 regulates thermotaxis. Genes Dev 19(4): 419–424
10. Viswanath V, Story GM, Peier AM, Petrus MJ, Lee VM, Hwang SW, Patapoutian A, Jegla T (2003) Opposite thermosensor in fruitfly and mouse. Nature 423(6942):822–823
11. Hamada FN, Rosenzweig M, Kang K, Pulver SR, Ghezzi A, Jegla TJ, Garrity PA (2008) An internal thermal sensor controlling temperature preference in drosophila. Nature 454(7201):217–220
12. Kwon Y, Shim HS, Wang X, Montell C (2008) Control of thermotactic behavior via coupling of a TRP channel to a phospholipase C signaling cascade. Nat Neurosci 11(8):871–873
13. Luo J, Shen WL, Montell C (2017) TRPA1 mediates sensation of the rate of temperature change in drosophila larvae. Nat Neurosci 20(1):34–41

14. Saito S, Nakatsuka K, Takahashi K, Fukuta N, Imagawa T, Ohta T, Tominaga M (2012) Analysis of transient receptor potential ankyrin 1 (TRPA1) in frogs and lizards illuminates both nociceptive heat and chemical sensitivities and coexpression with TRP vanilloid 1 (TRPV1) in ancestral vertebrates. J Biol Chem 287(36):30743–30754
15. Kurganov E, Zhou Y, Saito S, Tominaga M (2014) Heat and AITC activate green anole TRPA1 in a membrane-delimited manner. Pflugers Arch 466(10):1873–1884
16. Gracheva EO, Ingolia NT, Kelly YM, Cordero-Morales JF, Hollopeter G, Chesler AT, Sanchez EE, Perez JC, Weissman JS, Julius D (2010) Molecular basis of infrared detection by snakes. Nature 464(7291):1006–1011
17. Saito S, Banzawa N, Fukuta N, Saito CT, Takahashi K, Imagawa T, Ohta T, Tominaga M (2014) Heat and noxious chemical sensor, chicken TRPA1, as a target of bird repellents and identification of its structural determinants by multispecies functional comparison. Mol Biol Evol 31(3):708–722
18. Zhong L, Bellemer A, Yan H, Ken H, Jessica R, Hwang RY, Pitt GS, Tracey WD (2012) Thermosensory and nonthermosensory isoforms of Drosophila melanogaster TRPA1 reveal heat-sensor domains of a thermoTRP channel. Cell Rep 1(1):43–55
19. Gu P, Gong J, Shang Y, Wang F, Ruppell KT, Ma Z, Sheehan AE, Freeman MR, Xiang Y (2019) Polymodal nociception in drosophila requires alternative splicing of TrpA1. Curr Biol 29(23):3961–3973
20. Tracey WD Jr, Wilson RI, Laurent G, Benzer S (2003) Painless, a drosophila gene essential for nociception. Cell 113(2):261–273
21. Sokabe T, Tsujiuchi S, Kadowaki T, Tominaga M (2008) Drosophila painless is a Ca2+-=requiring channel activated by noxious heat. J Neurosci 28(40):9929–9938
22. Lee Y, Lee Y, Lee J, Bang S, Hyun S, Kang J, Hong ST, Bae E, Kaang BK, Kim J (2005) Pyrexia is a new thermal transient receptor potential channel endowing tolerance to high temperatures in Drosophila melanogaster. Nat Genet 37(3):305–310
23. Rosenzweig M, Kang K, Garrity PA (2008) Distinct TRP channels are required for warm and cool avoidance in Drosophila melanogaster. Proc Natl Acad Sci U S A 105(38): 14668–14673
24. Kwon Y, Shen WL, Shim HS, Montell C (2010) Fine thermotactic discrimination between the optimal and slightly cooler temperatures via a TRPV channel in chordotonal neurons. J Neurosci 30(31):10465–10471
25. Gallio M, Ofstad TA, Macpherson LJ, Wang JW, Zuker CS (2011) The coding of temperature in the drosophila brain. Cell 144(4): 614–624
26. Turner HN, Armengol K, Patel AA, Himmel NJ, Sullivan L, Iyer SC, Bhattacharya S, Iyer EPR, Landry C, Galko MJ, Cox DN (2016) The TRP channels Pkd2, NompC, and Trpm act in cold-sensing neurons to mediate unique aversive Behaviors to noxious cold in drosophila. Curr Biol 26(23):3116–3128
27. Ni L, Bronk P, Chang EC, Lowell AM, Flam JO, Panzano VC, Theobald DL, Griffith LC, Garrity PA (2013) A gustatory receptor paralogue controls rapid warmth avoidance in drosophila. Nature 500(7464):580–584
28. Ni L, Klein M, Svec KV, Budelli G, Chang EC, Ferrer AJ, Benton R, Samuel AD, Garrity PA (2016) The ionotropic receptors IR21a and IR25a mediate cool sensing in drosophila. elife 5. https://doi.org/10.7554/eLife.13254
29. Budelli G, Ni L, Berciu C, van Giesen L, Knecht ZA, Chang EC, Kaminski B, Silbering AF, Samuel A, Klein M, Benton R, Nicastro D, Garrity PA (2019) Ionotropic receptors specify the morphogenesis of phasic sensors controlling rapid thermal preference in drosophila. Neuron 101(4):738–747
30. Takeuchi K, Nakano Y, Kato U, Kaneda M, Aizu M, Awano W, Yonemura S, Kiyonaka S, Mori Y, Yamamoto D, Umeda M (2009) Changes in temperature preferences and energy homeostasis in dystroglycan mutants. Science 323(5922):1740–1743
31. Sokabe T, Chen HC, Luo J, Montell C (2016) A switch in thermal preference in drosophila larvae depends on multiple Rhodopsins. Cell Rep 17(2):336–344
32. Schindelin J, Arganda-Carreras I, Frise E, Kaynig V, Longair M, Pietzsch T, Preibisch S, Rueden C, Saalfeld S, Schmid B, Tinevez JY, White DJ, Hartenstein V, Eliceiri K, Tomancak P, Cardona A (2012) Fiji: an open-source platform for biological-image analysis. Nat Methods 9(7):676–682
33. R Core Team (2017) R: a language and environment for statistical computing. R Foundation for Statistical Computing, Vienna. https://www.R-project.org/
34. Harayama T, Riezman H (2018) Understanding the diversity of membrane lipid composition. Nat Rev Mol Cell Biol 19(5):281–296
35. Brand AH, Perrimon N (1993) Targeted gene-expression as a means of altering cell fates and

generating dominant phenotypes. Development 118(2):401–415

36. Suito T, Nagao K, Takeuchi K, Juni N, Hara Y, Umeda M (2020) Functional expression of Delta12 fatty acid desaturase modulates thermoregulatory behaviour in drosophila. Sci Rep 10. https://doi.org/10.1038/s41598-020-68601-2
37. Nagao K, Murakami A, Umeda M (2019) Structure and function of Delta9-fatty acid desaturase. Chem Pharm Bull (Tokyo) 67(4): 327–332
38. Klein M, Afonso B, Vonner AJ, Hernandez-Nunez L, Berck M, Tabone CJ, Kane EA, Pieribone VA, Nitabach MN, Cardona A, Zlatic M, Sprecher SG, Gershow M, Garrity PA, Samuel AD (2015) Sensory determinants of behavioral dynamics in drosophila thermotaxis. Proc Natl Acad Sci U S A 112(2):E220–E229
39. Yamaguchi S, Desplan C, Heisenberg M (2010) Contribution of photoreceptor subtypes to spectral wavelength preference in drosophila. Proc Natl Acad Sci U S A 107(12): 5634–5639
40. Branson K, Robie AA, Bender J, Perona P, Dickinson MH (2009) High-throughput ethomics in large groups of drosophila. Nat Methods 6(6):451–457
41. Romero-Ferrero F, Bergomi MG, Hinz RC, Heras FJH, de Polavieja GG (2019) Idtracker. Ai: tracking all individuals in small or large collectives of unmarked animals. Nat Methods 16(2):179–182
42. Mathis A, Mamidanna P, Cury KM, Abe T, Murthy VN, Mathis MW, Bethge M (2018) DeepLabCut: markerless pose estimation of user-defined body parts with deep learning. Nat Neurosci 21(9):1281–1289
43. Frank DD, Jouandet GC, Kearney PJ, Macpherson LJ, Gallio M (2015) Temperature representation in the drosophila brain. Nature 519(7543):358–361
44. Liu WW, Mazor O, Wilson RI (2015) Thermosensory processing in the drosophila brain. Nature 519(7543):353–357
45. Alpert MH, Frank DD, Kaspi E, Flourakis M, Zaharieva EE, Allada R, Para A, Gallio M (2020) A circuit encoding absolute cold temperature in drosophila. Curr Biol 30(12): 2275–2288
46. Marin EC, Buld L, Theiss M, Sarkissian T, Roberts RJV, Turnbull R, Tamimi IFM, Pleijzier MW, Laursen WJ, Drummond N, Schlegel P, Bates AS, Li F, Landgraf M, Costa M, Bock DD, Garrity PA, Jefferis G (2020) Connectomics analysis reveals first-, second-, and third-order Thermosensory and Hygrosensory neurons in the adult drosophila brain. Curr Biol 30(16):3167–3182
47. Umezaki Y, Hayley SE, Chu ML, Seo HW, Shah P, Hamada FN (2018) Feeding-state-dependent modulation of temperature preference requires insulin Signaling in drosophila warm-sensing neurons. Curr Biol 28(5): 779–787

Chapter 7

Quantifying Social Interactions in Medaka Fish

Saori Yokoi, Yasuko Isoe, Mu-Yun Wang, Masahiro Daimon, Teruhiro Okuyama, and Hideaki Takeuchi

Abstract

Zebrafish (*Danio rerio*) and medaka fish (*Oryzias latipes*) are popular vertebrate models for various biology fields. Medaka fish have recently attracted attention in the field of social neuroscience due to some of their unique advantages. First, medaka fish females have an extremely short ovarian cycle (24 h), which allows one to obtain a large number of female individuals of a similar physiological status according to their synchronized reproductive rhythms, facilitating quantitative analysis of female behaviors that are strongly influenced by the ovarian cycle; other experimental models such as mice and zebrafish have a longer ovarian cycle and/or unpredictable ovarian maturation. Second, medaka fish possess the visual ability to recognize conspecifics and make appropriate behavioral choices based on their social relationship. Third, the medaka fish are an established model in the field of molecular genetics (as are zebrafish) and state-of-the-art molecular-genetic methods are available. Here we describe several experimental procedures for behavioral paradigms to quantify medaka social behaviors, which are primarily mediated by visual cues.

Key words Social behavior, Medaka, Visual recognition, Molecular genetics, Social vision, Neuromodulator

1 Introduction

Various kinds of fish species are able to distinguish conspecific individuals, enabling them to choose the appropriate behavioral response and achieve social adaptation [1, 2]. For example, in the African cichlid fish *Astatotilapia burtoni*, males escape from higher-ranking individuals but approach lower-ranking individuals, demonstrating an ability to discriminate between group members and identify their social rank [3–5]. Guppies (*Poecilia reticulata*) can recognize familiar mates after social association and prefer to mate with unfamiliar conspecifics [6, 7]. Comparative studies among vertebrates could elucidate the evolutionary roots and basic neural mechanism underlying our social cognition and

Authors Saori Yokoi, Yasuko Isoe, Mu-Yun Wang have equally contributed to this chapter.

Daisuke Yamamoto (ed.), *Behavioral Neurogenetics*, Neuromethods, vol. 181, https://doi.org/10.1007/978-1-0716-2321-3_7,

behavioral choices. The neural/molecular basis of these mechanisms in fish species, however, remains poorly understood [8]. Medaka fish constitute an attractive model in the field of social neuroscience for further exploration of these issues [9].

Medaka (*Oryzias latipes*) is a small freshwater fish native to East Asia. The body length of sexually mature fish is approximately 3–4 cm. This species diverged from the zebrafish approximately 314–332 million years ago [10]. While the zebrafish females exhibit an ovarian cycle of 5 days [11], the medaka female has an ovarian cycle of only 24 h and spawns every morning. This confers a strong advantage while using medaka fish to analyze the social behaviors of females with a similar physiological status [12]. Different types of behavioral paradigms are available for investigating social interactions among medaka fish, such as mate preference, mate guarding, hierarchy formation, and schooling behaviors [13–17]. Importantly, medaka fish can distinguish familiar conspecific individuals on the basis of visual information. Medaka females discriminate between individual males and prefer to mate with socially familiar males (female mating preference) [18]. Furthermore, medaka use visual cues from the face (head part) for individual recognition. Females can discriminate between two male faces and two objects, but discrimination of upside-down faces is more difficult [19]. The face inversion effect in medaka fish implies that faces are particularly important for social cognition in medaka, as in humans [20]. Medaka males also show a visually mediated mate preference for females on the basis of body color, which requires color perception mediated by opsin genes in the retina [16]. Interestingly, the gene expression patterns of these opsin genes change according to seasonal reproduction, which could influence both color perception and mate preference [21].

Another advantage of medaka fish is that these fish are an established model organism for molecular genetics [22]. Genome editing using the CRISPR/Cas9 system [23–25] has contributed enormously to studies on the molecular basis of medaka social behaviors [26–28]. For example, behavioral analysis of a series of oxytocin ligand/receptor mutants revealed that these mutations influence mate preference in a sex-specific manner. In wild-type medaka fish, females prefer to choose familiar mates, whereas males court indiscriminately and irrespective of familiarity. The oxytocin signaling pathway is essential for eliciting female mate preference for familiar males, while it is also required for the unrestricted and indiscriminate preference of males. In addition, several efficient methods for targeted gene knock-in using the CRISPR/Cas9 system have been established in medaka fish [25, 29, 30], where a reporter gene of interest can mimic the expression of endogenous target genes in transgenic strains. Cre-loxP site-specific recombination is also used to manipulate gene expression in a spatially specific manner in medaka fish. For example, Cre/loxP site-specific recombination using an artificial heat-shock promoter with an infrared laser-evoked gene operator system allows for inducing spatiotemporal Cre expression in the neural precursor cells of medaka embryos [31–34]. More recently, it was discovered that the doxycycline (Dox)-inducible Tet-ON system [35], a robust tool for inducible and reversible transgene expression under a specific promoter, can be applied to medaka fish (personal communication). These characteristics of medaka fish make this species an excellent model for studies of the neuronal/molecular mechanisms that underlie visually mediated social recognition, using advanced genetic/molecular-genetic tools. Here we introduce experimental procedures for behavioral paradigms to quantify three types of social behaviors: (1) female mate preference for visually familiarized males, (2) mate-guarding behaviors, and (3) visually mediated social attraction.

1.1 Female Mate Preference for Visually-Familiarized Male

Medaka mating behaviors comprise a series of stereotyped steps (please see our video: https://www.youtube.com/watch?v=7HB5UaNrTSQ) [36]. In the first step, the male approaches the female and swims underneath her (approaching). When the female stops swimming, the male exhibits a courtship display by swimming rapidly in a circle in front of the female (quick-circle display); he then wraps her with the anal and dorsal fins (crossing). If the female is receptive, the pair simultaneously releases their gametes. If the female is not receptive, the female tends to exhibit head-up movements (rejection) and escapes from the male [37]. Thus, female receptiveness toward males can be judged based on female behavior.

Two behavioral paradigms are used for quantifying female mating preference. In the first paradigm, "latency to mate," which is defined as the interval from the first courtship display the "quick-circle display," to the beginning of "crossing" followed by gamete release, is measured when a pair of fish encounters each other in a small tank in the morning (Fig. 7.1a and b). "Latency to mate" negatively correlates with the degree of female receptiveness to a focal male [18, 19, 38, 39]. Visual familiarization, such as allowing fish to see each other through a transparent wall before a mating test, significantly reduces the latency of the female to mate with the familiar male (Fig. 7.1c). In the second paradigm, two males and one focal female (a triadic relationship) are placed together, and a paternity test is performed to determine which of the two males was selected by the female for mating [18] (Fig. 7.1d). The paternity test is conducted by evaluating the expression of fluorescent proteins in the progeny. We use one transgenic male expressing fluorescent proteins and one wild-type male in the triadic relationship. When either one of the two males is visually familiarized with the focal female, the female selects the

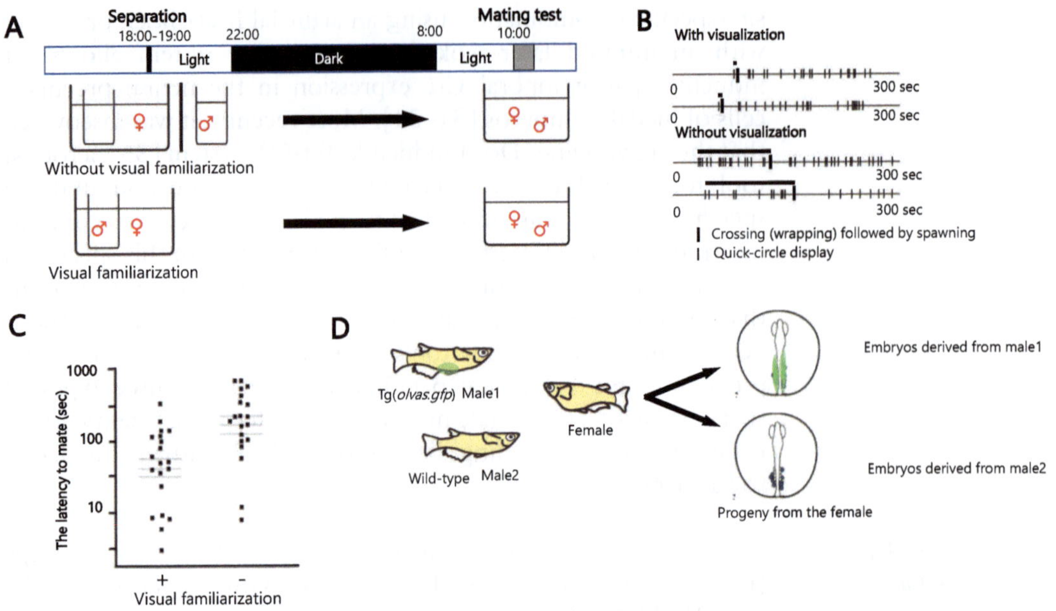

Fig. 7.1 Female mating preference test. (**a**) A procedure for mating tests to assess female receptiveness. A pair (female and male) were kept in a 14/10 h light/dark cycle. They were separated the day before the mating assay and placed together the following morning. To assess the effect of visual familiarization on female receptiveness, we compared two groups with and without visual familiarization. Previous studies show that visual familiarization could not enhance female receptiveness toward unfamiliar males [18, 19]. (**b**) Typical examples of the two groups with/without visual familiarization, respectively. We recorded mating behaviors for 300 sec and calculated "the latency to mate" as the interval between the first "quick-circle display" (thin horizontal lines) and "crossing following by spawning" (thick horizontal lines). (**c**) Comparison of "the latency to mate" between the two groups. We log-transformed "the latency to mate" to mate with/without visual familiarization, respectively. The latency was log-transformed to satisfy the assumption of normality. Grey lines indicate mean ± SEM time. This panel was modified from [19]. (**d**) Paternity test in a triad relationship (two males and one female). Heterozygote embryos derived from Tg(olvas:*gfp*) have GFP fluorescence in the abdomen (primordial germ cells). In most cases, the fertilized eggs are derived from only the first mating partner. A single brood (5–20 eggs) consists of only GFP-positive or GFP-negative embryos [18]

familiar male as the mating partner approximately 90% of the time [18].

1.2 Male Mate-Guarding Behaviors in Medaka Fish

When two males and one female medaka are placed in a small tank, the two males compete against each other for the female. In the male-male competition, the dominant male exhibits mate-guarding behavior: he maintains a dominant position near the female and interferes with the approach of the rival male toward the female [38–40]. To quantify the degree of mate-guarding behavior, we assess the guarding index of the focal male (frequency of locating between the female and rival male) by calculating the frequency at which he is located within the "guarding circle" between the female and the rival male (Fig. 7.2a). Mate-guarding behaviors can also be elicited using a three-chamber tank (Fig. 7.2b). The female is

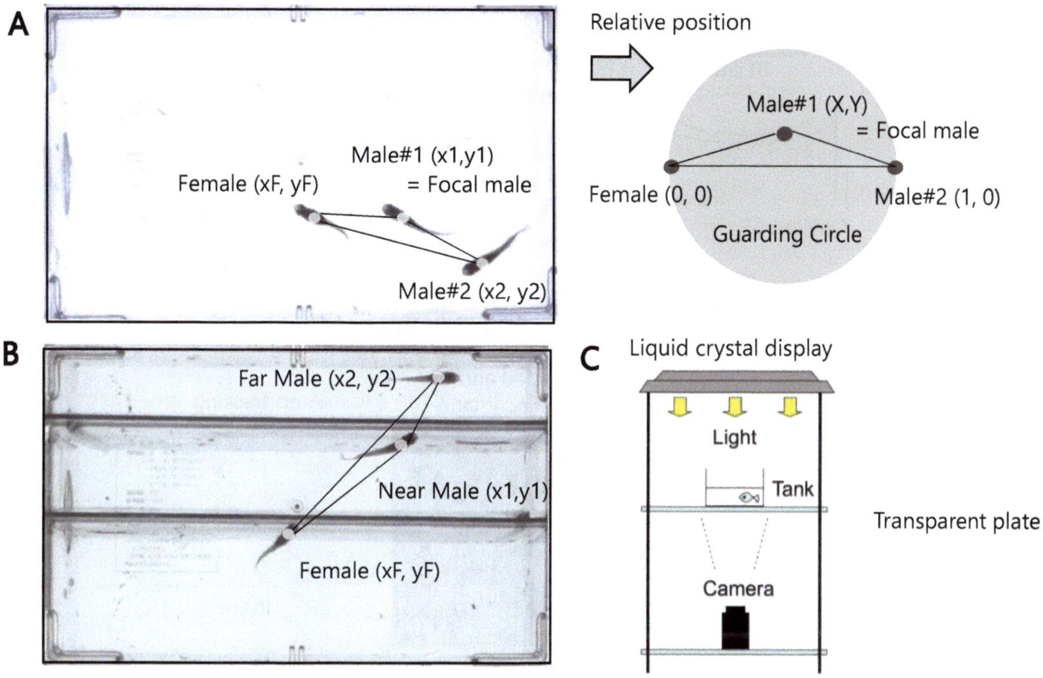

Fig. 7.2 Mate-guarding test in the triadic relationship (two males and one female)
(**a**) Based on the positions of the female (xF, yF), male #1 (x1, y1), and the other male #2 (x2, y2), the relative positions of the focal male (male #1) (X, Y) were calculated when the positions of the female, and the other male #2 were defined as (0, 0) and (1, 0), respectively. The "guarding circle" is defined as a circle with center (1/2, 0) and radius 1/2. When the focal male is present in the guarding circle, he remains near the female and interferes with the rival (the male #2). The probability of being in the guarding circle represents the degree of mate-guarding (guarding index) [38–40]. (**b**) Three-chamber test for mate-guarding. This condition could induce mate-guarding behaviors for only the "near male." Previously, we used this method to investigate the effect of mate-guarding behaviors on female mating receptivity [38]. (**c**) Setup for the behavior-recording system. The white screen of a liquid crystal display was fixed above the tank for a uniform illumination

placed in the larger chamber on one side, and then one male is placed in the "far" chamber and the other in the "near" chamber. Under these conditions, only the male in the "near" chamber exhibits mate-guarding behavior, maintaining a position between the female and the male in the "far" chamber.

1.3 Visually Mediated Social Attraction

Social attraction reflects the tendency to approach and interact with conspecifics, which is the first and an essential step in various social behaviors, such as schooling [14)] mating [18], mate-guarding [38–40], and aggressive behaviors [41–43]. Social attraction requires the ability to recognize conspecifics on the basis of cues derived from other individuals. In medaka fish, social attraction is primarily mediated by visual cues from conspecifics. Three-dimensional computer graphic animations of medaka fish are available for studying social attraction, and modifications of these animations reveal the types of morphological features and motion

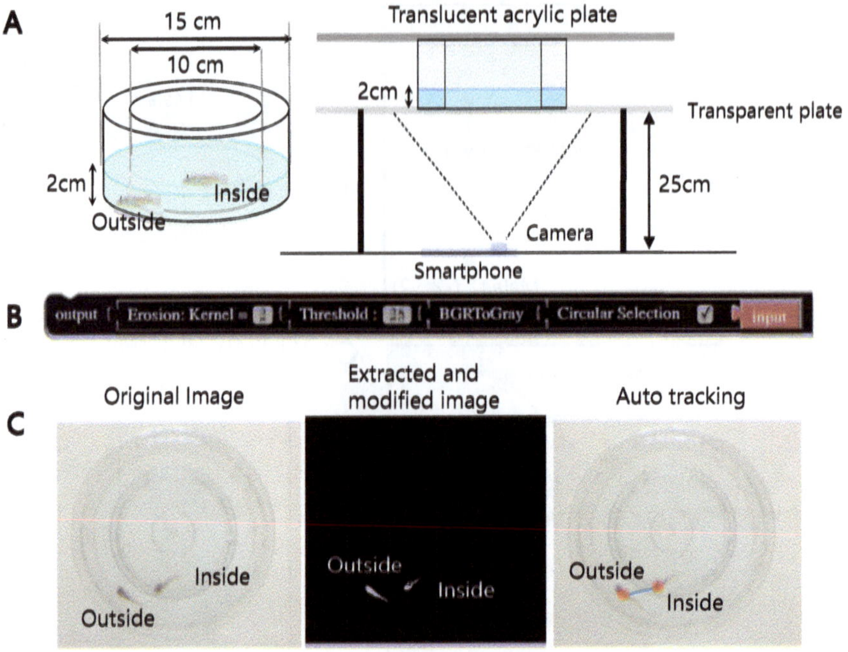

Fig. 7.3 Visually mediated social attraction. (**a**) Setup for behavior-recording system. (**b**) An example of a combination of block editors (UMATracker FilterGenerator): "Circular selection," "BGRToGRAY," "Threshold," and "Erosion kernel" to extract fish images from backgrounds. The value of "Threshold" and "Erosion kernel" should be adjusted to extract fish images. (**c**) After extraction and modification of fish images, the coordinate data of the gravity center of two fish were automatically measured using UMATracker-Tracking

characteristics that are necessary for the social attraction [44, 45]. We established a simple behavioral paradigm to assess visually mediated social attraction using real medaka fish [46]. When two fish were placed separately in a cylindrical tank with a concentric transparent wall, they maintained close proximity to each other (Supplemental Movie S1, Fig. 7.3a). An intact fish in the innermost chamber of the tank maintains proximity to a blind fish (optic nerve cut) in the outermost chamber of the tank, and not vice versa [46]. Thus, this behavioral paradigm allows for assessing the visually mediated social attraction of a single fish.

2 Materials

2.1 Preparation for Medaka Fish

The Medaka National BioResource Project (http://www.shigen.nig.ac.jp/medaka/) provides several experimental strains of medaka fish (*Oryzias latipes*; d-rR strain, Cab strain, and transgenic medaka Tg(olvas:*gfp*) for paternity tests. All fish are hatched and bred under laboratory conditions following a standard protocol [47]. In our laboratory, we use closed water recirculation systems (Meito Suien Co., Ltd.) that are placed indoors with a pump and a filter to maintain the water quality. The water temperature is maintained at ~28 °C, and light is provided via standard fluorescent/LED lamps for 14 h/day (08:00 to 22:00). Water quality deterioration disrupts the female reproduction cycle. To prevent infectious diseases, fish obtained from pet shops should not be brought into the laboratory facility. Sexually mature male and female fish (body length > 25 mm) are each kept in a plastic tank (22.5 cm × 15 mm × 16.5 cm [height]). The fish are fed twice per day with brine shrimp (*Artemia*) larvae and artificial powdered food (Otohime B2, 360–650 μm grain powder: Nisshin Marubeni Food Co. Ltd., Tokyo). The dried cyst-like eggs of brine shrimp (Brine Shrimp Eggs, Nichido, Japan) hatch after 18–24 h incubation in 2% (w/v) salt water with aeration at 25–28 °C.

3 Methods

3.1 Mating Test to Confirm Sexual Maturation

To confirm that the fish are sexually mature and ready to spawn, we always perform a mating test prior to performing the behavioral experiments. During the mating test, fish are fed brine shrimp (once per day) and artificial powdered food (twice per day). Male and female adult medaka (age > 2 months and body length > 25 mm) are selected and kept in a plastic aquarium (12 cm × 19.5 cm × 15.0 cm [height], water depth 7.5 cm). Female and male fish are separated the evening before the mating test. The male is separated from the female and placed in a columnar plastic container with a radius of 3.5–5 cm and a height of 9.0 cm (CE-300, Kenis, Japan), wrapped with opaque white paper. The following morning (9:00–12:00), the male is gently placed into the tank with the female along with all of the water from the plastic columnar container. After 10 min, we check whether females have fertilized eggs on their abdomen. Unfertilized eggs become white and turbid after 24 h.

3.2 Female Preference for Visual Familiarization

The female preference test is basically the same as the mating test described in Sect. 1.1. Males and females are separated during the evening (18:00–19:00) before the day of the assay (Fig.7.1a), kept under either of two conditions (with visual familiarization or without visual familiarization). Before separation, all debris is removed from the aquarium, and water flow from the water recirculation systems is suspended. For visual familiarization, the female is allowed to see the male prior to mating; the female is kept in the plastic aquarium (12 cm × 19.5 cm × 15.0 cm [height], water depth 7.5 cm), while the male is separated from the female in a transparent cylinder (glass/plastic) with a radius of 4–5 cm and a height of 15 cm. Clear transparent cylinders without any scratches should be used to allow the pair to see each other well. The container holding the male is placed in the plastic aquarium with the female for visual familiarization. For a negative control without familiarization, the cylinder is wrapped with opaque white paper and placed outside the tank that contains the female. The water temperature of the columnar container is kept the same as that in the female tank. The next morning (10:00), the male is quietly introduced into the aquarium with the female for the mating behavior assay (10:00–12:00). The behavior of the pair is recorded for 5 min using a wide-angle camera (BSW200MBK, Buffalo, Japan), which is placed on a shelf on the water recirculation systems just above the test tank 10 min before releasing the male. The recording is controlled using a personal computer running the Windows operating system with a USB video camera recording program (NVCap_R, Net Vision, Japan). The timing of the male courtship display (quick-circle display) and crossing followed by spawning is determined manually based on the recording. "Latency to mate" is defined as the interval between the first quick-circle display and the beginning of crossing followed by spawning.

3.3 Paternity Test Using Transgenic Fish

To determine which male is selected as the mating partner in the triadic relationship (two males and one female), a paternity assay is conducted (Fig.7.1d). The paternity test is basically the same as the mating test described in 2.2, except two males are used; one male is of the wild-type strain (the d-rR strain), while the other male is of the homozygote transgenic strain, Tg(olvas:*gfp*), in which green fluorescent protein (GFP) expression in primordial germ cells (PGCs) can be observed from the embryonic stage to an adult. If the female selects the Tg(olvas:*gfp*) male as her mating partner, the progeny (embryos) should show the GFP fluorescence in their PGCs, visible in the embryonic stage at 3 days after fertilization using fluorescence microscopy (MZ16F, Leica, Tokyo, Japan).

3.4 Mate-Guarding Behaviors

The setup for recording mate-guarding behavior is shown in Fig. 7.2c. Fish behaviors are recorded from below, as the bottom of the tank is transparent. Medaka fish seem to pay no attention to objects below them. To our knowledge, the size and the shape of the tank do not influence mate-guarding behavior. We tested two rectangular tanks (small 9.5 cm × 13 cm × 12 cm [height] and large 21 cm × 30 cm × 10 cm [height]) and 1 circular tank (15 cm diameter) [40]. The day before the behavioral test, the two males are separated into opaque cylinders (glass/plastic) with a radius of 4–5 cm and a height of 15 cm, while the female is kept in the test tank. Then they are placed into the same tank for the behavioral test. As far as we tested in the morning (before/after spawning) and the evening, the experimental time for the behavioral tests did not influence male mate-guarding activity [40].

After 1 min acclimation, the behaviors are recorded. To calculate the guarding index, time-series coordinate data of the three individuals are recorded for 100 s to calculate the relative locations of the three fish. Time-series coordinate data are measured with low time resolution (0.2 fps). The 100 s movie is then converted into 21 image sequences per 5 s and the head and tail positions of the 3 medaka fish are manually spotted using ImageJ (NIH, Bethesda, MD, USA) with 2 plugins ("Stack Sorter and "Cell counter"). You may use an auto-tracking system (UMATracker; please see Sect. 3.6) to measure coordinate data with high resolution (30 fps). To assess the frequency of the focal male locating between the other male and the female, we defined the "guarding circle," the zone between the other rival male and the female, as a circle with center (1/2, 0) and radius 1/2, when the female and the other rival male were defined as (0, 0) and (1, 0), respectively (Fig.7.2a). The presence of the focal male in the guarding circle indicates that the focal male occupies a dominant position to remain near the female as well as to interfere with the rival. The relative positions of the three fish and the probability of the focal male being within the "guarding circle" are calculated. The probability is defined as the "guarding index," which represents the degree of mate-guarding by the focal male.

The guarding index is used for two purposes. The first purpose is to evaluate how mate-guarding behavior emerges in the triadic relationship compared to a negative control. For negative control groups, we place and record one of the three fish, respectively, in the tank for 100 s and virtually merge the data of the three fish (one female and two males), which represents the condition without any inter-individual interaction among the three fish. If the "guarding index" of either of the two males significantly higher in the experimental groups than in the merged groups, this indicates that the focal males in the experimental groups exhibit mate-guarding. The second purpose is to determine which male is dominant based on the guarding index in a triadic relationship (two males, one female). The guarding indices between two males with different genotypes (heterozygous mutant and homozygous mutant) can be directly compared and the effect of mutation on mate-guarding activity evaluated [39, 40].

3.5 Mate-Guarding Behaviors in a Three-Chamber Tank

To assess the mate-guarding behaviors of a single focal male, a three-chamber tank is used [38]. An aquarium is divided into three chambers using transparent walls ("female zone": 6.5 cm × 19 cm × 12 cm [height], "near male zone," and "far male zone": each 3.25 cm × 19 cm × 12 cm [height]; Fig. 7.2b). A triad of size-matched fish (one female and two males) is selected, and each fish is placed separately into one of the three chambers. The procedure of recording and calculating the "guarding index" is the same as that described in Sect. 3.3. Under this condition, only the male in the "near zone" exhibits mate-guarding behaviors by maintaining a position between the far male and the female.

3.6 Social Attraction

The detailed procedure is explained in Fig. 7.3a. Two round plastic transparent tanks are used: one with a 15 cm diameter (Container φ150mm, 150-1, AS ONE Corporation) and the other with a 10 cm diameter (Round V Type V-9, φ99.5 × φ94 × 57 mm, AS ONE Corporation) (Fig. 7.3a). The water depth is maintained at ~2 cm. After placing a fish in each part ("outside" and "inside" in Fig. 7.3a), the water tanks are covered with a translucent acrylic plate (300 mm × 300 mm, thickness 2 mm) to provide a clear visual background for assessing fish movement (Fig.7.3a). A video(30fps) is recorded from below the water tanks through a transparent acrylic plate (300 mm x300mm, thickness 2 mm) using smartphones (Android or iOS devices). Just after the two fish began to move, a 1 min video was recorded. If neither of the two fish moves for 3 min under this condition, it may be that the fish condition is not appropriate for the test. In this case, fish in a more optimal condition should be used, and the water temperature should be confirmed to be 25–28 °C. For negative controls, we used opaque inner tanks wrapped with white vinyl tape (Supplemental Movie S2). The behavioral tests for the negative control should be conducted using the same fish pair at the same time zone when the experimental groups are used. To judge whether a focal fish in the inside exhibits social attraction behaviors toward fish in the outside, the distance between the two fish in the experimental groups is measured to determine whether it is statistically shorter than that in the negative control groups. The video data are analyzed with an open-source quantitative analysis toolkit for animal experiments, UMATracker [48]. To automatically extract shapes of the fish body from the background, a filter (UMATracker FilterGenerator) is created with a combination of four block editors: "Circular selection" to extract images within the outer tank, "BGRToGray" to converted them to gray scale, "Threshold" to enhance the contrast in extracted medaka body images, and "Erosion kernel" to remove images from debris (Fig.7.3b). UMATracker-Tracking is used to track coordinate data of the two medaka fish in the inner and outer tanks, respectively (Fig.7.3c). The distance between the two fish is calculated in each frame.

4 Notes

4.1 Breeding Condition Strongly Influence Ovarian Cycle and Female Behaviors

The most important point of behavioral tests for assessing intersex interaction is to prepare sexually mature male and female fish that are ready to spawn. It is particularly important to confirm that females spawn eggs at least continuously for 3 days just before the behavioral tests. The feeding frequency using brine shrimp can be increased to three to four times per day to improve the nutritional condition and promote egg spawning. No feeding for 24–48 h usually disrupts the female reproductive cycle for several days.

Because too much feeding may lead to water quality deterioration, the absence of leftover food should be confirmed. When the food stays unconsumed, algae can also grow on the tank walls. A thin layer of algae (< 5 mm deep) may improve fish conditions, although excessive algae can result in water deterioration.

For the mating test (female preference for visually familiarized male), we recommend that the tanks used for the behavioral experiment be fixed on the shelf of the breeding system, as tank movement may sometimes disrupt mating behaviors. A decrease in the water temperature will decrease the overall behavioral activity of medaka fish during behavioral tests. Thus, if the temperature of the rooms used for behavioral tests is lower than 25 °C, the water temperature should be checked regularly and maintained at 25–28 °C. In addition, in methods 3.1–3.5, females should not be kept in small tanks such as a cylinder with a radius of 4–5 cm for more than a few hours, as it may stress the females and disrupt the ovarian cycle.

4.2 Recording Conditions for the Automatic Tracking

To use the automatic tracking software (UMATracker), it is necessary to record videos with a uniform and bright background. A liquid crystal display should be used as an illuminator in methods 3.4 and 3.5 (Fig.7.2c) and a translucent acrylic plate to cover the experimental tanks in method 3.6 (Fig.7.3a). If the fish images cannot be extracted from the background, consider using brighter and more uniform illumination. All debris must be removed from the aquarium as it is often detected as a target object by UMATracker -Tracking.

Acknowledgments

We thank developers of UMATracker (Rito Takeuchi and Osamu Yamanaka) for giving us valuable comments. This work was supported by the National Institute for Basic Biology Priority Collaborative Research Project 10-104 and Cooperative Research Project 19-347; Joint Research Grant 01111904 by the National Institutes of Natural Sciences; Japan Society for the Promotion of Science (JSPS) KAKENHI Grants 21H04773, 20H04925 (to H.T.), 18H02479 (to H.T.), 19 K16247 (to S.Y.), 19H04889 (to S.Y.), Astellas Foundation for Research on Metabolic Disorders (to S.Y.), Takeda Science Foundation (to H.T. and S.Y.), and the natural science grant of the Mitsubishi Foundation.

References

1. Bshary R, Brown C (2014) Fish cognition. Curr Biol 24:R947–R950
2. Brown C, Laland K, Krause J Fish cognition and behaviour. pp 1–9. https://doi.org/10.1002/9781444342536.ch1
3. Maruska KP, Fernald RD (2018) *Astatotilapia burtoni*: a model system for analyzing the neurobiology of behavior. ACS Chem Neurosci 9: 1951–1962
4. Fernald RD (2017) Cognitive skills and the evolution of social systems. J Exp Biol 220: 103–113
5. Grosenick L, Clement TS, Fernald RD (2007) Fish can infer social rank by observation alone. Nature 445:429–432
6. Hughes KA et al (1999) Familiarity leads to female mate preference for novel males in the guppy, *Poecilia reticulata.* Anim Behav 58: 907–916
7. Mariette MM et al (2010) The effects of familiarity and group size on mating preferences in the guppy, *Poecilia reticulata.* J Evol Biol 23: 1772–1782
8. Bshary R, Gingins S, Vail AL (2014) Social cognition in fishes. Trends Cogni Sci 18: 465–471
9. Okuyama T, Yokoi S, Takeuchi H (2017) Molecular basis of social competence in medaka fish. Develop Growth Differ 59: 211–218
10. Kasahara M et al (2007) The medaka draft genome and insights into vertebrate genome evolution. Nature 447:714–719
11. Hisaoka KK, Firlit C (1962) Ovarian cycle and egg production in the zebrafish, *Brachydanio rerio.* Copeia 1962:788
12. Karigo T et al (2012) Time-of-day-dependent changes in GnRH1 neuronal activities and gonadotropin mRNA expression in a daily spawning fish, medaka. Endocrinol 153: 3394–3404
13. Airi O et al (2020) Immediate early gene expression in brain regions associated with the social behavioral network after male competition in medaka fish. Zool Sci 37:391–398
14. Imada H et al (2010) Coordinated and cohesive movement of two small conspecific fish induced by eliciting a simultaneous optomotor response. PLoS One 5:e11248
15. Ochiai T et al (2013) A new data-mining method to search for behavioral properties that induce alignment and their involvement in social learning in medaka fish (*Oryzias latipes*). PLoS One 8:e71685
16. Fukamachi S et al (2009) Dual control by a single gene of secondary sexual characters and mating preferences in medaka. BMC Biol 7:64
17. Kamijo M, Kawamura M, Fukamachi S (2018) Loss of red opsin genes relaxes sexual isolation between skin-colour variants of medaka. Behav Process 150:25–28
18. Okuyama T et al (2014) A neural mechanism underlying mating preferences for familiar individuals in medaka fish. Science 343:91–94
19. Wang MY, Takeuchi H (2017) Individual recognition and the 'face inversion effect' in medaka fish (*Oryzias latipes*). elife 6
20. Valentine T (1988) Upside-down faces: a review of the effect of inversion upon face recognition. Br J Psychol 79:471–491
21. Shimmura T et al (2017) Dynamic plasticity in phototransduction regulates seasonal changes in color perception. Nat Commun 8:412
22. Furutani-Seiki M, Wittbrodt J (2004) Medaka and zebrafish, an evolutionary twin study. Mech Dev 121:629–637
23. Ansai S, Kinoshita M (2014) Targeted mutagenesis using CRISPR/Cas system in medaka. Biol Open 3:362–371
24. Ansai S, Kinoshita M (2017) Genome editing of Medaka. Methods Mol Biol 1630:175–188
25. Watakabe I et al (2018) Highly efficient generation of knock-in transgenic medaka by CRISPR/Cas9-mediated genome engineering. Zool Lett 4:3
26. Nishiike Y et al (2021) Estrogen receptor 2b is the major determinant of sex-typical mating behavior and sexual preference in medaka. Curr Biol 31:1699–1710.e1696
27. Kayo D et al (2019) Gene knockout analysis reveals essentiality of estrogen receptor β1 (Esr2a) for female reproduction in medaka. Sci Rep 9:8868
28. Kanazawa N et al (2020) Changes in a cone opsin repertoire affect color-dependent social behavior in medaka but not behavioral photosensitivity. Front Genet 11:801–801
29. Murakami Y et al (2017) An efficient system for homology-dependent targeted gene integration in medaka (*Oryzias latipes*). Zool Lett 3: 10
30. Murakami Y et al (2020) CRISPR/Cas9 nickase-mediated efficient and seamless knock-in of lethal genes in the medaka fish Oryzias latipes. Develop Growth Differ 62: 554–567

31. Kobayashi K et al (2013) A heat-inducible CRE/LOXP gene induction system in medaka. Genesis 51:59–67
32. Deguchi T et al (2009) Infrared laser-mediated local gene induction in medaka, zebrafish and *Arabidopsis thaliana*. Develop Growth Differ 51:769–775
33. Kamei Y et al (2009) Infrared laser-mediated gene induction in targeted single cells in vivo. Nat Methods 6:79–81
34. Okuyama T et al (2013) Controlled Cre/loxP site-specific recombination in the developing brain in medaka fish, *Oryzias latipes*. PLoS One 8:e66597
35. Gossen M et al (1995) Transcriptional activation by tetracyclines in mammalian cells. Science 268:1766
36. Ono Y, Uematsu T (1957) Mating ethogram in *Oryzias latipes*. J Fac Sci Hokkaido Univ 13: 197–202
37. Walter RO, Hamilton JB (1970) Head-up movements as an indicator of sexual unreceptivity in female medaka, *Oryzias latipes*. Anim Behav 18:125–127
38. Yokoi S et al (2016) Mate-guarding behavior enhances male reproductive success via familiarization with mating partners in medaka fish. Front Zool 13:21
39. Yokoi S et al (2020) Sexually dimorphic role of oxytocin in medaka mate choice. Proc Natl Acad Sci U S A 117:4802–4808
40. Yokoi S et al (2015) An essential role of the arginine vasotocin system in mate-guarding behaviors in triadic relationships of medaka fish (*Oryzias latipes*). PLoS Genet 11: e1005009
41. Kagawa N et al (2017) Social rank-dependent expression of gonadotropin-releasing hormones and kisspeptin in the medaka brain. Gen Comp Endocrinol 249:48–54
42. Kagawa N (2014) Comparison of aggressive behaviors between two wild populations of Japanese medaka, *Oryzias latipes* and *O. sakaizumii*. Zool Sci 31:116–121
43. Kagawa N (2013) Social rank-dependent expression of arginine vasotocin in distinct preoptic regions in male Oryzias latipes. J Fish Biol 82:354–363
44. Nakayasu T, Watanabe E (2014) Biological motion stimuli are attractive to medaka fish. Anim Cogn 17:559–575
45. Nakayasu T et al (2017) Three-dimensional computer graphic animations for studying social approach behaviour in medaka fish: effects of systematic manipulation of morphological and motion cues. PLoS One 12: e0175059
46. Isoe Y et al (2016) Ontogeny and sexual differences in swimming proximity to conspecifics in response to visual cues in medaka fish. Zool Sci 33:246–254
47. Medaka Management. In: Medaka. pp 31–66. https://doi.org/10.1002/9780813818849.ch2
48. Yamanaka O, Takeuchi R (2018) UMA-Tracker: an intuitive image-based tracking platform. J Exp Biol 221

Chapter 8

A Method for Selective Breeding to Domesticate Mice

Bhim B. Biswa and Tsuyoshi Koide

Abstract

Domestication of animals plays an important role in human life by providing food and supporting work. A number of livestocks have been developed by selecting animals that exhibit an appropriate phenotype in the process of breeding in human history. Several experimental approaches have been applied to domesticated animals from wild stock in several species. Tameness, a major factor in domestication, is composed of two components—active and passive tameness. In a previous study, we established three behavioral tests (active tameness, passive tameness, and stay-on-hand tests) to measure these two components separately. For further studies, we applied selective breeding as a behavioral parameter of active tameness. Wild-derived heterogeneous stock (WHS), which was established from eight wild strains originating from different countries, was used as a founder stock for selective breeding. Given that WHS was derived from eight genetically different wild strains, the founder stock was highly diverse in terms of the domestication phenotype. The results of selection showed that the value of active tameness increased through the generation of selective breeding. Animals established by selective breeding are highly useful for not only studying the mechanism of domestication but also other behavioral studies. In this chapter, we have described the methods for measuring tameness in mice, developing WHS, and conducting selective breeding based on the score of active tameness.

Key words Tameness, Mouse, Domestication, Selective breeding, Heterogeneous stock, Behavior, Active tameness

1 Introduction

1.1 Domestication

Domestication, an age-old process, is defined as changing an organism phenotype for human benefits. Various livestocks, such as cows, pigs, goats, and sheep, and companion animals, such as dogs and cats, have been successfully domesticated. In animals, the primary phenotype of domestication is tameness or tolerance toward human handling and presence [1]. It is likely that many domesticated animals were first selected for tameness, which helps in rearing animals in captivity. In a famous domestication experiment of foxes (*Vulpes vulpes*) conducted at the Institute of Cytology and Genetics, Novosibirsk, Russia [2, 3]. Dmitry K. Belyaev and his

Daisuke Yamamoto (ed.), *Behavioral Neurogenetics*, Neuromethods, vol. 181, https://doi.org/10.1007/978-1-0716-2321-3_8,

colleagues used 30 male and 100 female foxes, mostly obtained from a commercial fur farm, as founder animals for the experiment.

For this experiment, an evaluation method of tameness was developed. Using the criterion of tameness, they crossed the top 4–5% tamest females with the top 20% tamest males by maintaining genetic heterogeneity to some extent. The proportion of male and female foxes showing a high level of tameness increased from 18% to 35% from the 10th to the 20th generation by selective breeding [4]. Similar selective breeding was conducted on wild-caught grey rats (*Rattus norvegicus*), in which rats were selected for tameness and increased defensive aggression toward humans [1, 5–7]. Thirty percent of the animals that showed the highest or lowest scores of aggressions toward a human hand were chosen and used for cross-breeding to produce successive generations. This selective breeding resulted in a behavioral difference by the 10th–12th generation and generated two lines, one of which was aggression and the other was tameness. Successful results of selective breeding in different animals indicate that tameness is influenced by genetic factors and may have been used for selection during the domestication of animals in human history.

However, these experiments did not report on the genetic history of the founder animals; hence, it is impossible to repeat selective breeding from the same founder stock. In addition, it is difficult to know how many alleles existed at the beginning and how many alleles were lost during breeding. To address the issue of unknown genetic profiles in the founder stock, a new heterogeneous stock established from multiple inbred strains was used for the selective breeding experiment.

1.2 Tameness

The establishment of an ideal behavior test and a clear definition of tameness are major issues in domestication studies. Albert et al. used 13 different parameters in their study [1], while Blanchard et al. used 10 parameters [6]. The methods used in these experiments were very different, and it is likely that tameness is composed of various factors. Therefore, it is necessary to clearly define tameness and establish a behavioral test that measures the various factors of tameness.

Tameness can be categorized as active and passive tameness [8, 9]. Active tameness is the motivation to approach humans, and passive tameness is a reluctance to avoid humans. To study tameness, it is essential to quantify these two components separately. Previously, we have developed a tameness test that can quantify both active and passive tameness separately in mice [10]. A cluster analysis revealed that behavioral indices obtained from the test represent the difference between active and passive tameness [11]. We used this tameness test to develop two independent groups of tamed mice [10].

1.3 Development of Wild-Derived Heterogeneous Stock

To perform selective breeding efficiently, it is critical to use mice that have a large genetic heterogeneity as the founder stock. Mice from eight wild strains—BFM/2Ms (BFM/2), BLG2/Ms. (BLG2), CHD/Ms. (CHD), HMI/Ms. (HMI), KJR/Ms. (KJR), MSM/Ms. (MSM), NJL/Ms. (NJL), and PGN2/Ms. (PGN2)—were crossed to develop a wild-derived heterogeneous stock (WHS) [10]. These mice were from different countries and represented three subspecies of *Mus musculus* [12–15]. A new outbred WHS was established by crossing the wild strains. To determine the appropriate size of the breeding pairs, a simulation study of the distribution of allele frequencies and the number of lost alleles for a heterogeneous stock descended from 16 breeding pairs with the assumption of no selection for 100,000 realizations of a simple mating scheme was conducted. This analysis concluded that heterogeneous stock at the population level could be maintained using 16 breeding pairs [16].

2 Materials

2.1 Mice

The mice were maintained according to the National Institute of Genetics (NIG) guidelines; all procedures were approved by the Institutional Committee for Animal Care and Use. Eight wild strains (BFM/2, BLG2, CHD, HMI, KJR, MSM, NJL, and PGN2) were used as founder strains (Table 8.1). These strains were derived from wild mice [12–15] and kept at the NIG as inbred strains. In the case of cage exchange and behavioral tests, each mouse was gently caught by its tail using large tweezers covered with silicon tubing to reduce pain. Using tweezers may also reduce the animal's fear response to a hand that approaches the body to capture the mouse. After crossbreeding, the hereditary histories of all animals were recorded.

2.2 Breeding Equipment and Environment

All animals were maintained at NIG under a 12/12-h light/dark cycle (light on from 6:00 to 18:00) in a temperature-controlled room (23 °C ± 2 °C). The mice were housed in standard-sized transparent polycarbonate cages with wood chips as bedding material. All mice were fed a standard laboratory diet (CE-2; CLEA Japan, Tokyo, Japan). Food and water were provided ad libitum. For the breeding pairs, the Aspen Enrichment Tunnel (Tapvei Estonia OÜ, Harjumaa, Estonia) was provided in the cage to allow the animals to gnaw as they like. For handling of mice, large tweezers, with tips covered with silicon tubes to reduce pain, were used. The mice were weaned at approximately 3–4 weeks of age and housed in same-sex groups until the tameness tests. All tests were conducted from 14:00 to 18:00 h.

Table 8.1
Information regarding different strains of wild mice used to establish wild heterogenous stock

Strain	Subspecies Group	Subspecies	Place of collection	Year of capture	Current generations
PGN2/Ms	Domesticus	*M. m. domesticus*	Ontario, Canada	1979	F95
BFM/2Ms	Domesticus	*M. m. brevirostris*	Montpellier, France	1976	F130
HMI/Ms	Castaneus	*M. m. castaneus*	Heimei, Taiwan	1986	F88
NJL/Ms	Musculus	*M. m. musculus*	Northern Jutland, Denmark	1980	F100
BLG2/Ms	Musculus	*M. m. musculus*	General Toshevo, Bulgaria	1980	F114
CHD/Ms	Musculus	*M. m. gansuensis*	Chendu, China	1981	F102
KJR/Ms	Musculus	*M. m. yamasinai*	Kojuri, Korea	1984	F107
MSM/Ms	Musculus	*M. m. molossinus*	Mishima, Japan	1978	F121

2.3 Preparation for the Tameness Test

A square open-field box (size: 400 × 400 × 400 mm) made from gray polyvinyl chloride (O'Hara & Co. Ltd., Tokyo, Japan) was used for behavioral tests (Fig. 8.1). The field was illuminated with 100 lux at the center of the field using a neutral white bulb fixed at 450 mm above the basement of the field. The floor of open-field box was evenly lit in all corners with a compact fluorescent lamp (CFL). A commercially available compact digital camera (CX-5; Ricoh Company, Ltd., Tokyo, Japan) was fixed to a tripod 850 mm above the open field. The behavior of mice was recorded with high quality in the AVI format. Before commencing the tameness test, the open-field box was cleaned and dried to avoid the influence of the previous animal's odor. The experimenter wore cloth gloves under latex rubber gloves to protect hands from being bitten during handling and tameness tests. A stopwatch was used to measure the time; a beeping sound was produced at the end of a 1-min tameness test.

2.4 Preparation for Video Analysis

An experienced person who was blind to the test conditions analyzed all the videos using tanaMove software, which can be freely downloaded from http://www.mgrl-lab.jp/MGRL_software.html. The data obtained were summarized using an Excel macro summarizer, which is also available from the same site.

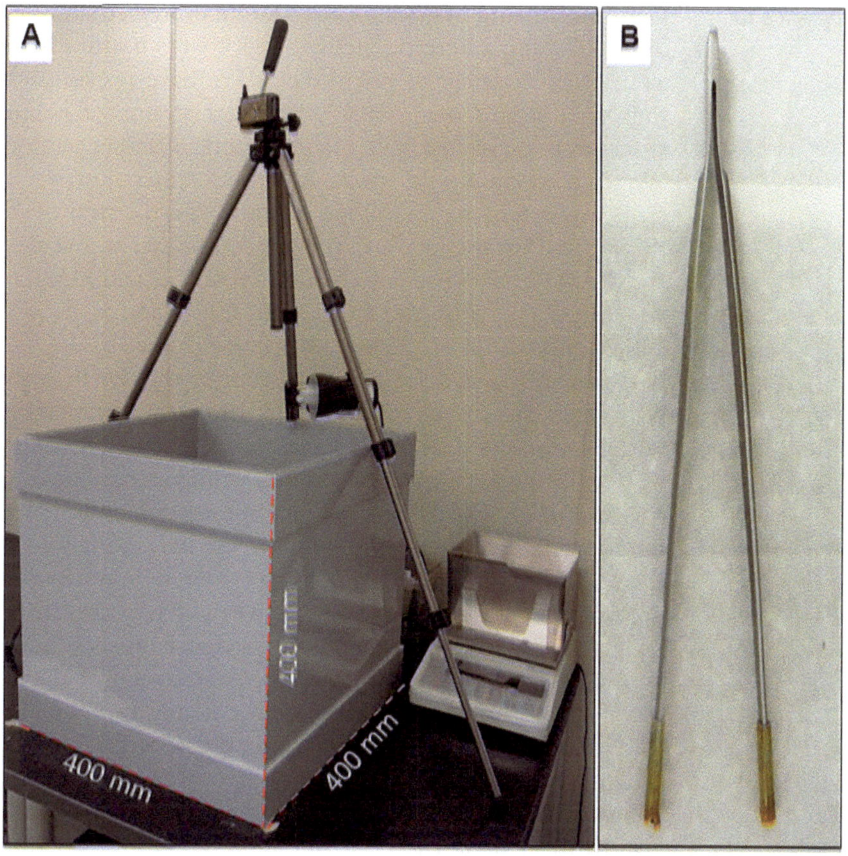

Fig. 8.1 Tameness test setup. (**a**) 400 × 400 × 400-mm monocolor square open-field box for the tameness test with optimum camera and light positioning. (**b**) Silicon tubes covered tweezers for mice handling

2.5 Breeding Records

Breeding of the selection and control groups was conducted based on a precise mating scheme. Given that breeding records are essential for genetic analysis, all animal information used for breeding and behavioral tests was collected and saved/stored meticulously.

3 Methods

3.1 Handling of Mice

All mice were handled using large tweezers with tips covered with silicon tubes to reduce pain when capturing the tail. It was important to use large tweezers to catch wild-derived strains and WHS mice as these mice exhibit higher wildness. The use of large tweezers was also essential to avoid habituation of animals with human hands. When mice were younger (<2 weeks), they were captured using hands. Before the tameness test, which was conducted at 6 weeks of age, mice were grouped and housed with their same-sex littermates to avoid the influence of single housing.

3.2 Mating Mice to Establish Heterogeneous Stock

Genomic mixing of genetically different inbred strains is a key point in establishing a heterogeneous stock that is useful for the genetic analysis of a variety of phenotypes. We have previously proposed that wild strains are useful as founder strains of heterogeneous stocks as these strains have a large genetic difference among strains. We selected eight wild strains from different countries. The initial cross was conducted as follows (in each pair, the strain of the female is indicated on the left, and the family number is given): (1) MSM and BFM/2, (2) HMI and MSM, (3) BLG2 and HMI, (4) PGN2 and BLG2, (5) KJR and PGN2, (6) CHD and KJR, (7) NJL and CHD, and (8) BFM/2 and NJL (Fig. 8.2) [10]. The family numbers in the next generations are given for the female lineage; thus, both male and female progeny were given the same family number as their mothers. In the next generation, the female of each family was crossed with a male of another family according to a circular rotation rule. From G3, the mating scheme was changed to a random mating equal contribution rule because it was difficult to maintain a precise circular rotation rule as some pairs may not produce progeny, and we may have had to substitute with the progeny of another pair. The number of families doubled at G2, with family numbers from 9 to 16. Each group of the WHS consisted of 16 pairs of mice. At G3, the genomes of all eight strains were mixed randomly in each mouse; the genetic diversity based on the combination of polymorphic genes was expanded extensively compared to that in the eight founder strains. Because of the high genetic diversity in the WHS, a higher level of phenotypic diversity is expected in the WHS.

3.3 Animal Preparation

All mice were weaned at 3 weeks of age and housed in same-sex groups with littermates using standard-sized plastic cages containing wood chips until the test. The mice were group housed until all the behavioral tests were completed. Large tweezers were used to separate them from their parents at weaning and move them during the experiments. The mice were not directly touched by human hands until the tameness test, except when they were still receiving milk from their mother for the purpose of cage change. At 42–48 days (6 weeks) of age, mice were tested for their tameness levels. On the day of the test, the mice were transferred to the experimental room prior to conducting the test.

3.4 Tameness Test

The tameness test was performed to measure the degree of tameness in the mice. A series of tameness tests is an approximately 3-min procedure consisting of active tameness for 1 min, passive tameness for 1 min, and stay-on-hand tests. The active tameness test measures animal heading or approaching a human hand. The passive tameness test measures an animal's tolerance to be touched by a human hand. The stay-on-hand test measures an animal's tolerance when stroked by a human hand [17].

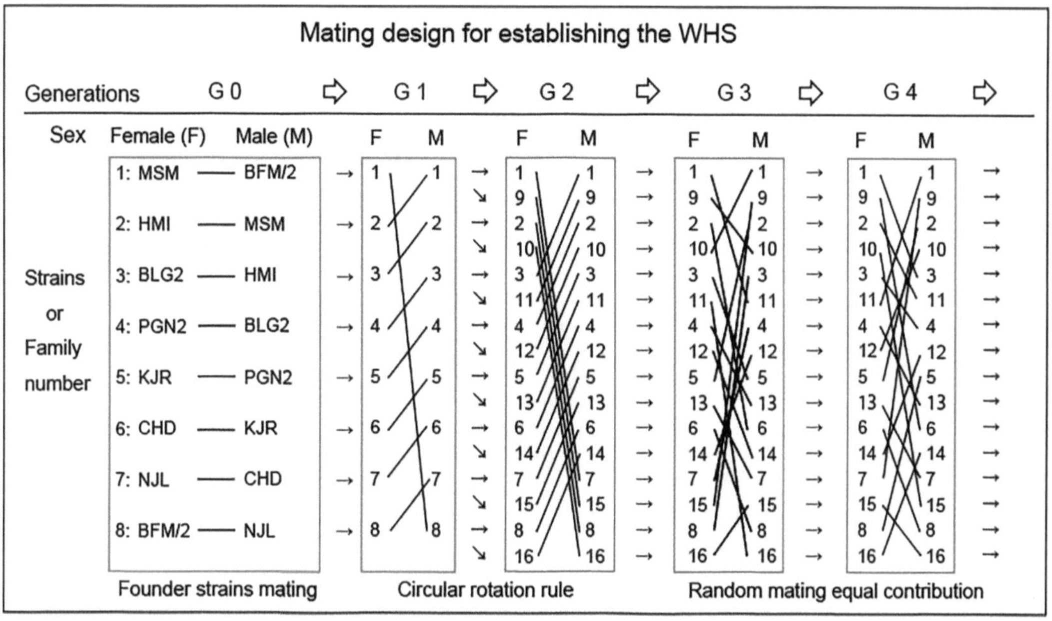

Fig. 8.2 A scheme for mating to establish the wild-derived heterogeneous stock (WHS). At generation zero (G0), eight wild strains were crossed with different strain each other. Numbers indicate family numbers to consist the stock. Solid lines between the founder strains and numbers indicate crosses to produce offspring. The family numbers for the next generations are given based on the family number of the dam; hence, both male and female progenies were given the same family numbers as their dam. Crosses at G1 and G2 were designed based on the circular rotation rule. For example, at G1, a female produced by family number 1 was mated with a male produced by family number 8, a female produced by family number 2 was mated with a male produced by family number 1, and so on. At the G2 generation, the number of families was expanded from 8 to 16, and the mice were mated based on the circular rotation rule. From G3, mating of families was designed based on the random mating rule to avoid inbreeding

The active tameness test was performed according to the following protocol:

1. Start video recording.
2. Pick up a mouse gently by the tail using large tweezers from the home cage, and gently place at the center of the open-field apparatus.
3. Start the stopwatch, with the time set at 1 min.
4. Place the left hand on the lower-left corner of the apparatus.
5. Keep the palm facing up, and move fingers slowly and rhythmically from the little finger to the index finger to ensure that the mouse recognizes the hand as a human hand.
6. Move the hand toward the mouse slowly, stopping at 10 cm from the side of the mouse's body (Fig. 8.3).
7. Maintain the same distance if the mouse attempts to move away. If the mouse approaches, keep the hand in the same position, and continue to move their fingers slightly. If the mouse attempts to climb up the arm, return it to the field using the other hand.

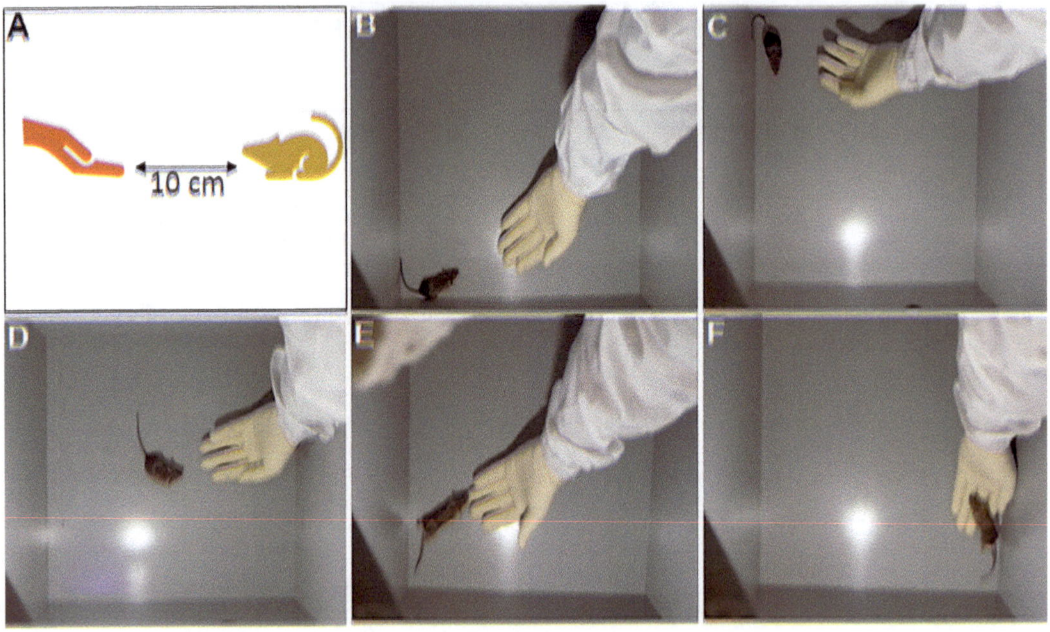

Fig. 8.3 Different behavioral indices of active tameness test. (**a**) Illustration of active tameness test. (**b**) Maintaining 10-cm distance during active tameness test. (**c**) Locomotion, (**d**) heading. (**e** and **f**) Contacting

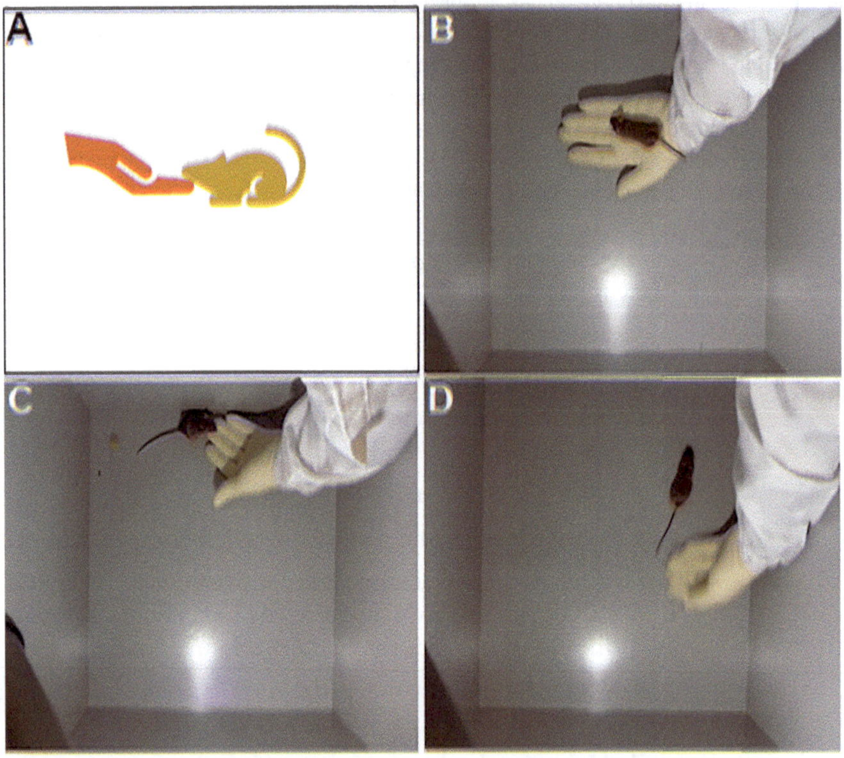

Fig. 8.4 Different behavioral indices of passive tameness test. (**a**) Illustration of passive tameness test. (**b**) Accepting by mice. (**c**) Heading. (**d**) Locomotion

8. After the stopwatch beeps, remove the hand from the box and conclude the active tameness test.

The passive tameness test was performed according to the following protocol:

1. Start the stopwatch, with the time set at 1 min.
2. Place the left hand on the lower-left corner of the apparatus.
3. Keep the palm facing up, and continuously pursue the mouse's body without moving the fingers until touching the body during the 1-min trial (Fig. 8.4).
4. Leave the left hand in the same position in contact with the body as long as possible until the mouse moves away. When the mouse attempts to move away, slowly follow the mouse, and reestablish the touch. If the mouse climbs up the arm, return it to the field in the same way as that used in the active tameness test.
5. After the stopwatch beeps, conclude the passive tameness test, and immediately commence the stay-on-hand test.

The stay-on-hand test was performed according to the following protocol:

1. Start the stopwatch, with the time set at 1 min.
2. Lift the mouse by its tail using tweezers, and gently place it on the left palm (Fig. 8.5). The hand is kept close to the floor of the open-field apparatus.
3. Stroke the mouse's back gently at a frequency of once a second using the left thumb until the mouse gets off from the hand.
4. Repeat steps 2 and 3 three times.
5. After the third trial, remove the hand from the box, and stop the video recording.

After all the tameness tests are completed, the following procedures should be followed:

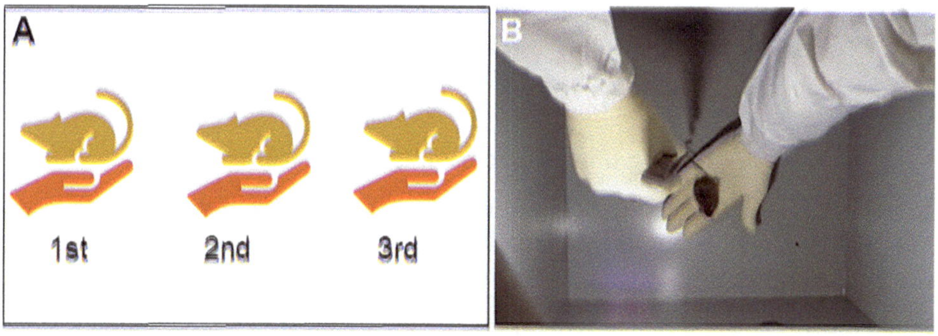

Fig. 8.5 Stay-on-hand test. (**a**) Illustration of stay-on-hand test. (**b**) Representative image of keeping mice on the hand using silicon covered tweezers for the stay-on-hand test

1. Count the number of feces and urine in the open field.
2. Weigh and mark the mouse individually by the ear punch method.
3. Place the mouse back into the cage. Males are kept in individual housing conditions until mating as some animals may show high aggression due to their wild origin. Females can be kept in group housing conditions.
4. Clean the apparatus and experimenter gloves with a paper towel soaked in 0.2% Iodophor sanitizer.
5. Clean the apparatus with a dry paper towel.

3.5 Scoring Tameness Using Video Data

A person who was blind to the sample information scored tameness using the video data. All videos were analyzed using the event recording software tanaMove (version 0.01). Durations of all behavioral indices were scored in 0.1 second units. During video analysis, the appropriate keyboard letters were pressed when observing a given behavior. Nine behavioral indices were scored using video data as follows: keyboard letters assigned to each behavioral index are indicated in parentheses.

1. Active tameness test.

Heading (A): heading toward or approaching the hand, including standing to sniff.

Locomotion (R): traveling without heading toward the hand.

Contacting (T): voluntary contact behavior, except accidental contacts such as landing on a palm after jumping.

Jumping (J).

2. Passive tameness test.

Heading (A): heading toward or approaching the hand.

Locomotion (R): traveling without heading toward the hand.

Accepting (T): allowing the experimenter to touch its body.

Jumping (J).

3. Stay-on-hand test.

Staying time (T): remaining on the hand.

Nine behavioral indices were analyzed during three rounds of video analyses, restarting from the beginning of each session and generating a separate .csv file for each round.

The following procedure was followed:

1. Double click the tanaMove icon and start the software.
2. Drag and drop the video file that needs to be analyzed in the window of tanaMove.

First Round of Video Analysis

Clicking the **Ready** button opens a new window where we can enter the assigned keys. The **play** button can be used to start the video, and the **stop** button can be used to pause.

Click the play button and start the analysis. Initiate the analysis of the active tameness test. Press key **A** when the mouse is observed heading toward or approaching the hand, including standing to a sniff. Press key **R** when the mouse shows locomotion without heading toward the hand.

Press key **W** for 0.3 s when the tameness test finishes at 60-s intervals and the timer rings.

Immediately initiate the analysis of the passive tameness test. Initiate analysis of the passive tameness test after the left hand is seen in the video. Press key **A** when the mouse is observed heading toward or approaching the hand, and press key **R** when the mouse is observed traveling without heading toward the hand during the passive tameness test.

Press key **W** for 0.3 s when the tameness test finishes at 60-s intervals and the timer rings.

Analyze the stay-on-hand test immediately. Press key **T** when the mouse is observed standing on the hand.

After finishing the stay-on-hand test, click the **stop** button to end the video. Click the **back** button and save data in the CSV file format (.csv).

Second Round of Video Analysis

Clicking the **Ready** button opens a new window, where we can enter the assigned keys.

Click the play button, start the analysis, and initiate analysis of the active tameness test. Start from the beginning of the second session of analysis. Press key **T** when the mouse exhibits contact with the hand, which includes all voluntary contacting behaviors, but not accidental contacts, such as landing on a palm after jumping.

Press key **W** for 0.3 s when the tameness test finishes at 60-s intervals and the timer rings.

Immediately initiate the analysis of the passive tameness test. Press key **T** when the mouse allows the experimenter to touch its body.

Press key **W** for 0.3 s when the tameness test finishes at 60-s intervals and the timer rings.

After finishing the passive tameness test, click the **stop** button to stop the video. Click the **back** button and save data in the CSV file format (.csv).

Clicking the **Ready** button opens a new window, where we can enter the assigned keys.

Third Round of Video Analysis

Clicking the **Ready** button opens a new window, where we can enter the assigned keys. Click the play button, start the analysis, and initiate analysis of the active tameness test. Start from the beginning of the third session of analysis. Press key **J** when the mouse exhibits jumping during the active tameness test.

Press key **W** for 0.3 s when the tameness test finishes at 60-s intervals and the timer rings.

Immediately initiate the analysis of the passive tameness test. Press key **J** when the mouse exhibits jumping during the passive tameness test.

Press key **W** for 0.3 s when the tameness test finishes at 60-s intervals and the timer rings.

After finishing the passive tameness test, click the **stop** button to end the video. Click the **back** button and save data in the CSV file format (.csv).

Keep all the .csv file generated in a single folder and start the Summarizer macro to summarize the data.

3.6 Selective Breeding

In a previous study, we established a selection group, selection 1 (S1), and a non-selection (control) group, control 1 (C1), from G3 (Fig. 8.6) [10]. Another selection group, S2, was created from the C1 group, and selection for higher active tameness was initiated at G5. In addition, another non-selection control group, C2, was split from the S1 group at G5 after selection for two generations and bred without any selection thereafter (Fig. 8.6). Further splitting into two groups was conducted for the following reasons. First, the analyses to detect a selected signature in the genome might be affected by genetic drift. Therefore, to distinguish a selected region from the potentially obscuring factor of genetic drift, we attempted to replicate the results by duplicating the groups. Second, this increased the chance of detecting selected loci considering that each group consisted of a relatively small number of mice.

The selection was applied only to selection groups S1 and S2. To perform selective breeding, animals that exhibited higher levels of tameness were chosen for mating. Five females and five males were obtained from each pair and underwent the three tameness tests. In the case where the number of females and males in the litter was smaller than five for each, we added subsequent litters

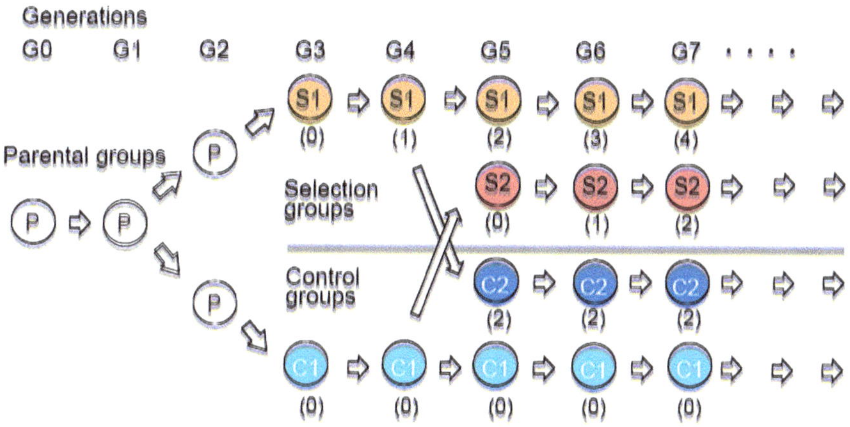

Fig. 8.6 Design of selective breeding. A process to establish the two selection groups (S1 and S2) and the two control groups (C1 and C2). Each number in parentheses indicates the number of selections applied to the group of mice. The orange and red circles indicate selection groups, and blue and light blue indicate control groups, respectively

from the same pair. In some cases, we used a third litter for the tameness test to increase the total number of animals to five. In the case where the number of animals for each sex in each pair was less than five, even with multiple litters or due to poor reproduction, we used a smaller number of animals to select the best animals for the next cross.

The mice used for selective breeding were chosen based on the highest scores for contact with the human hand in the active tameness test among five females and five males. When more than two animals exhibited the same highest score for contact within the progeny of each sex in the same pair, the animals exhibited a higher heading score in the active tameness test. Crosses within each family were not applied to avoid inbreeding. Each selected male and female from one family was mated to a selected female and male from another family, respectively, to produce the next generation. For the control groups, randomly selected mice from five males and five females in each family were chosen and mated with a similarly chosen female and male, respectively, from another randomly selected family (Fig. 8.1). This method prevents inbreeding and helps continue selective breeding while maintaining genetic diversity as much as possible.

Changes in the behavior indices for tameness across generations are shown in Fig. 8.7. The behavioral index, used for the selection of tameness, in the active tameness test was clearly changed across generations. The results showed that selective breeding dramatically influenced the WHS phenotype. Some other behavioral indices, such as heading in passive tameness test and staying on the hand in stay on hand test, were co-changed with contacting and showed differences between selected and control

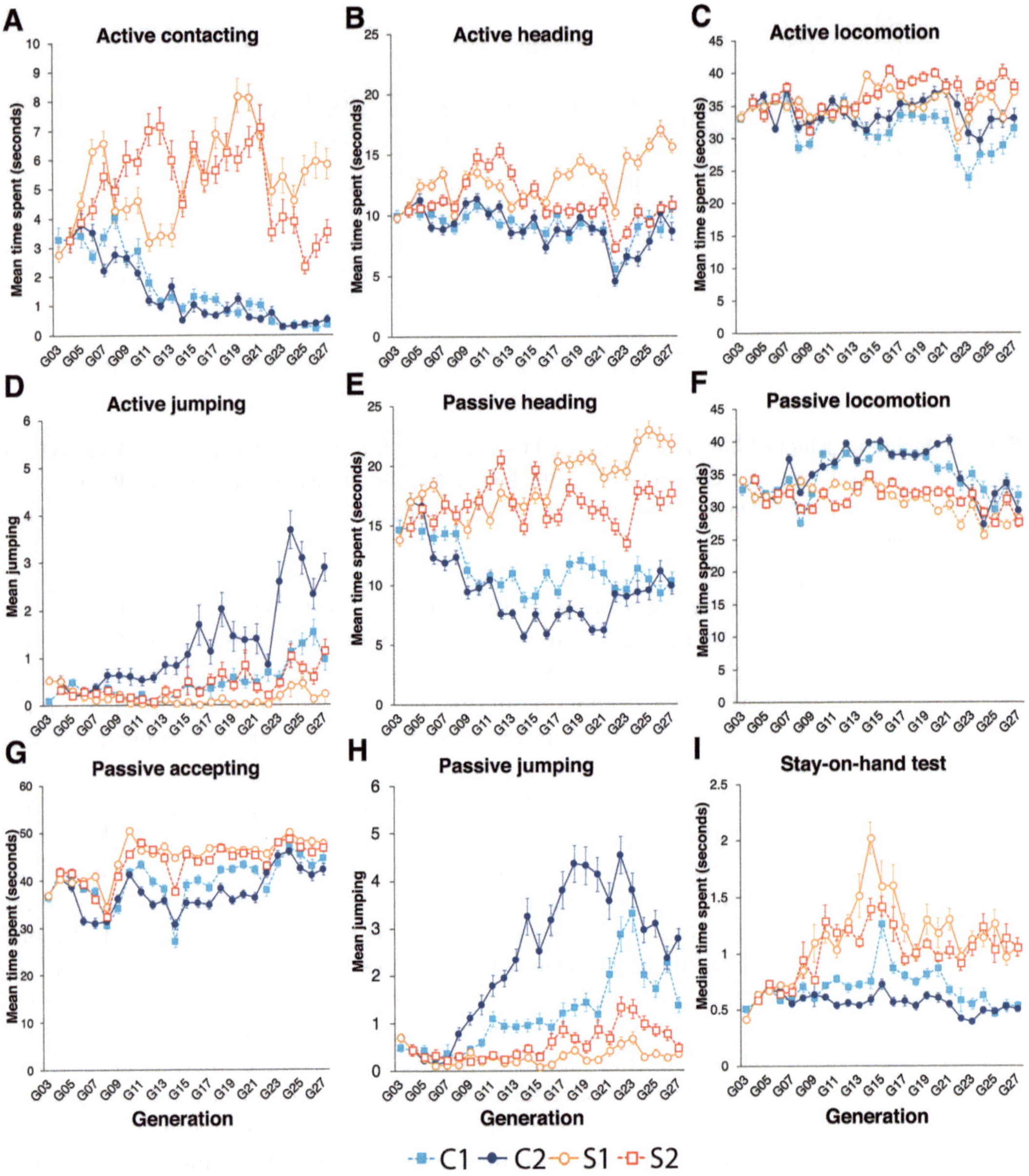

Fig. 8.7 Generational effects of selective breeding on tameness in mice. Filled squares (light blue) and filled circles (blue) indicate control groups, C1 and C2. Open circles (orange) and open squares (red) indicate selected groups, S1 and S2. Different indices obtained from three tameness tests are indicated in graphs. (**a–d**) Active tameness test, revealing contacting (**a**), heading (**b**), locomotion (**c**), and jumping (**d**). (**e–h**) Passive tameness test, revealing heading (**e**), locomotion (**f**), accepting (**g**), and jumping (**h**). (**i**) Stay-on-hand test. A single point shows mean ± standard error of the mean of the score in the population at each generation

groups. The results showed that there was some overlap of genetic mechanisms among these behavioral indices.

Animals established by selective breeding are highly useful for not only studying the mechanism of domestication but also other behavioral studies. The tamed mice exhibited higher sociability than the control mice. Many behavioral studies that require animal sociability may have the advantage of using these tamed mice. In addition, these mice are highly useful for understanding the genetic and neural mechanisms of domestication in animals. Furthermore, applying this method to other animal species is expected to be a very effective method in the future.

4 Notes

1. Detailed information on breeding and behavioral data are essential for genetic analysis. All breeding records and behavioral data should be kept appropriately.
2. Before the tameness tests, it is important to avoid any contact or interaction between the human hand and mice, except during the lactation period. Therefore, all transfers from one place to another should be performed using large tweezers covered with silicon tubes.
3. It is recommended to perform all the tameness tests in an isolated room from other experiments and away from other mice.
4. To clean the apparatus after the tameness tests, a disinfectant that does not have a strong odor should be used, that is, the disinfectant should be diluted with an appropriate amount of water. We used 0.2% Iodophor sanitizer Mikro Klene (Ecolab G.K., Tokyo, Japan). The test arena should be wiped a cloth soaked with the sanitizer and then wiped with a dry paper towel to clean and dry the test field.
5. During the rhythmic movement of the finger, the rubbing noise of the experimenter's gloved fingers may scare the mice. Hence, as little rubbing noise as possible should be made.
6. Some odors may affect the response of the mice to the tameness tests. The experimenter should avoid any type of perfume, deodorant, or any strong odor-producing substance on the day of the experiment.
7. Mice usually do not show aggressive biting to the human hand during tameness tests, as the procedure does not include any restraint situation. However, it is recommended to wear cloth gloves inside latex gloves to protect the hand from mice biting.
8. In our study, no sex effect was observed in the tameness tests. However, it is better to examine the sex effect when an analysis of tameness tests is conducted.
9. When multiple groups are compared, Bonferroni correction is recommended for statistical analysis.

10. Breeding wild-derived mice is often difficult because of high anxiety. Therefore, it is advisable to use wooden boxes to provide environmental enrichment and hiding places. In particular, Aspen products have less odor; therefore, they seem to be less stressful for mice.

References

1. Albert FW et al (2008) Phenotypic differences in behavior, physiology and neurochemistry between rats selected for tameness and for defensive aggression towards humans. Horm Behav 53(3):413–421
2. Belyaev D. K., The Wilhelmine E. Key (1979) Invitational lecture. Destabilizing selection as a factor in domestication. J Hered 70(5), 301–308
3. Trut LN (1999) Early canid domestication: the farm-fox experiment. Am Sci 87:160–169. http://wwwjstororg/stable/27857815. Accessed 15 Oct 2020
4. Belyaev DK, Borodin PM (1982) The influence of stress on variation and its role in evolution. Biol Zentralbl 101(6):705–714
5. Albert FW et al (2009) Genetic architecture of tameness in a rat model of animal domestication. Genetics 182(2):541–554
6. Blanchard DC et al (1994) Defensive reactions of "wild-type" and "domesticated" wild rats to approach and contact by a threat stimulus. Aggr Behav 20:387–397
7. Plyusnina IZ et al (2009) Cross-fostering effects on weight, exploratory activity, acoustic startle reflex and corticosterone stress response in Norway gray rats selected for elimination and for enhancement of aggressiveness towards human. Behav Genet 39(2):202–212
8. Price EO (2002) Animal domestication and behavior. Cabi, Wallingford
9. Goto T et al (2013) Selection for reluctance to avoid humans during the domestication of mice. Genes Brain Behav 12(8):760–770
10. Matsumoto Y et al (2017) Selective breeding and selection mapping using a novel wild-derived heterogeneous stock of mice revealed two closely-linked loci for tameness. Sci Rep 7: 4607
11. Matsumoto Y et al (2021) Combined change of behavioral traits for domestication and gene-networks in mice selectively bred for active tameness. Genes Brain Behav 20(3):e12721
12. Ogasawara M et al (2005) Length variation of CAG/CAA triplet repeats in 50 genes among 16 inbred mouse strains. Gene 349:107–119
13. Koide T et al (2000) Multi-phenotype behavioral characterization of inbred strains derived from wild stocks of *Mus musculus*. Mamm Genome 11:664–670
14. Furuse T et al (2002) Genetic diversity underlying capsaicin intake in the Mishima battery of mouse strains. Brain Res Bull 57(1):49–55
15. Moriwaki K et al (2009) Unique inbred strain MSM/Ms established from the Japanese wild mouse. Exp Anim 58(2):123–134
16. Koide T, Goto T, Takano-Shimizu T (2012) Genomic mixing to elucidate the genetic system of complex traits. Exp Anim 61(5): 503–509
17. Nagayama H et al (2018) Measuring active and passive tameness separately in mice. J Vis Exp 138:e58048

Chapter 9

Bioassaying the Function of Pheromones in *Drosophila melanogaster's* Social Behavior

Thomas A. Verschut, Philip Kohlmeier, and Jean-Christophe Billeter

Abstract

Social interactions are generally regulated by pheromones that convey information about the identity, physiological state, and location of an individual. The fruit fly, *Drosophila melanogaster*, offers a powerful model system to study the mechanisms through which pheromones modulate social interactions. Most of the fruit fly's social behavior is demonstrably modulated by pheromones, and many of the chemical compounds composing its pheromonal profile have been characterized. This chapter describes several behavioral bioassays that can be used to determine the function of contact and short-range volatile pheromones in *D. melanogaster's* social behavior. The chapter first provides instructions on how to rear flies for pheromonal experimentation and how to generate flies that cannot produce cuticular hydrocarbons. Afterward, protocols on how to determine the function of pheromones in courtship behavior and mate choice are provided, followed by protocols to determine whether pheromones function as volatile or contact cues during oviposition site selection. Finally, the last section of the chapter gives general advice on how to work with pheromones in the laboratory.

Key words Aggregation, Courtship, *Drosophila melanogaster*, Mate choice, Olfaction, Oviposition, Pheromones, Social behavior

1 Introduction

In many species, social interactions are orchestrated by pheromones that are released by an individual to convey information about its identity, physiological state, and location [1–3]. This information can, in turn, influence the physiological state and behavior of conspecifics that are equipped to detect it. Understanding the mechanisms through which pheromones function is a central goal for several scientific disciplines ranging from neurosciences that seek to understand how pheromones influence brain function [4, 5] to pest management, making use of pheromones to control pest populations [6, 7]. Determining the function of pheromones typically begins with analytical chemistry to identify which chemicals –

Daisuke Yamamoto (ed.), *Behavioral Neurogenetics*, Neuromethods, vol. 181, https://doi.org/10.1007/978-1-0716-2321-3_9,

extracted from the body surface, the headspace, or from a substrate visited by an individual – cause a response in conspecifics. Afterwards, the identified chemicals are synthesized and tested as single compounds, or combinations thereof, in behavioral experiments to determine whether they trigger a response. Finally, the receptors for these compounds are identified by electrophysiological recordings of candidate sensory receptor neurons. This now classical approach has been used to identify pheromones, and their mode of action, in various insect orders [8–10] as well as in several animal taxa [11–13].

The biology and genetic traceability of the fruit fly, *Drosophila melanogaster* (Diptera: Drosophilidae), makes it a powerful model system for studying the mechanisms through which pheromones modulate social interactions, such as courtship, egg-laying, and aggregation [1, 14–17]. Moreover, its pheromonal profile has been described in detail [14, 18–23] and individual pheromones can be synthesized or purchased [18, 22–26]. The pheromonal profile of *D. melanogaster* predominantly consists of fatty acid-derived cuticular hydrocarbons present on the body surface. These compounds include male-dominant monoalkenes, like (*Z*)-7-tricosene (7-T) and (*Z*)-9-tricosene (9-T), and female-specific dienes like (7*Z*,11*Z*)-heptacosadiene (7,11-HD), and (7*Z*,11*Z*)-nonacosadiene (7,11-ND) [15]. Additionally, males also produce the aggregation pheromone, *cis*-11-vaccenyl acetate (cVA), in the ejaculatory bulb, and both sexes produce several other fatty acids with demonstrable pheromonal roles [21–23, 27]. The cells producing the cuticular hydrocarbons have also been identified and can be genetically manipulated to change the type of pheromones produced [28, 29] or even ablated to test the necessity of specific pheromones [26, 30].

Identifying the receptor that detects a pheromone is an important step in understanding the mechanism underlying social interactions. Functionally, pheromones can be classified by the distance at which they trigger a response [4], with volatile pheromones such as cVA being sensed over several meters [31, 32], and cuticular hydrocarbons usually only being detected at short ranges through contact receptors [33–36]. However, several studies have suggested that cuticular hydrocarbons can also be detected over small distances as volatiles [20, 37–39], and a recent study showed that the oxidation of 7,11-HD produces the volatile attractant (*Z*)-4-undecenal [25]. Functional genetic screens, in which individual sensory neurons have been systematically silenced, are a powerful neurogenetic tool for identifying the chemosensory receptor detecting a compound. Such screens have, for example, revealed the importance of *Or47b* olfactory sensory neurons (ORN) in detecting the aphrodisiac pheromone palmitoleic acid [22, 23] and the involvement of *Or65a* and *Or67d* in responding to cVA [40–43]. Subsequent recordings of neuronal activity through physiological recordings have successfully been used to test whether changes in the neuronal responsiveness of those receptors modulate the respective social behavior. An exciting recent development is that once these behavioral mechanisms have been identified in *D. melanogaster*, comparative studies in other species allow the reconstruction of how these interactions might have evolved [44, 45].

This chapter describes various behavioral bioassays that can be used to determine the function of contact and short-range volatile pheromones that fulfill diverse social functions in *D. melanogaster*. This includes detailed protocols on how to rear flies and how to generate flies unable to produce cuticular hydrocarbons (Sect. 2), how to bioassay the function of pheromones in the courtship and mate choice (Sect. 3), how to determine whether pheromones function as volatile or contact cues during oviposition site selection (Sect. 4), and general advice on how to work with pheromones in a laboratory (Sect. 5). Complementary information about the production of *D. melanogaster* pheromones [46, 47] and the analytical chemistry methods typically used to identify [22, 23, 38, 48] or quantify pheromones [49, 50] can be consulted elsewhere.

2 How to Rear *Drosophila melanogaster* for Behavioral Experiments with Pheromones

2.1 Rearing Flies for Behavioral Experiments

The behavior, pheromonal profile, and physiology of *D. melanogaster* are influenced by numerous factors including nutrition, social environment, age, and mating status [1, 51, 52]. Therefore, increasing the reproducibility of pheromonal bioassays requires flies that were reared under standardized conditions, that did not experience resource limitation due to overcrowding, and that are of known age and mating status. The section below describes how to generate such flies. It is recommended to maintain constant temperature and light conditions as this will affect the length of the life cycle. Generally, the life cycle takes about 10 days at 25 °C and a 12:12 h light-dark cycle.

Materials

170 mL polypropylene rearing bottles (734-1249, VWR)

37 mm cellulose plugs (734-2673, VWR)

Food medium ingredients:

Water

Molasses

Agar

Sucrose

Glucose

Cornmeal

Wheat germ

Soy flour

Dried yeast

Propionic acid (CAS# 79-09-4)

Tegosept (CAS# 99-76-3)

Ethanol (CAS# 64-17-5)

Laboratory utensils for food preparation

Stereoscope with CO2 pad and utensils for fly sorting

Step 1. Prepare Fly Food for Optimal Larval Developmental Conditions First prepare the following ingredients: water (1000 mL), molasses (30 g/L), agar (10 g/L), sucrose (15 g/L), glucose (30 g/L), cornmeal (15 g/L), wheat germ (10 g/L), soy flour (10 g/L), dried yeast (35 g/L), propionic acid (5 mL/L), tegosept (2 g/L), and ethanol (10 mL/L). Start by mixing the water and molasses, and add the other ingredients (except the preservatives propionic acid, tegosept, and ethanol) once it has started to boil. Keep stirring and let it boil for 10 min. Afterwards, let it cool down to approximately 55 °C and add the three preservatives.

Step 2. Pour 50 mL of Liquid Fly Food into the Rearing Bottles Let the medium solidify and cool down to room temperature. Afterwards, store the bottles in the refrigerator until needed.

Step 3. Anesthetize Flies from an Ongoing Rearing on a CO_2 Pad Gently knock a rearing bottle on the lab bench to remove flies from the lid. Open the bottle and place it upside down in the CO_2 pad. Gently tap the bottle to distribute the flies on the CO_2 pad.

Step 4. Collect 30 Virgin Females and 30 Males and Transfer Them into the Bottle Virgin flies can be recognized up to 6 h after eclosion, by their lighter colors and the dark greenish meconium on the underside of the abdomen. If the flies are still anesthetized, place the new bottle horizontally to prevent the flies from getting stuck on the food while anesthetized.

Step 5. Transfer the Flies to a Fresh Rearing Bottle After 3 Days Removing adults from the rearing bottle after 3 days will limit the number of eggs deposited in the bottle and will reduce the risk of larval overcrowding.

Step 6. Transfer the Flies to a Fresh Rearing Bottle After 2 Days Most of the flies have already mated and will immediately resume laying eggs in the fresh bottle. Therefore, they will need less time in the rearing bottle.

Step 7. Discard the Adult Flies and Let the Offspring Develop The flies should not be transferred a third time as the fecundity and egg quality decreases with age [53].

Step 8. Collect Virgin Flies Upon Eclosion and Keep Them in Vials Separated by Sex Only collect during the first couple of days as flies that hatched later may have already experienced increased larval competition due to overcrowding. Only collect the flies as virgins to have control over the age and mating status.

Step 9. Keep the Flies Separated by Sex in Groups of Ten, and Age Them for a Minimum of 3 Days The flies need this maturation period to start producing their mature pheromonal profile [54]. Place the vials horizontally to prevent flies from getting stuck in the medium.

2.2 Controlling the Mating Status of Flies

Because mating induces changes in *D. melanogaster's* behavior and chemical profile [20, 55, 56], it is essential to control their mating status for behavioral experiments. An essential tool for gently transferring flies between containers is the mouth aspirator, which can be easily constructed by placing a disposable plastic pipette tip and a fine soft mesh on a piece of tubing. The narrow opening of the pipette tip can be truncated to a wider diameter allowing the flies to enter the pipette tip. The piece of mesh will ensure that the collected flies are not sucked into the tubing. The virgin flies can be reared and collected following the protocol of Sect. 2.1. Once the flies have aged for 3 days, they can be mated by transferring groups of virgin females and males into a fresh rearing vial containing food. It is possible to check whether a female has recently mated by shining ultraviolet light on the ovipositor to inspect for the presence of the fluorescent mating plug [57]. Once the females have mated, it is important not to anaesthetize them with CO_2 prior to using them in a behavioral experiment, as this will impact nervous system functioning [58, 59]. Instead, close the rearing vials with soft polyurethane foam plugs that will allow a mouth aspirator to be inserted into the vial to remove the flies.

Materials

25 mm × 95 mm polystyrene rearing vials (734-2255, VWR)

25 mm polyurethane foam plugs

Mouth aspirator made of:

Disposable pipette tips

Fine mesh

Soft tubing

Step 1. Collect Virgin Flies Upon Eclosion, and Age for 5–7 Days Separated by Sex Age the flies in groups of ten individuals in rearing vials containing a food medium in an incubator with stable temperature and light cycle (25 °C, 12:12 h light-dark cycle). Place the vials horizontally to prevent flies from getting stuck in the medium.

Step 2. Transfer 10 Aged Males and 10 Aged Females into a Fresh Rearing Vial Containing Food Close the vial with a soft foam plug that will allow a mouth aspirator to be inserted later on to remove the mated females without anesthetizing them.

Step 3. Give the Males and Females 3 h to Interact to Ensure that Most of Them Have Mated Place the vials in an incubator with stable conditions (25 °C, 12:12 h light-dark cycle), and let the flies mate. When the females are to be used in oviposition experiments and start to lay eggs within those 3 h, it is advised to shorten the mating time to ensure that they retain their eggs for further experiments.

Step 4. Remove the Females Individually from the Rearing Vials A mouth aspirator can be easily inserted past the soft foam plug to remove the mated females. Depending on the experimental purpose, these females can either be directly used or transferred into an insolation vial to boost the egg development prior to using them in behavioral experiments (see Sect. 3.2).

2.3 Generating Oenocyte Ablated Flies

A powerful tool for investigating the function of specific pheromonal compounds are flies that cannot produce cuticular hydrocarbons. The following protocol describes how to generate flies in which the cells that produce cuticular hydrocarbons – the oenocytes located at the ventral and dorsal surface of the abdomen of adult flies – have been ablated (Fig. 9.1). It is important to note that pheromones that are not produced in the oenocytes, including cVA, are still present on oenocyte ablated flies [26]. These flies can then be used to test both the necessity of cuticular hydrocarbons and the sufficiency of single pheromonal compounds for modulating social interactions (e.g., perfuming them with a pheromone as described in Sect. 2.1).

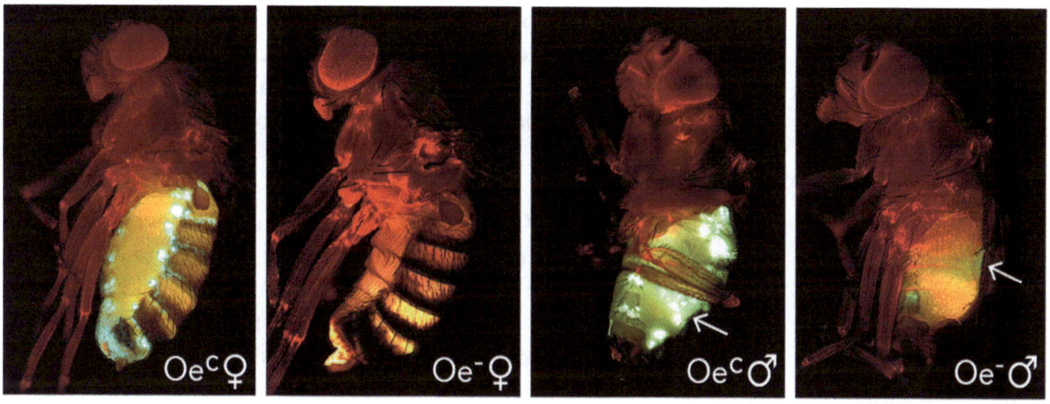

Fig. 9.1 Green fluorescent protein (GFP) expression in 6-day-old adult male and female flies. The oenocyte control flies (Oe^{C}) are transgenic flies in which PromE(800)-Gal4 drives the expression of UAS-nuclearGFP. The oenocyte ablated flies (Oe^{-}) are transgenic flies in which PromE(800)-Gal4 simultaneously drives the expression of UAS-nuclearGFP and the cell-death inducer *head involution defective* (*hid*). Both female and male Oe^{C} flies specifically express GFP in the oenocytes. Oe^{C} males have additional expression in the accessory glands (indicated by white arrow). Oe^{-} flies are devoid of GFP expression in the oenocytes because those cells have been ablated by *hid* expression. Oe^{-} males however keep faint fluorescent in their abdomen (white arrow) coming from the accessory gland, which are not grossly disrupted by *hid* expression

Materials

170 mL polypropylene rearing bottles (734-1249, VWR)

37 mm cellulose plugs (734-2673, VWR)

25 mm x 95 mm polystyrene rearing vials (734-2255, VWR)

23 mm cellulose plugs (734-2875, VWR)

Stereoscope with CO_2 pad and utensils for fly sorting

Microscope with GFP scope

Polyurethane ice bucket

Lines for oenocyte ablated flies:

P{w[+mC] = *Desat1-GAL4.E800}2 M, P{w[+mC]* = *tubP-GAL80 [ts]}20* [BDSC# 65406].

P{w[+mC] = *UAS-Stinger}2, P{w[+mC]* = *UAS-hid.Z}2/CyO* [BDSC# 65408].

Lines for oenocyte control flies:

P{w[+mC] = *Desat1-GAL4.E800}2 M, P{w[+mC]* = *tubP-GAL80 [ts]}20* [BDSC# 65406].

w$^{[1118]}$*; P{w[+mC]* = *UAS-Stinger}2* [BDSC# 84277].

Step 1. Setting Up the Experimental and Control Crosses Cross +: *PromE(800)-Gal4 [2 M], Tub:Gal80*^{ts};+ to +: *UAS-StingerII, UAS-hid/CyO*";+ to generate oenocyte ablated flies and +: *PromE(800)-Gal4 [2 M], Tub:Gal80*^{ts};+ to *w*$^{[1118]}$:

UAS-StingerII";+ to generate control flies. The direction of the cross, from which line the female or the male is taken, does not have an effect on the success of the oenocyte ablation. However, the crosses generally produce less offspring if the females are from the *Gal4*-driver lines.

Step 2. Keep the Crosses at 18 °C During Larval Development A temperature sensitive Gal80 coding sequence (*Gal80^{ts}*) expressed under the ubiquitous *Tubulin* promoter (*Tub:Gal80^{ts}*) prevents Gal4 from binding to the UAS promoter [60], which suppresses the expression of the *head involution defective* (*hid*) gene that encodes a cell death inducer. This is crucial since oenocyte ablation during development is lethal. *Gal80^{ts}* is a temperature sensitive *Gal80* mutant allele acting as a negative regulator of the Gal4 transcriptional activator at the permissive temperature of 18 °C. At restrictive temperatures above 25 °C, Gal80^{ts} can no longer bind Gal4, and its repressive function is lost, leading to *hid* expression and oenocytes ablation.

Step 3. Collect Virgin Flies Upon Eclosion and Keep Them in Vials Separated by Sex Collect adult progeny at room temperature then return crosses to 18 °C. Ensure to select against *CyO*.

Step 4. Transfer Adult Progeny to +25 °C and Age for a Minimum of 5 days At this temperature, *Gal4* is expressed in the oenocytes and initiates the expression of *hid* which ablates the oenocytes. These flies are sensitive to getting stuck in the food medium, so it is recommended to transfer the flies to a fresh vial after 3 days.

Step 5. Confirm the Successful Ablation of the Oenocytes The experimental crosses will express GFP in the oenocytes if they are not properly ablated. On day 6, check male flies under a GFP stereomicroscope to ensure that no GFP expression is visible in the oenocytes. Note that male flies will still show GFP expression in the accessory glands, but should have no visible oenocytes (Fig. 9.1). Because oenocyte ablated flies have a lowered desiccation resistance [52], it is important to anaesthetize them on ice, as they cannot retain their body water content on a CO_2 pad.

3 Bioassaying the Role of Single or Combinations of Pheromones in Courtship Behavior and Mate Choice

3.1 Perfuming *Drosophila Melanogaster* Using the Tumbling Method

This section describes protocols to test the function of specific pheromones in courtship behavior and mate choice [26, 56]. The protocols below rely on applying synthetic cuticular hydrocarbons onto flies before using them in behavioral experiments. This is achieved by placing flies in a glass vial coated with the compound and gently vortexing the vial to transfer the compound onto the cuticle of the flies. While the amount of synthetic compound used to coat the glass vial should be determined empirically – as not all molecules will be transferred in equal ratios onto the cuticle – by quantification with gas chromatography, it is possible to roughly estimate how much to treat flies with based on transfer efficiency. The protocol below is based on transferring pheromones on groups of seven flies at once with a transfer efficiency of approximately 25%. As an example, a vial coated with 100 μg of a compound will transfer 25 μg of the compound onto to the group of seven flies. As a result, each individual fly will be perfumed with approximately 3.6 μg. Although the transfer efficiency may vary slightly between individual flies, the method is replicable for testing precise amounts of pheromonal compounds and determining dose-response curves. It is highly recommended to check perfuming efficiency by quantifying the pheromonal profile of the perfumed flies using gas chromatography coupled with flame ionization detection (GC-FID), as described in [26, 56].

Materials

25 mm x 95 mm polystyrene rearing vials (734-2255, VWR)

23 mm cellulose plugs (734-2875, VWR)

≥ 99% n-Hexane (CAS# 110-54-3)

2 mL glass screw cap vials (29378-U, Supelco)

Synthetic compounds

Nitrogen flow

Polyurethane ice bucket

Petri dishes (glass or disposable plastic)

Fine tipped forceps (#5 Inox, Dumont)

20 mL glass bottles with urea caps and metal foil liner (218-0000, VWR)

Vortex

Step 1. Determine the Amount of Compound to be Transferred onto Each Fly The amount is based on experimental quantification of natural amounts of a cuticular hydrocarbon exhibited by *D. melanogaster*. Although these amounts can be found in published work [26, 50, 56], it is recommend that they are checked by GC-FID using the actual experimental flies. Multiply the quantity of compound that will eventually coat 1 fly by 28 to perfume 7 flies with a transfer efficiency of 25%.

Step 2. Coat a 2 mL Glass Vial with the Hydrocarbon of Interest Pipette the diluted synthetic compound into a clean 2 mL glass vial, and briefly vortex the vial. Place the vial under a nitrogen gas flow and wait until the hexane is evaporated. Visually inspecting the vial will reveal an opaque oily precipitate at the bottom of the vial.

Step 3. Clean the Forceps Used for Handling Flies Residual chemicals on the forceps can be removed by first immersing the tips of the forceps into a 20 mL glass bottle containing n-Hexane and wiping it with a clean disposable tissue. Next, the tips should be immersed into a 20 mL bottle containing methanol and wiped again. Finally, the forceps should be immersed into a second 20 mL bottle containing n-Hexane and let to air dry (see Sect. 4.4).

Step 4. Anesthetize the Flies on Ice, and Transfer Them into the Vial Coated with the Compound First transfer the flies in a vial without food and anesthetize them on ice. Empty the vial onto a glass or unused disposable plastic Petri dish, and use the cleaned forceps to gently transfer seven flies by their wings into the coated vial.

Step 5. Tumble the Flies in the Vials Using a Vortex Vortex the vial three times at the lowest speed setting for 20 s with 20 s pauses between each pulse. Keep the speed and length of these pulses constant to standardize the transfer rate of the compounds.

Step 6 (Optional). Perfume the Fly with Additional Compounds This can be done by repeating steps 1 to 5 with another compound.

Step 7. Transfer the Treated Flies to a Fresh Rearing Vial Containing Food Either directly tap the treated flies from the glass vial into a fresh food vial or first anesthetize the flies by briefly putting the glass vial on ice. It is essential to give the flies an hour recovery period before using them in behavioral experiments. During the recovery time, flies will start grooming themselves and spread the pheromones uniformly over their bodies. The time between the perfuming and the experiment should therefore be kept constant as the pheromone concentration decreases over time (see supplemental information of [26]).

Step 8 (Optional). Quantify the Transfer Efficiency of the Compound It is recommended that the transferred amounts of hydrocarbons are measured by gas chromatography by sampling a fraction of the treated flies (see [26, 56]). Of the seven perfumed flies, three can be used for quantification, and the four remaining flies can be tested in the behavioral assays. Quantification is strongly recommended to determine if the flies were perfumed with the predetermined amount of compound or that the starting quantity has to be adjusted of compound added to the perfuming vial.

3.2 Assaying Courtship Behavior

Pheromones play a central role in the recognition of species, sex, and mating status of fruit flies [1, 51]. The role of individual pheromones can be assessed by measuring the courtship intensity of males toward perfumed females or males (see Sect. 2.1). Courtship behavior can be observed or video recorded by containing the flies in courtship arenas. Because flies shed cuticular hydrocarbons including those that can impact other flies' behavior in their environment [61–63], it is important to use single-use courtship arenas made of disposable plastic Petri dishes. Furthermore, it is important to standardize the environmental conditions to a stable temperature around 25 °C, a relative humidity around 50%, and to run the experiments in a darkened room where external light conditions cannot affect the behavior of the flies.

Materials

Temperature-controlled room

35 mm disposable Petri dishes (627,102, Greiner Bio-One)

Soldering iron

Mouth aspirator made of:

Disposable pipette tips

Fine mesh

Soft tubing

Video camera

Step 1. Prepare the Courtship Arenas Melt a 3 mm hole through the side of both lids of a Petri dish using a soldering iron. Afterward, pour a 2 mm layer of fly food in the bottom lid of the Petri dish, and let it solidify. Close the hole with a small plug made of Parafilm (100 mm), and store the dishes overnight at +4 °C. Let the dishes warm up to room temperature, and remove any water droplets from the lids before using them.

Step 2. Transfer the Tester Fly in Mating Arenas Transfer the fly whose courtship response to a pheromone is tested to a mating arena using a mouth aspirator. It is recommended to exchange the tip and the mesh of the mouth aspirator after each experimental treatment to prevent cross-contaminations of pheromones. Let the flies acclimatize for 5 min.

Step 3. Place the Dish Under a Video Camera and Start Recording Adjust the light conditions to reduce the light glare on the surface of the Petri dish to ensure that all flies can be observed in detail.

Step 4. Add a Perfumed Target Fly Transfer a female or male perfumed as described in Sect. 2.1 into the mating arena using a mouth aspirator.

Step 5. Record the Pair of Flies for 10 min The usual courtship observation is 10 min, but this can be adjusted to longer observation times to meet the needs of a specific experimental question.

Step 6. Score the Courtship Index The courtship index is the fraction of the observation period spent by the male exhibiting courtship steps such as following the female, tapping, wing extension, licking of the female's abdominal tip, and attempting copulation. The wing extension index represents the fraction of a 10 min observation period during which the male performed the discrete courtship step of extending one of its wings. Attempted copulations represent the number of occurrences over a 10 min observation period of a male bending its abdomen at an angle greater than 90° according to his anterior-posterior body axis toward the courted subject and failing to initiate copulation. These steps are described in [64].

3.3 Assaying Mate Choice Behavior

The effect of a specific pheromone can depend on the mating state of the individual detecting it. For instance, mated females are less sensitive to certain pheromones than virgins, resulting in higher choosiness when selecting their next mate [65]. The following section describes how to determine whether females that are offered a choice between two different males, prefer one over the other for mating and whether this choice differs when the same female has to re-mate. In this protocol, flies are labelled with a fluorescent powder to distinguish the two males from each other, rather than using invasive techniques such as wing clipping. Ensure to randomly assign colors to genotypes to avoid confounding effects of color selection. It is important to note that the outcome of mate choice assays is sensitive to the conditions the flies experienced during development and that small batches of flies from several independent crosses should be tested over several days to prevent behavioral bias. Furthermore, ensure to standardize the environmental conditions to a stable temperature around 25 °C, a relative humidity around 50%, and to run the experiments in a darkened room where external light conditions cannot affect the behavior of the flies.

Materials

25 mm x 95 mm polystyrene rearing vials (734-2255, VWR)

23 mm cellulose plugs (734-2875 - VWR)

Two colors of UV fluorescent powder (Slice of Moon)

Laboratory utensils

Parafilm (100 mm)

Fly food (see Sect. 1.1)

35 mm disposable Petri dishes (627,102, Greiner Bio-One)

Soldering iron

Mouth aspirator made of:

Disposable pipette tips

Fine mesh

Soft tubing

UV flashlight

Step 1. Prepare the Mating Arenas Melt a 3 mm hole through the side of both lids of a Petri dish using a soldering iron. Afterwards, pour a 2 mm layer of fly food in the bottom lid of the Petri dish, and let it solidify. Close the hole with a small plug made of Parafilm, and store the dished overnight at 4 °C. Let the dishes warm up to room temperature, and remove any water droplets from the lids before using them.

Step 2. Prepare Vials with Fluorescent Powder for Coloring - Transfer 0.5 g of fluorescent powder into an empty rearing vial and rotate the vial, while holding it horizontally, until its inner surface is fully covered by the powder. Tap the vials upside down to remove residual powder.

Step 3. Transfer 15 Males into the Coloration Vial, and Rotate the Vial for 3 s to Color Them The males are then visibly covered with fluorescent powder (Fig. 9.2b and f). The flies will groom off residual powder during the experiment (Fig. 9.2c and g), but the coloration will remain visible under UV light for at least 30 h (Fig. 9.2d and h).

Step 4. Color the Second Group of Males with a Different Color Repeat step 2 using a different fluorescent color to

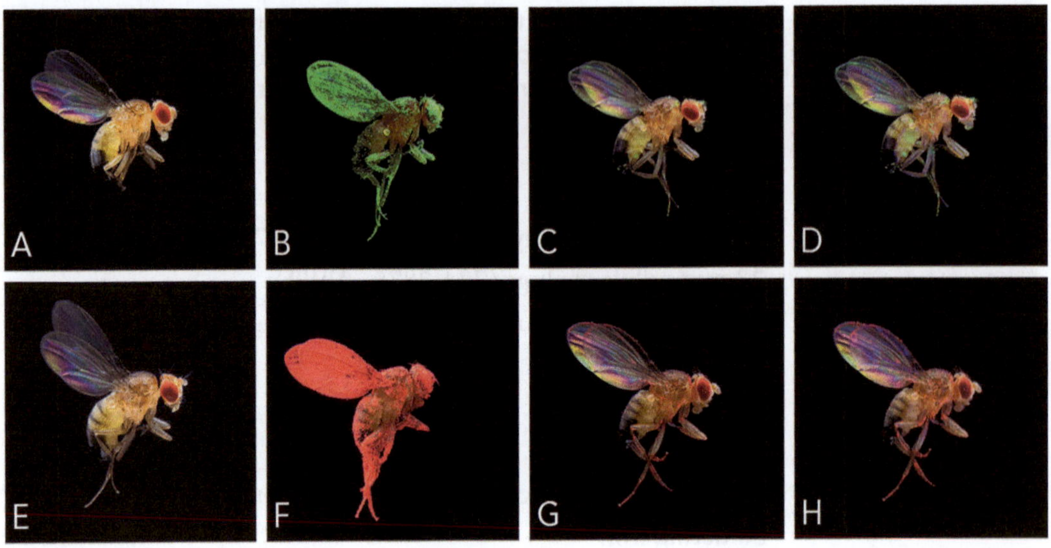

Fig. 9.2 Untreated *Drosophila melanogaster* males (**a** & **e**) can be labelled with different colors using fluorescent powders to distinguish them during mate choice assays. Immediately after being transferred to the coloration vials, males are entirely covered with the green (**b**) or pink (**f**) powder. Most of the powder will shed from the flies following the first 2 h after coloration (**c** & **g**). Using a UV flashlight, the coloration will remain visible for at least 30 h after powdering the flies as show for the green (**d**) and pink (**h**) powder

distinguish males from both groups. Ensure to randomly assign colors to genotypes to avoid confounding effects of color selection.

Step 5. Transfer the Colored Males into the Mating Arenas - Transfer one of each colored males into a mating arena using a mouth aspirator. It is recommended to place an additional filter in the tube to block the fluorescent powder from entering the tube. Exchange the tip and the mesh of the mouth aspirator after each genotype to prevent cross-contaminations of pheromones and coloration. Let the flies acclimatize for 30 min after which most of the powder will have been groomed off but flies remain distinguishable under UV light (Fig. 9.2d and h).

Step 6. Add a Female to the Mating Arena and Start the Experiment Check the dishes every 3–5 min for copulations. Replicates with virgin females that did not mate within 2 h should be discarded as this is likely caused by one or more flies being physically damaged.

Step 7. Confirm Mating and Remove Males Assess the color of the male upon copulation using a UV flashlight, and check for the presence of a mating plug in the female once the copulation has ended. The mating plug is visible as a small fluorescent spot when shining ultraviolet light on the ovipositor of a female [57]. When the mate choice of mated female will also be assessed, remove the males, and keep the female in the dish at 25 °C overnight. This allows females to display the post-mating response before retesting females and prevents females from re-mating unobserved during the night.

Step 8 (Optional). Prepare New Dishes for Re-mating Repeat steps 1 till 5 to test re-mating choice 24 h after the virgin mating choice assay. Check the dishes every 3–5 min, and record re-mating latency and the selected male. As mated females are less receptive than virgin females, the observation time for re-mating experiments has to be extended to 10 h. Some genetic manipulations may result in very low receptivity and may require even longer observation times.

4 Assaying Volatile and Contact Pheromones for Oviposition Site Selection

4.1 Preparing the Oviposition Assays

Flies visiting potential oviposition sites deposit volatile pheromones including cVA and with several nonvolatile cuticular hydrocarbons, to regulate the oviposition choice of other females visiting those sites [61–63]. While the difference in volatility dictates that both classes of pheromones are detected through different sensory modalities [4], some cuticular hydrocarbons appear to act as volatiles over a small distance [20, 37–39]. Furthermore, the oxidation of 7,11-HD produces the volatile attractant (Z)-4-undecenal detected by *Or69a* [25], showing that cuticular hydrocarbons can indirectly operate through the olfactory system. The protocol below describes how to prepare disposable two-choice oviposition assay to determine whether a specific compound functions as a volatile or nonvolatile cue during oviposition site selection. The assay consists of a 57 × 38 × 17 mm rectangular transparent polystyrene dish accommodating two 10 × 38 × 4 mm oviposition zones filled with 0.75% agar containing 100 mM of sucrose as a nutrient source to instigate oviposition [66–70]. The two zones are separated by a 37 × 38 × 4 mm middle zone of 3% agar that is unsuitable for oviposition [71]. The steps below describe the preparation of two alternative versions of the assay: one in which the pheromones are enclosed in an inaccessible cup that prevents direct contact with the pheromones, ensuring that the pheromones can only be sensed through volatiles (e.g., olfactory assay), and a second version in which the pheromones are accessibly placed on top of the oviposition zones, allowing direct contact with the pheromones while selecting an oviposition site (e.g., contact assay).

Materials

57 × 38 × 17 mm transparent polystyrene dishes (31,170, Bodemschat, the Netherlands)

Bacteriological agar (CAS # 9002-18-0)

D-sucrose (CAS # 57-50-1)

0.2 mL flat-capped PCR vial strips (72.991.002, Sarstedt)

5 mL volumetric pipet (606,107, Greiner Bio-One)

25 mL volumetric pipet (760,180, Greiner Bio-One)

Beakers and laboratory utensils

Large filter paper sheets or fine cloth

4.1.1 Preparing the Olfactory Oviposition Assay

Step 1. Prepare 3% Agar for the Middle Zone Add 3 gr of agar powder in a glass beaker, and fill it with distilled water until 100 mL is reached. Boil for at least 1 min in a microwave or on a heating plate. Afterward, cover the beaker with aluminum foil, to prevent evaporation and contamination, and let it cool down for several minutes before handling it.

Step 2. Dispense 8 mL of 3% Agar into the Dishes with a 25 mL Volumetric Pipette The resulting 4-mm-thick layer of 3% agar should cover the bottom of the dish evenly and should be free of air bubbles (Fig. 9.3a).

Step 3. Let the Agar Solidify at Room Temperature for 30–45 min Cover the dishes with filter paper sheets or fine mesh to prevent contamination of the agar while solidifying. Agar that has solidified for less than 30 min may be too soft for removing the sections in the following step.

Step 4. Remove 10-mm-wide Sections from the 3% Agar at Both Ends of the Dish The agar can easily be cut and removed with the flat end of a stainless steel spatula. It is advisable to place the dish on a printed template corresponding to the size of the sections that have to be removed for size standardization of the zones.

Step 5. Prepare Cups that Will Contain the Odor Treatment - Cut the caps of 0.2 mL flat-capped PCR vials at the base of the vial while keeping the plastic connection strip attached to the cap. It is recommended to wear gloves to prevent contaminations and to use PCR vials that come in strips of eight as all caps can be cut in one continuous movement.

Step 6. Place a Detached Cap with the Opening Upward in Both Empty Sections of the Assay Use forceps to slide the

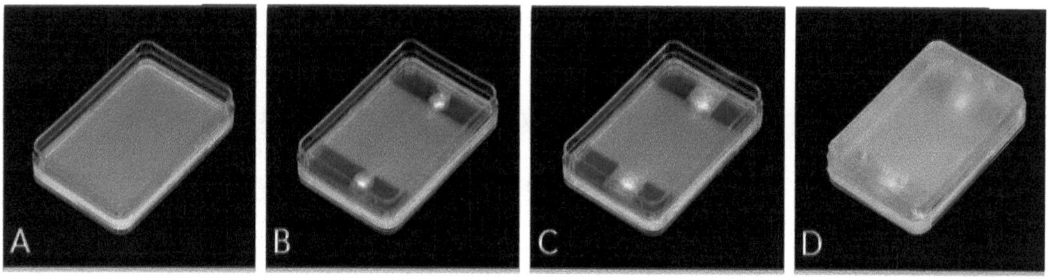

Fig. 9.3 The olfactory oviposition assay consists of a 57 × 38 x 17 mm rectangular transparent polystyrene dish accommodating two 10 × 38 × 4 mm oviposition zones filled with 0.75% agar containing 100 mM of sucrose. The two zones are separated by a 37 × 38 × 4 mm middle zone of 3% agar that is unsuitable for oviposition. The figure shows (**a**) step 2 of Sect. 3.1.1 in which 8 mL of 3% agar is dispensed into the dishes. (**b**) shows the cumulative result of steps 4 till 6 of Sect. 3.1 and step 5 of Sect. 3.5, in which the cellulose paper discs are placed in the cups. (**c**) shows how the pieces of mesh are placed on top of the odor containing cups in step 8 of Sect. 3.5. Finally, (**d**) shows how the assay is covered with Parafilm and punctured as explained in steps 9 till 11 of Sect. 3.5. In the contact oviposition assay, the oviposition zones do not contain the cups shown in (**b**). Instead, the discs containing the pheromones are directly placed on top of the oviposition zones as explained in steps 4 and 5 of Sect. 3.6

connection strip of the cap under the remaining 3% agar, and position them in the center of the open sections (Fig. 9.3b). The connection strip under the 3% agar ensures that the cap is secured in place when the oviposition zones are filed with agar (step 8).

Step 7. Prepare 0.75% Agar Containing 100 mM D-sucrose for the Oviposition Zones Add 0.75 g of agar powder and 3.42 g of D-sucrose in a glass beaker, and fill it with distilled water until 100 mL is reached. Heat the mixture and let it boil for at least 1 min. Cover the beaker with aluminum foil, and let it cool down for several minutes before handling it.

Step 8. Dispense 1.5 mL of the 0.75% Agar and Sugar Mixture into the Open Sections Use a 5 mL volumetric pipette to dispense the agar around the odor cups. Ensure not to spill agar over the 3% agar middle zone or into the odor cups (Fig. 9.3b).

Step 9. Let the Agar Solidify at Room Temperature for at Least 1 h Cover the dishes with filter paper sheets to prevent contamination of the assays. The assays can be stored in a fridge for up to 2 days. Ensure to place them with the lid facing downward to prevent condensation dropping on to the surface of the agar.

4.1.2 Preparing the Contact Oviposition Assay

Step 1. Prepare 3% Agar for the Middle Zone Add 3 gr of agar powder in a glass beaker, and fill it with distilled water until 100 mL is reached. Boil for at least a minute in a microwave or on a heating plate. Afterward, cover the beaker with aluminum foil to prevent evaporation and contamination, and let it cool down for several minutes before handling it.

Step 2. Dispense 8 mL of 3% Agar into the Dishes with a 25 mL Volumetric Pipette The resulting 4-mm-thick layer of 3% agar should cover the bottom of the dish evenly and should be free of air bubbles.

Step 3. Let the Agar Solidify at Room Temperature for 30–45 min Cover the dishes with filter paper sheets or fine mesh to prevent contamination of the agar while solidifying. Agar that has solidified for less than 30 min may be too soft for removing the sections in the following step.

Step 4. Remove 10-mm-wide Sections from the 3% Agar at Both Ends of the Dish The agar can easily be cut and removed with the flat end of a stainless steel spatula. It is advisable to place the dish on a printed template corresponding to the size of the sections that have to be removed for size standardization of the zones.

Step 5. Prepare 0.75% Agar Containing 100 mM D-sucrose for the Oviposition Zones Add 0.75 g of agar powder and 3.42 g of D-sucrose in a glass beaker, and fill it with distilled water until 100 mL is reached. Heat the mixture and let it boil for at least a minute. Cover the beaker with aluminum foil, and let it cool down for several minutes before handling it.

Step 6. Dispense 1.7 mL of the 0.75% Agar and Sugar Mixture into the Open Sections Using a 5 mL volumetric pipette with a narrow opening will help preventing the agar from spilling over the 3% agar middle zone.

Step 7. Let the Agar Solidify at Room Temperature for at Least 1 h Cover the dishes with filter paper sheets to prevent contamination of the assays. The assays can be stored in a fridge for up to 2 days. Ensure to place them with the lid facing downward to prevent condensation dropping on to the surface of the agar.

4.2 Isolating Females for Oviposition Assays

After mating, the number of matured eggs in the ovaries starts to increase over a period of several hours [72, 73]. Therefore, it is advisable to standardize the time given between mating and testing oviposition behavior to ensure the production of a reliable number of eggs for quantification in the experiment. This can be done by isolating mated females for 18–24 h prior to the experiment in 2 mL screw cap vials, of which the vial contains 300 μL 3%

bacteriological agar to provide the flies with a humid medium, and of which the screw caps contain 150 μL of yeast extract paste to provide the flies with nutrients (from here onward isolation vials). Ensure to only use 3% agar and yeast extract paste as they do not provide suitable oviposition substrates during the isolation time. An additional advantage of the 2 mL screw cap vials is that they offer a space efficient way to isolate large number of females prior to an experiment. The steps below describe how to prepare the vials and how to transfer the females into the isolation vials after mating them (see Sect. 1.2).

Materials

2 mL screw cap microvials (72.609, Sarstedt)

Screw cap for microvials (65.716, Sarstedt)

Bacteriological agar (CAS # 9002-18-0)

Yeast extract (CAS # 8013-01-02)

5 mL volumetric pipet (606,107, Greiner Bio-One)

10 mL syringe (473,202, Greiner Bio-One)

Beakers and laboratory utensils

Four well plates (96,077,307, Greiner Bio-One)

94 mm disposable Petri dishes (633,180, Greiner Bio-One)

Step 1. Prepare 3% Agar for the Isolation Vials Add 3 g of agar powder in a glass beaker, and fill it with distilled water until 100 mL is reached. Larger amounts can be prepared by multiplying these volumes. Heat the mixture and let it boil for at least a minute. Cover the beaker with aluminum foil, and let it cool down for several minutes before handling it.

Step 2. Dispense 300 μL of 3% Agar into the Vials Use a 5 mL volumetric pipette with a narrow opening to reduce the risk of coating the inside of the vial with small agar drops that may offer suitable oviposition substrates during the isolation period.

Step 3. Prepare Yeast Extract Paste for the Screw Caps Mix yeast extract powder with small amounts of water until it becomes a foamy liquid. A mixture of approximately 20 g of yeast extract powder and 8–10 mL of distilled water is needed to fill 150–200 caps with yeast extract paste.

Step 4. Fill the Openings of the Screw Caps with 150 μL of Yeast Extract Paste Place the caps with the opening upward in 4 well plates or Petri dishes, and dispense the yeast extract paste with a 10 mL syringe into the screw caps. Ensure to keep the screw threads of the cap free of yeast extract paste.

Step 5. Store the Vials and Screw Caps Overnight In a Fridge to Solidify Keep them in closed containers to prevent contamination and dehydration of the materials. Hold several of the yeast extract paste caps upside down before using them to check if any liquid runs out. Discard the caps if liquid runs out and repeat steps 3 and 4 with thicker yeast extract paste.

Step 6. Transfer the Females into the Isolation Vials with a Mouth Aspirator Cap the vials loosely with the yeast extract paste caps to allow for some air circulation. See Sect. 1.2 on how to control the mating state of female flies.

Step 7. Isolate the Females for 18 h Prior to the Experiment - Place the isolation vials vertically in a microvial container, without closing the lid, and label them according to genotype or experimental treatment. It is advised to keep the vials at the location where the experiment is run for proper acclimatization.

4.3 Extracting Cuticular Pheromones for the Oviposition Assays

The oviposition assay can be used to test oviposition site selection in response to full pheromone profiles or to any predetermined compounds. The pheromonal profile of *D. melanogaster* predominantly consists of cuticular hydrocarbons, like the male-dominant 7-T and 9-T and the female-specific 7,11-HD and 7,11-ND, along with cVA and other fatty acids produced in tissues other than the oenocytes [14, 15, 21]. The cuticular hydrocarbons, along with cVA and other nonpolar compounds, can be extracted by immersing adult flies into n-Hexane. It should be noted that this method is designed for extracting cuticular hydrocarbons and that it will underperform in extracting cVA and other pheromones stored internally [56]. The protocol can be applied to single or groups of flies by adjusting the amount of n-Hexane used to extract the pheromones.

Materials

2 mL glass screw cap vials (29378-U, Supelco)

≥ 99% n-Hexane (CAS# 110-54-3)

≥ 99% Methanol (CAS # 67-56-1)

Fine tipped forceps (#5 Inox, Dumont)

Polyurethane ice bucket

25 mm × 95 mm polystyrene rearing vials (734-2255, VWR)

23 mm cellulose plugs (734-2875, VWR)

50 mm Petri dishes (glass or disposable plastic)

20 mL glass bottles with urea caps and metal foil liner (218-0000P, VWR).

Step 1. Transfer the Flies into an Empty Rearing Vial Use a rearing vial without food to prevent the flies from getting stuck in the food medium while anesthetized, and use hard cellulose plugs to close the vials.

Step 2. Anesthetize the Flies on Ice Push the rearing vial into the ice and tap the cellulose plug after several minutes. The flies are properly anesthetized when they fall to the bottom of the vial and stop moving.

Step 3. Clean the Forceps Used for Handling Flies Contaminants can be removed by first immersing the tips of the forceps into a 20 mL glass bottle containing n-Hexane and wiping it with a clean disposable tissue. Next, the tips should be immersed into a 20 mL bottle containing methanol and wiped again. Finally, the forceps should be immersed into a second 20 mL bottle containing n-Hexane and let to air dry (see Sect. 4.4).

Step 4. Transfer the Anesthetized Flies onto a Petri Dish Placed on Ice Use either glass Petri dishes that are properly cleaned (see Sect. 4.3) or unused disposable plastic Petri dishes. Exclude any physically damaged flies from the extraction. Also, the sexes can be separated on the Petri dish when extracting pheromones originating from mixed groups of flies.

Step 5. Transfer the Flies into a 2 mL Screw Cap Glass Vial Kept on Ice Use the cleaned forceps to pick individual flies up by the appendages, and transfer them in a 2 mL glass screw cap vial. Ensure to count the number of flies transferred into the vial (in preparation of step 6), and do not exceed 80 flies per vial to allow for space to vortex the flies in step 7.

Step 6. Pipet 12 μL n-Hexane per Fly into the 2 mL Screw Cap Vial The total volume of n-Hexane can be calculated by multiplying the number of flies counted in step 5 by 12. Ensure to keep all materials on ice and to close the vial immediately after pipetting the hexane to limit evaporation.

Step 7. Vortex the Vial with the Flies for 3 min Check after 30 s if any flies have gotten stuck in the lid of the vial. If this happened, tap the flies down and vortex at a lower speed.

Step 8. Transfer the Supernatant into a Clean 2 mL Vial Kept on Ice Ensure that no body parts have contaminated the supernatant. The extract will comprise of approximately 10 μL per fly, but this may differ depending on the amount of evaporation and absorption of the hexane during the previous steps.

Step 9. Prepare the Pheromone Extract for Storage at −20 °C Label the vial with its contents and the date it was prepared. Afterward, it is advisable to flush the vials with nitrogen and to seal the screw caps with Teflon tape or Parafilm to prevent deterioration of the extracts during long-term storage at −20 °C.

4.4 Preparing Pheromone Dilutions for the Oviposition Assay

For some experiments it may be more relevant to test single pheromones instead of full pheromone profiles. The steps below describe how to dilute synthetic compounds for the oviposition assay. Keep in mind that the dilutions should be prepared ten times lower than the test concentration, to account for the 10 μL of solvent that is used to load the pheromones onto the filter paper disc in the oviposition assay. Considering that many of the solvents used in behavioral experiments are hydrocarbons themselves, it is important to only use solvents of analytical grade purity (e.g., purity ≥99% n-Hexane) to lower the change of physiological or behavioral responses caused by contaminants [74]. For the preparation of pheromones, it is also recommended to equip air displacement pipettes with filter barrier tips, or to use positive displacement pipettes (e.g., tip with a piston), to prevent any contaminations from being transferred through the pipette.

Materials

2 mL glass screw cap vials (29378-U, Supelco)

20 μL Pipette filter tips (773 26, Greiner Bio-One)

1250 μL Pipette filter tips (750 26, Greiner Bio-One)

≥ 99% n-Hexane (CAS# 110-54-3)

Vortex

Nitrogen flow

Step 1. Determine the Pheromone Concentration for the Experiment Concentrations can be based on chemical analysis or on previously published results (see [49, 50, 56]).

Step 2. Prepare 1 mL of a 1:100 Stock Solution in Hexane in a 2 mL Glass Screw Cap Vial The stock solution will be used for making dilutions of the predetermined concentrations. When using commercially available products, the initial density is reported in g/mL on the bottle. First convert this into ng/μL before continuing with the following steps. Next, pipette 990 μL of n-Hexane into the screw cap vial, and add 10 μL of the compound to it. Finally, divide the initial density in ng/μL by a hundred, and note the density on the vial.

Step 3. Calculate the Volumes Needed for Diluting the Pheromones The volumes used to make the dilution can be calculated through $\text{Concentration}_{(\text{start})} \times \text{Volume}_{(\text{start})} = \text{Concentration}_{(\text{final})} \times \text{Volume}_{(\text{final})}$. Use a maximum final volume of 1000 μL to ensure that the dilution can be prepared in a 2 mL vial.

Step 4. Dilute the 1:100 Stock Solution to the Desired Concentration in a 2 mL Glass Screw Cap Vial Pipette the calculated volume of the 1:100 stock solution into the vial and add n-Hexane to reach the final volume of 1000 μL. Vortex the solution for several seconds to mix the two chemicals.

Step 5. Prepare the Pheromone Dilutions for Storage at −20 °C Clearly label the vial with its content and the date it was prepared. It is advisable to flush the vials with nitrogen and to seal the screw caps with Teflon tape or Parafilm to prevent deterioration of the pheromones during long-term storage at −20 °C.

4.5 Running the Olfactory Oviposition Assays

The oviposition behavior of *D. melanogaster* females is an intricate process in which multiple sensory pathways, including olfaction, gustation, vision, and mechanosensation are employed to choose an appropriate egg-laying site [75–77]. When interested in testing the function of pheromones in oviposition site selection, it is important to standardize the environmental conditions to a stable temperature around 25 °C, a 12:12 h light-dark cycle, and a relative humidity around 50% and to run the experiments in a darkened room where external light conditions cannot affect the behavior of the flies. The steps below describe how to run the olfactory oviposition assay in which the pheromones are enclosed in an inaccessible cup that prevents direct contact with them, ensuring that only volatile pheromones can only be sensed. This assay is useful in determining whether the tested pheromones can attract flies through volatiles alone or whether direct contact between the fly and the pheromones is needed for a behavioral response.

Materials

Prepared oviposition assays (Sect. 3.1)

Grade 1 Chr cellulose chromatography paper (10 mm × 100 m roll, Whatman)

Polyamide mesh (mesh 0.4–0.6 mm/threads 0.15–0.25 mm)

Parafilm (100 mm)

Straight dissecting needle

Curved tip forceps

Nitrogen flow

3 mm diameter one-hole paper puncher

50 mm Petri dishes (glass or disposable plastic)

Mouth aspirator made of:

Disposable pipette tips

Fine mesh

Soft tubing

Step 1. Prepare 3 mm Chromatography Paper Discs for the Odor Treatments The discs can be punched out of cellulose chromatography paper rolls (Grade 1 Chr, 0.18-mm-thick) using a 3 mm diameter one-hole paper puncher or purchased.

Step 2. Cut 6 × 6 mm Square Pieces of Mesh to Cover the Odor Treatments Use a product with a mesh size between 0.4 and 0.6 mm made out of rigid polyamide fibers of 0.15–0.25 mm. Discard any pieces of mesh that are slightly curved as they cannot attach to the agar properly.

Step 3. Cut 100 × 50 mm Strips of Parafilm These strips will be used to cover the oviposition assay during the experiment instead of the polystyrene lids of the dishes. The benefit of using Parafilm to cover the assays is that it offers an odorless, colorless, and opaque seal that prevents bright light from disturbing the flies while limiting the loss of moisture in the assay.

Step 4. Mark on the Outside of the Assay Which Side Receives the Solvent Control and the Pheromone Treatment Use a permanent marker to prevent the marking from accidently being removed during the preparation of the assay.

Step 5. Place Two 3 mm Chromatography Paper Discs on Top of Each Other in Each Cup Placing two discs on top of each other will ensure that the 10 μL of n-Hexane will be properly absorbed without spilling over the edge of the cup (Fig. 9.3b). It is easiest to pick the discs up with a curved tip stainless steel forceps.

Step 6. Pipette 10 μL of the Solvent Control in the Cup at the Side Marked to Receive the Solvent Once the solvent has been loaded into the cup, briefly place a nitrogen flow above the cup until the hexane has evaporated.

Step 7. Pipette 10 μL of the Pheromone Treatment in the Cup at the Side Marked to Receive the Pheromones Once the pheromones have been loaded, briefly place a nitrogen flow above the cup until the hexane has evaporated. When interested in testing a combination of pheromones, follow the same procedure to add a second pheromone once the hexane of the first pheromone has evaporated.

Step 8. Cover Both Cups with the 6 × 6 mm Pieces of Mesh - Press the mesh gently with the blunt curved end of a curved tip stainless steel forceps onto the agar in the oviposition zone to secure it into place (Fig. 9.3c). Use separate forceps for the solvent side and pheromone side to prevent contamination.

Step 9. Cover the Assay with a 100 × 50 mm Parafilm Strip - Stretch the Parafilm tightly over the assay, and keep one of sides open to insert the fly with a mouth aspirator.

Step 10. Introduce a Mated Female into the Assay and Close It with the Parafilm Use a mouth aspirator to transfer a mated female from the isolation vial into the assay. Immediately close the open side of the assay with the Parafilm to prevent the fly from escaping.

Step 11. Puncture Three Small Holes in the Parafilm Above the Oviposition Zones These holes will prevent odor saturation in the assay. Use a fine tipped dissecting needle with a diameter smaller than the body size of the flies to prevent them from escaping (Fig. 9.3d).

Step 12. Let the Oviposition Experiment Run for 24 h Keep the experimental conditions stable over the entire run of the experiment, and place the solvent and pheromone treatments in random directions to prevent bias from environmental cues.

Step 13: Count the Eggs Laid at Both Sides of the Assay Under a Stereomicroscope, and Calculate the Oviposition Index Eggs are visible through the agar and can be easily counted through the bottom of the assay without having to remove the opaque Parafilm from the top. The oviposition index can be calculated per replicate as (eggs pheromones side – eggs solvent side)/(eggs pheromones side + eggs solvent side).

4.6 Running the Contact Oviposition Assays

Several gustatory receptors [16, 33, 36] and *pickpocket* ion channels [35, 78] are involved in detecting cuticular hydrocarbons. Therefore, it may not suffice to only test the involvement of volatile compounds in oviposition site selection. The steps below describe the preparation of the contact oviposition assay, in which the female fly is allowed to make direct contact with the pheromones while selecting an oviposition site, allowing for the additional involvement of gustation. While most of the experimental procedure is the same between both types of assays, the biggest difference is the application of the pheromones. In the contact oviposition assay, the pheromones are loaded onto cellulose discs that are coated with a barrier of Parafilm to prevent the disc from absorbing liquid from the oviposition zones. These discs should have a larger diameter compared to those of the olfactory assay to allow for proper absorption of the pheromone solution.

Step 1. Prepare 5 mm Chromatography Paper Discs Coated with Parafilm for the Odor Treatments Place a strip of cellulose chromatography paper (Grade 1 Chr, 0.18-mm-thick) on top of a strip of Parafilm, and punch a hole with a 5 mm diameter one-hole paper puncher. When executed properly the chromatography paper will stick onto the Parafilm.

Step 2. Cut 100 × 50 mm Strips of Parafilm These strips will be used to cover the oviposition assay during the experiment.

Step 3. Mark on the Outside of the Assay Which Side Receives the Solvent Control and the Pheromone Treatment Use a permanent marker to prevent the marking from accidently being removed during the preparation of the assay.

Step 4. Pipette the Solvent Control on a Disc, and Place it on the Side Marked to Receive the Solvent Place the filter paper disc that has been coated with Parafilm in a glass Petri dish, and once the solvent has been loaded, briefly place a nitrogen flow above the dish until the hexane has evaporated. Once the hexane has evaporated, place the disc on the side of the assay marked to receive the solvent treatment with a curved tip stainless steel forceps. Use separate forceps for the solvent side and pheromone side to prevent contamination.

Step 5. Pipette the Pheromone Treatment on a Disc, and Place it on the Side Marked to Receive the Pheromones Repeat step 4 using the pheromone treatment instead, and place the disc on the side of the assay marked to receive the pheromone treatment. When interested in testing a combination of pheromones, follow the same procedure to add a second pheromone once the hexane of the first pheromone has evaporated.

Step 6. Cover the Assay with a 100 × 50 mm Parafilm Strip - Stretch the Parafilm tightly over the assay, and keep one of sides open to insert the fly with a mouth aspirator.

Step 7. Introduce a Mated Female into the Assay and Close it with the Parafilm Use a mouth aspirator to transfer a mated female from the isolation vial into the assay. Immediately close the open side of the assay with the Parafilm to prevent the fly from escaping.

Step 8. Puncture Three Small Holes in the Parafilm Above the Oviposition Zones These holes will prevent odor saturation in the assay. Use a fine-tipped dissecting needle with a diameter smaller than the body size of the flies to prevent them from escaping (Fig. 9.3d).

Step 9. Let the Oviposition Experiment Run for 24 h Keep the experimental conditions stable over the entire run of the experiment and to place the solvent and pheromone treatments in random directions to prevent bias from environmental cues.

Step 10: Count the Eggs Laid at Both Sides of the Assay Under a Stereomicroscope, and Calculate the Oviposition Index Eggs are visible through the agar and can be easily counted through the bottom of the assay without having to remove the opaque Parafilm from the top. The oviposition index can be calculated per replicate as (eggs pheromones side – eggs solvent side)/(eggs pheromones side + eggs solvent side).

5 Laboratory Practices When Working with Pheromones

5.1 General Laboratory Practices

Besides adhering to general laboratory safety guidelines and informing oneself with the potential toxicity and health-related hazards of the chemicals that are being worked with, it is important to take several additional measures into account to prevent contaminating samples or the working environment with pheromones. Firstly, wear nitrile gloves at all times as contaminants present on bare hands may share chemical similarity to fly cuticular hydrocarbons. Secondly, prevent contamination of working surfaces and laboratory equipment with pheromones as they can have a lasting effect on future experiments due to their stability and behavioral effect at low doses. This can be achieved by covering work surfaces with bench protector films to form a barrier when chemicals are spilled, and by using disposable products or products equipped with filter barriers. A good example of such products are micropipette tips that are equipped with filter barriers, which will help to prevent volatile contaminants from reaching the inside mechanisms of the micropipette. Another thing to keep in mind is that pheromones will stick to the pipette point and mesh of a mouth aspirator when transferring flies. Therefore, it is recommended to change these two parts of the mouth aspirator when running experiments with different genotypes, or experimental treatments, that differ in pheromonal profile. Thirdly, only run experiments in rooms that are well ventilated to prevent the saturation of the environment with volatile pheromones. When possible, it is strongly advised to equip the ventilation system with carbon filters to remove odors if the air is recycled. Finally, clean the working environment and laboratory equipment meticulously, and avoid reusing materials that cannot be properly cleaned. The sections below provide advice on how clean pheromones from work surfaces and several types of laboratory equipment.

5.2 Cleaning Work Surfaces

All the work surfaces need to be cleaned after using them to remove any chemicals left behind on them. The best cleaning agent depends on both the type of surface and chemical characteristics of the pheromones. In most cases, the first step is to clean the surface with a soap solution to remove the any dirt that may have collected on the surface. Most dishwashing soaps already remove small amounts of several types of pheromones. Afterward, technical-grade solvents can be used to remove specific pheromone compounds. Nonpolar solvents like pentane, hexane, and acetone can be used to clean glass and stone surfaces. However, for acrylic surfaces it is advisable to only use pentane and to avoid acetone and chlorinated solvents as they may damage the surface permanently. Finally, the surfaces should be wiped with ethanol to remove any polar phenomes, like cVA and fatty acids, and to expedite the drying process.

Materials

Technical-grade acetone (CAS# 67-64-1)

Technical-grade ethanol (CAS# 64-17-5)

Technical-grade n-Hexane (CAS# 110-54-3)

Technical-grade pentane (CAS# 109-66-0)

Dishwashing soap

5.3 Cleaning Glassware

The preparation of pheromones for behavioral experiments may require the use of several types of glassware. In many cases it is recommended to prefer glassware over disposable plastic products as the nonpolar solvents used as a vehicle for the pheromones in the experiments, as well as the cleaning solvents, may dissolve the plastic surfaces over time. It is advisable to clean glassware immediately after being used as it that may become difficult to remove certain residues after some time. The protocol below applies to glassware that has been in direct contact with pheromones, and the first step can be skipped for glassware that has only been used to prepare fly food or agar.

Materials

Technical-grade acetone (CAS# 67-64-1)

Technical-grade ethanol (CAS# 64-17-5)

Technical-grade n-Hexane (CAS# 110-54-3)

Dishwashing soap

Step 1. Rinse the Glassware with Hexane Hexane will remove most of the soluble pheromones from the glass surface. Ensure to take safety measures and discard the used hexane rinse into the organic solvent waste. This step can be omitted for glassware used to prepare fly food or agar, which should first be cleaned with warm water and dishwashing soap.

Step 2. Clean the Glassware with Warm Water and Detergent - Gently scrub the glass surface to remove any residues and pheromones from the surface. Rinse the glassware thoroughly with tap water to remove any remains of the soap.

Step 3. Rinse the Glassware with Ethanol Ethanol will remove any remains of the soap and also helps removing cVA from the glass surface.

Step 4. Rinse the Glassware with Distilled Water Rinsing the glassware with distilled water will prevent mineral water stains from forming while the glassware dries.

Step 5. Let the Glassware Air Dry Do not dry the glassware with a paper towel or forced air as this can leave behind fibers and impurities. Acetone can be used to expedite the drying process of glassware with narrow openings. Ensure to dispose the used acetone rinse into the organic solvent waste.

Step 6. Bake the Glassware in an Oven at 250 °C for 6–8 h Most volatile chemicals will be broken down and removed from the glass surface when it is baked at 250 °C. Once the glassware has been baked, cover any openings with a sheet of aluminum foil, and store it in a closed cabinet until further use.

5.4 Cleaning Laboratory Utensils

Several types of laboratory utensils are used during the preparation of behavioral experiments. Where applicable, it is highly recommended to use stainless steel utensils as most pheromones and other contaminants can be easily removed from stainless steel utensils using the protocol below. It is generally good practice to apply this cleaning protocol before and after using a utensil as this lowers the chance that contaminants are introduced into the experiment and ensures that no pheromones stay behind on the utensils.

Materials

Technical-grade n-Hexane (CAS# 110-54-3)

Technical-grade methanol (CAS # 67-56-1)

20 mL glass bottles with urea caps and metal foil liner (218-0000P, VWR)

Step 1. Immerse the Utensil in Hexane and Wipe it Clean with a Tissue Hexane is a nonpolar solvent that removes cuticular hydrocarbons from the utensil. For most utensils it is only necessary to immerse those parts into the hexane that have been in direct contact with pheromones. For practical and safety reasons, it is advisable to keep the hexane in a glass vial of maximum 20 mL that can be closed directly after using it with a screw cap.

Step 2. Immerse the Utensil in Methanol and Wipe it Clean with a Tissue Methanol is an alcohol that is particularly effective in removing cVA and fatty acids. Only immerse the parts that have been in contact with pheromones, and keep the methanol in a 20 mL glass vial that can be closed immediately after using it.

Step 3. Immerse the Utensil in Second Vial of Hexane and let it Dry in the Air This second 20 mL glass vial containing hexane removes the last contaminants that may be present on the utensil. It is important to let the utensil dry in the air and not to wipe it as this may leave behind fibers and impurities on the utensil. Once the utensil is dry, it is ready to be used.

References

1. Billeter JC, Wolfner MF (2018) Chemical cues that guide female reproduction in *Drosophila melanogaster*. J Chem Ecol 44:750–769. https://doi.org/10.1007/s10886-018-0947-z
2. Wyatt TD (2014) Pheromones and animal behavior: chemical signals and signatures. Cambridge University Press
3. Brennan PA, Zufall F (2006) Pheromonal communication in vertebrate. Nature:308–315. https://doi.org/10.1038/nature0540
4. Kohl J, Huoviala P, Jefferis GSXE (2015) Pheromone processing in *Drosophila*. Curr Opin Neurobiol 34:149–157. https://doi.org/10.1016/j.conb.2015.06.009
5. Li Y, Dulac C (2018) Neural coding of sex-specific social information in the mouse brain. Curr Opin Neurobiol 53:120–130. https://doi.org/10.1016/j.conb.2018.07.005
6. Cardé RT (2007) Using pheromones to disrupt mating of moth pests. In: Kogan M, Jepson P (eds) Perspectives in ecological theory and integrated pest management. Cambridge University Press, Cambridge, pp 122–169
7. Reddy GVP, Guerrero A (2004) Interactions of insect pheromones and plant semiochemicals. Trends Plant Sci 9:253–261. https://doi.org/10.1016/j.tplants.2004.03.009
8. Cardé RT, Millar JG (2009) Pheromones. In: Resh V, Cardé RT (eds) Encyclopedia of insects. Academic Press, pp 766–772
9. Wertheim B, van Baalen EJA, Dicke M, Vet LEM (2005) Pheromone-mediated aggregation in nonsocial arthropods: an evolutionary ecological perspective. Annu Rev Entomol 50: 321–346. https://doi.org/10.1146/annurev.ento.49.061802.123329

10. Byers JA (1991) Pheromones and chemical ecology of locusts. Biol Rev 66:347–378. https://doi.org/10.1111/j.1469-185x.1991.tb01146.x
11. Srinivasan J, Kaplan F, Ajredini R et al (2008) A blend of small molecules regulates both mating and development in *Caenorhabditis elegans*. Nature 454:1115–1118. https://doi.org/10.1038/nature07168
12. Liberles SD (2014) Mammalian pheromones. Annu Rev Physiol 76:151–175. https://doi.org/10.1146/annurev-physiol-021113-170334
13. Roberts SA, Simpson DM, Armstrong SD et al (2010) Darcin: a male pheromone that stimulates female memory and sexual attraction to an individual male's odour. BMC Biol 8. https://doi.org/10.1186/1741-7007-8-75
14. Jallon JM (1984) A few chemical words exchanged by Drosophila during courtship and mating. Behav Genet 14:441–478. https://doi.org/10.1007/BF01065444
15. Ferveur JF (2005) Cuticular hydrocarbons: their evolution and roles in Drosophila pheromonal communication. Behav Genet 35:279. https://doi.org/10.1007/s10519-005-3220-5
16. Sato K, Yamamoto D (2020) Contact-chemosensory evolution underlying reproductive isolation in *Drosophila* species. Front Behav Neurosci 14:1–13. https://doi.org/10.3389/fnbeh.2020.597428
17. Billeter JC, Levine JD (2014) Neurogenetics: sex and the female brain. Curr Biol 24:R812–R814. https://doi.org/10.1016/j.cub.2014.07.052
18. Bartelt RJ, Schaner AM, Jackson LL (1985) cis-Vaccenyl acetate as an aggregation pheromone in *Drosophila melanogaster*. J Chem Ecol 11:1747–1756. https://doi.org/10.1007/bf01012124
19. Yew JY, Dreisewerd K, Luftmann H et al (2009) A new male sex pheromone and novel cuticular cues for chemical communication in *Drosophila*. Curr Biol 19:1245–1254. https://doi.org/10.1016/j.cub.2009.06.037
20. Everaerts C, Farine JP, Cobb M, Ferveur JF (2010) *Drosophila* cuticular hydrocarbons revisited: mating status alters cuticular profiles. PLoS One 5:e9607. https://doi.org/10.1371/journal.pone.0009607
21. Butterworth FM (1969) Lipids of *Drosophila*: a newly detected lipid in the male. Science 163:1356–1357. https://doi.org/10.1126/science.163.3873.1356
22. Lin HH, Cao DS, Sethi S et al (2016) Hormonal modulation of pheromone detection enhances male courtship success. Neuron 90:1272–1285. https://doi.org/10.1016/j.neuron.2016.05.004
23. Dweck HHKM, Ebrahim SAM, Thoma M et al (2015) Pheromones mediating copulation and attraction in *Drosophila*. Proc Natl Acad Sci USA 112:E2829–E2835. https://doi.org/10.1073/pnas.1504527112
24. Antony C, Davis TL, Carlson DA et al (1985) Compared behavioral responses of male *Drosophila melanogaster* (Canton S) to natural and synthetic aphrodisiacs. J Chem Ecol 11:1617–1629. https://doi.org/10.1007/BF01012116
25. Lebreton S, Borrero-Echeverry F, Gonzalez F et al (2017) A *Drosophila* female pheromone elicits species-specific long-range attraction via an olfactory channel with dual specificity for sex and food. BMC Biol 15:88. https://doi.org/10.1186/s12915-017-0427-x
26. Billeter JC, Atallah J, Krupp JJ et al (2009) Specialized cells tag sexual and species identity in *Drosophila melanogaster*. Nature 461:987–991. https://doi.org/10.1038/nature08495
27. Guiraudie-Capraz G, Pho DB, Jallon JM (2007) Role of the ejaculatory bulb in biosynthesis of the male pheromone cis -vaccenyl acetate in *Drosophila melanogaster*. Integr Zool 2:89–99. https://doi.org/10.1111/j.1749-4877.2007.00047.x
28. Ferveur JF, Savarit F, O'Kane CJ et al (1997) Genetic feminization of pheromones and its behavioral consequences in *Drosophila* males. Science 276:1555–1558. https://doi.org/10.1126/science.276.5318.1555
29. Chertemps T, Duportets L, Labeur C et al (2007) A female-biased expressed elongase involved in long-chain hydrocarbon biosynthesis and courtship behavior in *Drosophila melanogaster*. Proc Natl Acad Sci USA 104:4273–4278. https://doi.org/10.1073/pnas.0608142104
30. Savarit F, Sureau G, Cobb M, Ferveur JF (1999) Genetic elimination of known pheromones reveals the fundamental chemical bases of mating and isolation in *Drosophila*. Proc Natl Acad Sci USA 96:9015–9020. https://doi.org/10.1073/pnas.96.16.9015
31. Lebreton S, Grabe V, Omondi AB et al (2014) Love makes smell blind: mating suppresses pheromone attraction in *Drosophila* females via Or65a olfactory neurons. Sci Rep 4:7199. https://doi.org/10.1038/srep07119
32. Wertheim B, Dicke M, Vet LEM (2002) Behavioural plasticity in support of a benefit for aggregation pheromone use in *Drosophila*

melanogaster. Entomol Exp Appl 103:61–71. https://doi.org/10.1046/j.1570-7458.2002.00954.x

33. Miyamoto T, Amrein H (2008) Suppression of male courtship by a *Drosophila* pheromone receptor. Nat Neurosci 11:874–876. https://doi.org/10.1038/nn.2161
34. Moon SJ, Lee Y, Jiao Y, Montell C (2009) A *Drosophila* gustatory receptor essential for aversive taste and inhibiting male-to-male courtship. Curr Biol 19:1623–1627. https://doi.org/10.1016/j.cub.2009.07.061
35. Thistle R, Cameron P, Ghorayshi A et al (2012) Contact chemoreceptors mediate male-male repulsion and male-female attraction during *Drosophila* courtship. Cell 149:1140–1151. https://doi.org/10.1016/j.cell.2012.03.045
36. Wang L, Han X, Mehren J et al (2011) Hierarchical chemosensory regulation of male-male social interactions in *Drosophila*. Nat Neurosci 14:757–762. https://doi.org/10.1038/nn.2800
37. Grillet M, Dartevelle L, Ferveur JF (2006) A *Drosophila* male pheromone affects female sexual receptivity. Proc R Soc B Biol Sci 273: 315–323. https://doi.org/10.1098/rspb.2005.3332
38. Farine JP, Ferveur JF, Everaerts C (2012) Volatile *Drosophila* cuticular pheromones are affected by social but not sexual experience. PLoS One 7:e40396. https://doi.org/10.1371/journal.pone.0040396
39. Lin CC, Prokop-Prigge KA, Preti G, Potter CJ (2015) Food odors trigger *Drosophila* males to deposit a pheromone that guides aggregation and female oviposition decisions. eLife 4: e08688. https://doi.org/10.7554/eLife.08688
40. Kurtovic A, Widmer A, Dickson BJ (2007) A single class of olfactory neurons mediates behavioural responses to a *Drosophila* sex pheromone. Nature 446:542–546. https://doi.org/10.1038/nature05672
41. Wang L, Anderson DJ (2010) Identification of an aggression-promoting pheromone and its receptor neurons in *Drosophila*. Nature 463: 227–231. https://doi.org/10.1038/nature08678
42. Ha TS, Smith DP (2006) A pheromone receptor mediates 11-cis-vaccenyl acetate-induced responses in *Drosophila*. J Neurosci 26: 8727–8733. https://doi.org/10.1523/JNEUROSCI.0876-06.2006
43. van der Goes van Naters W, Carlson JR (2007) Receptors and neurons for fly odors in *Drosophila*. Curr Biol 17:606–612. https://doi.org/10.1016/j.cub.2007.02.043
44. Seeholzer LF, Seppo M, Stern DL, Ruta V (2018) Evolution of a central neural circuit underlies *Drosophila* mate preferences. Nature 559:564–569. https://doi.org/10.1038/s41586-018-0322-9
45. Auer TO, Khallaf MA, Silbering AF et al (2020) Olfactory receptor and circuit evolution promote host specialization. Nature. https://doi.org/10.1038/s41586-020-2073-7
46. Montagne J, Wicker-Thomas C (2021) *Drosophila* pheromone production. In: Insect pheromone biochemistry and molecular biology. Elsevier, pp 163–181
47. Wicker-Thomas C, Chertemps T (2010) Molecular biology and genetics of hydrocarbon production. In: Insect hydrocarbons biology, biochemistry, and chemical ecology, pp 53–74. https://doi.org/10.1017/CBO9780511711909.005
48. de la Paz Fernández M, Chan YB, Yew JY et al (2010) Pheromonal and behavioral cues trigger male-to-female aggression in *Drosophila*. PLoS Biol 8:e1000541. https://doi.org/10.1371/journal.pbio.1000541
49. Kent C, Azanchi R, Smith B et al (2007) A model-based analysis of chemical and temporal patterns of cuticular hydrocarbons in male *Drosophila melanogaster*. PLoS One 2. https://doi.org/10.1371/journal.pone.0000962
50. Marcillac F, Grosjean Y, Ferveur JF (2005) A single mutation alters production and discrimination of *Drosophila* sex pheromones. Proc R Soc B Biol Sci 272:303–309. https://doi.org/10.1098/rspb.2004.2971
51. Billeter JC, Levine JD (2013) Who is he and what is he to you? Recognition in *Drosophila melanogaster*. Curr Opin Neurobiol 23:17–23. https://doi.org/10.1016/j.conb.2012.08.009
52. Krupp JJ, Nayal K, Wong A et al (2020) Desiccation resistance is an adaptive life-history trait dependent upon cuticular hydrocarbons, and influenced by mating status and temperature in *D. melanogaster*. J Insect Physiol 121: 103990. https://doi.org/10.1016/j.jinsphys.2019.103990
53. Mossman JA, Mabeza RMS, Blake E et al (2019) Age of both parents influences reproduction and egg dumping behavior in *Drosophila melanogaster*. J Hered 110:300–309. https://doi.org/10.1093/jhered/esz009
54. Arienti M, Antony C, Wicker-Thomas C et al (2010) Ontogeny of *Drosophila* melanogaster female sex-appeal and cuticular hydrocarbons. Integr Zool 5:272–282. https://doi.org/10.1111/j.1749-4877.2010.00213.x

55. Scott D (1986) Sexual mimicry regulates the attractiveness of mated *Drosophila melanogaster* females. Proc Natl Acad Sci USA 83: 8429–8433. https://doi.org/10.1073/pnas.83.21.8429
56. Laturney M, Billeter JC (2016) *Drosophila melanogaster* females restore their attractiveness after mating by removing male anti-aphrodisiac pheromones. Nat Commun 7:12322. https://doi.org/10.1038/ncomms12322
57. Lung O, Wolfner MF (2001) Identification and characterization of the major *Drosophila melanogaster* mating plug protein. Insect Biochem Mol Biol 31:543–551. https://doi.org/10.1016/S0965-1748(00)00154-5
58. Badre NH, Martin ME, Cooper RL (2005) The physiological and behavioral effects of carbon dioxide on *Drosophila melanogaster* larvae. Comp Biochem Physiol Mol Integr Physiol 140:363–376. https://doi.org/10.1016/j.cbpb.2005.01.019
59. Bartholomew NR, Burdett JM, Vandenbrooks JM, et al (2015) Impaired climbing and flight behaviour in *Drosophila melanogaster* following carbon dioxide anaesthesia. Sci Rep 5:1–10. https://doi.org/https://doi.org/10.1038/srep15298
60. McGuire SE, Le PT, Osborn AJ et al (2003) Spatiotemporal Rescue of Memory Dysfunction in *Drosophila*. Science 302:1765–1768. https://doi.org/10.1126/science.1089035
61. Duménil C, Woud D, Pinto F et al (2016) Pheromonal cues deposited by mated females convey social information about egg-laying sites in *Drosophila Melanogaster*. J Chem Ecol 42:259–269. https://doi.org/10.1007/s10886-016-0681-3
62. Keesey IW, Koerte S, Retzke T et al (2016) Adult frass provides a pheromone signature for *Drosophila* feeding and aggregation. J Chem Ecol 42:739–747. https://doi.org/10.1007/s10886-016-0737-4
63. Mercier D, Tsuchimoto Y, Ohta K, Kazama H (2018) Olfactory landmark-based communication in interacting *Drosophila*. Curr Biol 28: 2624–2631.e5. https://doi.org/10.1016/j.cub.2018.06.005
64. Billeter JC, Rideout EJ, Dornan AJ, Goodwin SF (2006) Control of male sexual behavior in *Drosophila* by the sex determination pathway. Curr Biol 16:R766–R776. https://doi.org/10.1016/j.cub.2006.08.025
65. Kohlmeier P, Zhang Y, Gorter JA et al (2021) Mating increases *Drosophila melanogaster* females' choosiness by reducing olfactory sensitivity to a male pheromone. Nat Ecol Evol 5: 1165–1173. https://doi.org/10.1038/s41559-021-01482-4
66. Schwartz NU, Zhong L, Bellemer A, Tracey WD (2012) Egg laying decisions in *Drosophila* are consistent with foraging costs of larval progeny. PLoS One 7:e37910. https://doi.org/10.1371/journal.pone.0037910
67. Yang CH, Belawat P, Hafen E et al (2008) *Drosophila* egg-laying site selection as a system to study simple decision-making processes. Science 319:1679–1683. https://doi.org/10.1126/science.1151842
68. Zhu EY, Guntur AR, He R et al (2014) Egg-laying demand induces aversion of UV light in *Drosophila* females. Curr Biol 24: 2797–2804. https://doi.org/10.1016/j.cub.2014.09.076
69. Yang CH, He R, Stern U (2015) Behavioral and circuit basis of sucrose rejection by *Drosophila* females in a simple decision-making task. J Neurosci 35:1396–1410. https://doi.org/10.1523/jneurosci.0992-14.2015
70. Nojima T, Rings A, Allen AM et al (2021) A sex-specific switch between visual and olfactory inputs underlies adaptive sex differences in behavior. Curr Biol 31:1175–1191.e6. https://doi.org/10.1016/j.cub.2020.12.047
71. Karageorgi M, Bräcker LB, Lebreton S et al (2017) Evolution of multiple sensory systems drives novel egg-laying behavior in the fruit pest *Drosophila suzukii*. Curr Biol 27: 847–853. https://doi.org/10.1016/j.cub.2017.01.055
72. Bloch Qazi MC, Heifetz Y, Wolfner MF (2003) The developments between gametogenesis and fertilization: ovulation and female sperm storage in *Drosophila melanogaster*. Dev Biol 256:195–211. https://doi.org/10.1016/S0012-1606(02)00125-2
73. Rubinstein CD, Wolfner MF (2013) *Drosophila* seminal protein ovulin mediates ovulation through female octopamine neuronal signaling. Proc Natl Acad Sci USA 110: 17420–17425. https://doi.org/10.1073/pnas.1220018110
74. Schorkopf DLP, Molnár BP, Solum M et al (2019) False positives from impurities result in incorrect functional characterization of receptors in chemosensory studies. Prog Neurobiol 181:101661. https://doi.org/10.1016/j.pneurobio.2019.101661
75. Laturney M, Billeter JC (2014) Neurogenetics of female reproductive behaviors in *Drosophila melanogaster*. In: Friedmann T, Dunlap JC, Goodwin SF (eds) Advances in genetics. Academic Press, San Diego, pp 1–108

76. Verschut TA, Carlsson MA, Anderson P, Hambäck PA (2017) Sensory mutations in *Drosophila melanogaster* influence associational effects between resources during oviposition. Sci Rep 7:9352. https://doi.org/10.1038/s41598-017-09728-7
77. Cury KM, Prudhomme B, Gompel N (2019) A short guide to insect oviposition: when, where and how to lay an egg. J Neurogenet 33:75–89. https://doi.org/10.1080/01677063.2019.1586898
78. Lu B, LaMora A, Sun Y et al (2012) ppk23-dependent chemosensory functions contribute to courtship behavior in *Drosophila melanogaster*. PLoS Genet 8:e1002587. https://doi.org/10.1371/journal.pgen.1002587

Chapter 10

Miniature Microscopy of Hippocampal CA1 to Identify Engram Cells and Record Calcium Transients for Analyses of Ensemble Activities

Noriaki Ohkawa and Khaled Ghandour

Abstract

In the brain, memories are thought to be stored by neurons termed engram cells that are activated during learning. Engram cells in the hippocampus can be targeted in c-Fos-tTA mice and have been identified as those that are activated during the acquisition and retrieval of a specific memory. Notably, the recent application of a genetically encoded calcium (Ca^{2+}) indicator (GECI) combined with miniature microscopy (i.e., Ca^{2+} imaging) enables us to observe neuronal activity within deep brain regions of rodents. Here, we explain the technical details of systems for engram cell targeting and Ca^{2+} imaging with a miniature microscope specifically applied for hippocampal CA1. In addition, we introduce a method to combine engram identification and Ca^{2+} imaging, which provides the ability to analyze the ensemble activity of engram cells during memory processes.

Key words Memory, Engram (cell), Ca^{2+} imaging, Hippocampus, CA1 region, GRIN lens, Photoconversion, Miniature microscope

1 Introduction

1.1 Engram Cells: Memory-Encoding Cells

To study memory, it is crucial to determine how memories are encoded in the brain and to identify its physiological nature. Richard Semon hypothesized that "a permanent record has been written or engraved on the irritable substance by a stimulus" during an event. In his book, *The Mneme*, the record is called an "engram," which has roles for "persisting effects of past events" in organisms [1, 2]. Currently, neuroscientists postulate that this writing or engraving on a substance is the phosphorylation or de novo expression of specific proteins in response to neuronal activation during a novel event or learning. On the other hand, Donald Hebb, who hypothesized long-term potentiation as synaptic model of learning and memory, proposed that "a network of neurons that is activated repeatedly during a certain mental process has strengthened synaptic connections" [3]. In this hypothesis, the neurons activated

Daisuke Yamamoto (ed.), *Behavioral Neurogenetics*, Neuromethods, vol. 181, https://doi.org/10.1007/978-1-0716-2321-3_10,

during a mental process are responsible for memory encoding. On the basis of the concepts posited by Semon and Hebb, neuroscientists postulated that memory is stored in the neurons preferentially activated during a learning event.

To investigate the concept of engram cells, c-Fos-tetracycline transactivator (tTA) transgenic mice [4] were established and became a decisive factor. Neuronal activation triggers promoter activity at the *Fos* gene. In these mice, tTA binds to a tetracycline-responsive element (TRE) and drives the expression of a downstream target gene upon the removal of doxycycline (Dox). In c-Fos-tTA mice injected with a recombinant adeno-associated virus (AAV) or lentivirus (LV) harboring the TRE promoter followed by preferred gene *X*, the subpopulation of cells activated during learning are specifically labeled with the X protein in the absence of Dox (OFF Dox). The cells activated during learning and are labeled by this c-Fos-tTA system are preferentially reactivated in "good learner" mice for the corresponding memory [4]. In addition, artificial activation and inhibition of the cells labeled by the c-Fos-tTA system can drive [5–10] and attenuate [11, 12] memory retrieval, respectively, such as by using optogenetics [13, 14] and chemogenetics [15, 16]. These sophisticated studies using the c-Fos-tTA system provide strong support for the engram concept of memory encoding.

1.2 Visualization of Neuronal Activity with a Miniature Microscope

Although the c-Fos-tTA system enabled us to identify and manipulate engram cells, there was no approach to investigate their patterns of activity. When a neuron is activated, Ca^{2+} flows into the soma. This neuronal activity can be translated into green fluorescence of a Ca^{2+} indicator, such as GCaMP [17, 18]. GCaMP is a genetically encoded Ca^{2+} indicator (GECI) composed of circularly permuted green fluorescent protein (GFP), a Ca^{2+}-binding domain (calmodulin domain), and an M13 domain, a peptide sequence from myosin light-chain kinase, as a binding target. The use of GECI for the visualization of neuronal activity is called Ca^{2+} imaging. The application of Ca^{2+} imaging is rapidly expanding among neuroscientists with a background in molecular biology, providing an opportunity to observe and track the activity of a large number of neurons simultaneously [18]. In support of the broadening molecular biological applications, optical systems have also been undergoing rapid improvements, including the development of miniature microscopes. Currently, two major miniature microscopes are typically utilized by neuroscientists [19, 20]. Both systems are similar and work as fluorescence microscopes for use in small animals, as they weigh around 2–3 g (Fig. 10.1). Moreover, these microscopes can be applied to freely moving animals, including mice [20, 21], rats [22], songbirds [23], and monkeys [24], to visualize events occurring within their brains. Ca^{2+} imaging was expected to be highly adaptive to engram identification, because

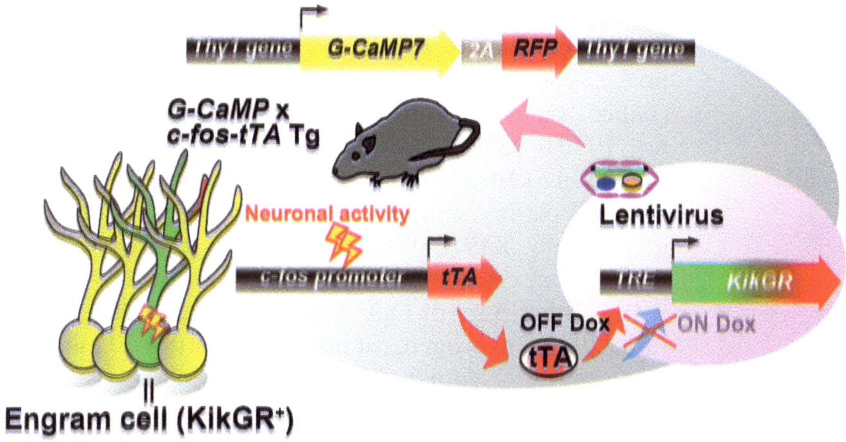

Fig. 10.1 Schematic diagram showing labeling and visualization of engram and nonengram cells in Thy1-GCaMP7 × c-Fos-tTA double transgenic mice injected with TRE-KikGR LV. Modified from Ghandour et al. (2019) [26]

both Ca^{2+} imaging and the c-Fos-tTA system are genetic techniques for the visualization of neurons. However, it is critical to discriminate the signals from Ca^{2+} transients and those form engram cells via the miniature microscope, which contains optics compatible only for green fluorescence.

2 Materials

2.1 LV for Gene Transfer

Photoconvertible fluorescent proteins change their spectrum in response to a specific trigger. Kikume Green-Red (KikGR) emits bright green fluorescence that can be irreversibly photoconverted to red by UV light exposure [25]. KikGR provides an opportunity to overcome the restrictions of optics designed for only one-color fluorescence, enabling the investigation of several targets. GCaMP and KikGR are detected at the same time by the miniature microscope adapted for GFP; however, KikGR fluorescence can be erased when it is no longer required. Therefore, photoconversion of KikGR before Ca^{2+} imaging will prevent the obstruction of Ca^{2+} signals during miniature microscopy. The combined use of KikGR with Ca^{2+} imaging using GCaMP was proposed for engram identification [26]. Thus, we utilized LV infection for gene transfer of TRE-KikGR into mice expressing the c-Fos-tTA system (Fig. 10.1). AAV is also useful for gene transfer; however, LV can restrict infection of neurons to a specific area because of suppressed diffusion of LV by its larger diameter. The LV we utilized was prepared essentially according to the protocol from the laboratory of K. Deisseroth (http://www.stanford.edu/group/dlab/optogenetics/expression_systems.html) [27]. 293FT cells (Invitrogen) were maintained in maintenance medium (DMEM;

Gibco, catalog number [cat. no.] 11,995) containing 10% heat-inactivated fetal bovine serum, 2 mM L-glutamine (Gibco, cat no. 25030–149), 0.1 mM MEM nonessential amino acids (Gibco, cat no. 11140–050), 500 μg/ml Geneticin (Gibco, cat no. 10131–035), and 1% 100× penicillin/streptomycin (Gibco, cat no. 15140–148). Twenty micrograms of the pLenti-TRE-hKiKGR plasmid and 48 μg of ViraPower packaging mix (Invitrogen, cat no. K4975–00) were combined in 6 ml of Opti-MEM (Invitrogen, cat no. 31985). The DNA mixture was mixed gently with another 6 ml of Opti-MEM containing 144 μl of Lipofectamine 2000 (Invitrogen, cat no. 11668). After 20 min, the transfection mix was added to a 225 cm^2 flask (BD Falcon, cat. no. 353138) containing 293FT cells at 95% confluence cultured in 28 ml of maintenance medium without Geneticin. After 16 h, spent medium was replaced with 30 ml of the virus production medium, Ultraculture (Lonza, cat. no. 12-725F), supplemented with 4 mM L-glutamine, 2 mM GlutaMAX-I (Gibco, cat. no. 35050–061), 0.1 mM MEM nonessential amino acids, 1 mM sodium pyruvate (Gibco, cat. no. 11360–070), and penicillin/streptomycin. Seventy-two hours after transfection, the culture medium was collected, filtered with a Millex-HV filter unit (0.45 μm) (Millipore), and transferred to a centrifuge tube. Three milliliters of 20% sucrose in phosphate-buffered saline (PBS; Gibco, cat. no. 14190) was gently added to the bottom of each centrifuge tube; the tubes were then centrifuged in a swing rotor, e.g., Beckman SW-28 or Hitachi P28S, for 2 h at 50,000 × *g*. The pelleted TRE-hKiKGR-LV was resuspended in 12.5 μl (1/2400 original volume of harvested medium) of cold PBS per tube to achieve a sufficient amount of fluorescent proteins with detectable native fluorescence; viral titers were approximately 5×10^9 to 1×10^{10} IU/ml.

2.2 Miniature Microscope

The miniature fluorescence microscope developed by Schnitzer's group (Fig. 10.2a) [19] is commercially available from Inscopix Inc. as "nVista." Golshani and Silva's group developed another small fluorescence microscope (UCLA miniscope; Fig. 10.2b) [20] and provided public access to the manufacturing method (http://miniscope.org/index.php/Main_Page). UCLA miniscope kits are now available from LabMaker (https://www.labmaker.org/products/miniscope-complete-set-of-components) and Open Ephys (https://open-ephys.org/miniscope-v4), and their use is becoming widespread among neuroscientists around the world. To relay fluorescence signals from deep brain regions to the external miniature microscope, a gradient refractive index (GRIN) relay lens (Fig. 10.2c) is applied as an endoscope [28]. In our case, we select lenses according to their working distance and ability to collimate a point source from an image (Fig. 10.2d).

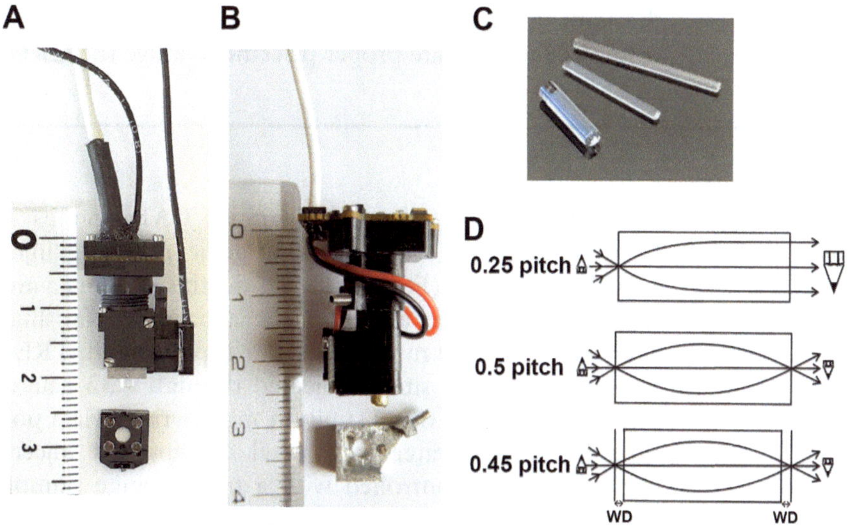

Fig. 10.2 Miniature microscopes and GRIN lenses
(**a**) Miniature microscope nVista V2 and its baseplate supplied from Inscopix Inc.
(**b**) UCLA miniscope V3 and its baseplate.
(**c**) Three GRIN lenses: from left to right, 1-mm-diameter 4-mm-length, 500-μm-diameter 6-mm-length, and 500-μm-diameter 8.4-mm-length. For CA1 imaging, the 1-mm-diameter 4-mm-length lens was chosen.
(**d**) Pitch of GRIN lens. A quarter-pitch lens and a half-pitch lens relay a point of an object on the entrance surface into infinity and inverted to the exit surface, respectively. Slightly less than a half pitch is applied to a lens to provide the working distance (WD) for an endoscope. Modified from a web page of GoPhoton Co., Ltd. with consent.

2.3 Animals

c-Fos-tTA mice were purchased from the Mutant Mouse Regional Resource Center (stock number 031756-MU). Thy1-G-CaMP7-T2A-DsRed2 mice have been described previously [29]. Other transgenic mice are available that express Ca^{2+} indicators, e.g., Thy1-GCaMP6f mice from the Jackson Laboratory [30]. We crossed c-Fos-tTA and Thy1-GCaMP7 mice to generate double transgenic mice for combined engram identification and Ca^{2+} imaging. However, GCaMP expression can also be induced by AAV-mediated gene transfer, such as by the commercially available CaMKII-GCaMP AAV (Addgene), but the expression of the Ca^{2+} indicator is less stable with an AAV-mediated system than in a transgenic mouse. Therefore, appropriate timing of Ca^{2+} imaging within a narrow time window should be considered when utilizing AAV-mediated expression.

2.4 LED for Photoconversion

For photoconversion, Black-LED-365 (Prizmatix Ltd., Israel) was used as a light source, and a SH4001 Super Eska polyethylene jacketed optical fiber cord (numerical aperture, 0.5; core diameter, 1000 μm; Mitsubishi Rayon Co., Ltd., Japan) was used to target 365 nm light to the CA1 region through the GRIN lens. The tip of the light illumination end was cut to form a blunt end with no connector to facilitate proper placement above the GRIN lens.

3 Methods

3.1 Animal Preparations

For our method, the hippocampal CA1 regions in Thy1-GCaMP7/c-Fos-tTA double transgenic mice are injected with TRE-hKiKGR-LV (Fig. 10.1). To do this, the mice are anesthetized and placed in a stereotactic apparatus (Narishige), and a craniotomy of ~1.0 mm in diameter is performed. TRE-hKiKGR-LV (1 μl/injection site) is injected through a 32 gauge injection cannula connected to a Hamilton microsyringe via a polyethylene tube filled with water or mineral oil, and the injection speed (0.1 μl/min) is controlled with a microsyringe pump (Harvard Apparatus, cat. no. CMA 400). The injection cannula is targeted to the right CA1 (anterior-posterior, 2.0 mm; medial-lateral, 1.4 mm; dorsal-ventral, 1.5 mm from bregma) and is left in place for 10 min before being slowly withdrawn. The craniotomy is closed with dental cement, and the skin is closed.

Around 3 weeks after LV injection, complete recovery of the mice from the LV injection is confirmed, and surgery is performed to implant a GRIN relay lens over hippocampal CA1, similarly as described in a previous report [28]. The following includes some tips for this procedure commonly adapted to the nVista and UCLA miniscope. Mice are anesthetized and placed in a stereotactic apparatus (Narishige). The GRIN lens is targeted to the center of the right hemisphere (anterior-posterior, 2.0 mm; medial-lateral, 1.5 mm). A 1.5-mm-diameter craniotomy is made in the skull to provide access for the GRIN lens (1.0 mm outer diameter and ~ 4.0 mm length; Inscopix Inc. or CLH lens from GoFoton). A cylindrical column of neocortex and corpus callosum above the alveus covering the dorsal hippocampus should be aspirated with continuously dripping saline using a 27 gauge blunt drawing-up needle. The corpus callosum, the thick bundle of fibers that appears white running in the medial-lateral direction (Fig. 10.3a), prevents transmission of light for imaging; therefore, the corpus callosum should be removed from the region between the GRIN lens and CA1. By contrast, the alveus, composed of thin, more fragile fibers running at ~45° from the medial-lateral direction (Fig. 10.3b), contains axonal fibers derived from CA1 and other hippocampal subregions. Thus, researchers should proceed with aspiration carefully after the appearance of the corpus callosum to avoid damaging the alveus. The GRIN lens is then gently placed on the alveus and fixed to the edge of the burr hole with bone wax melted by low-temperature cautery (e.g., https://www.boviemedical.com/lowtempcauteries/) (Fig. 10.3c). After positioning approximately eight anchor screws in the skull and confirming there is no bleeding

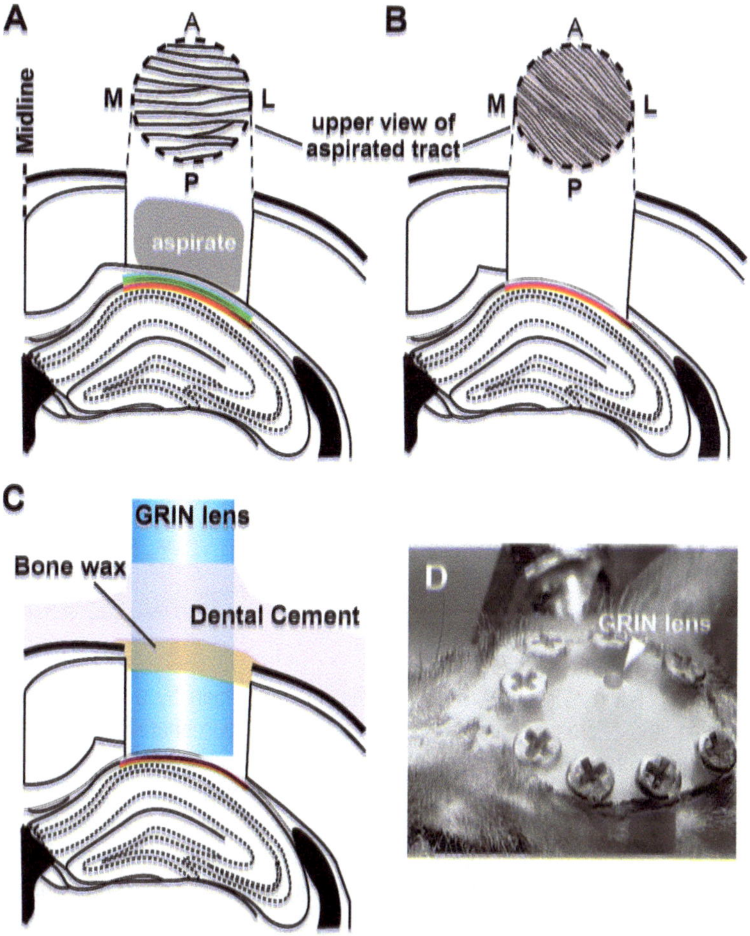

Fig. 10.3 GRIN lens setting on CA1

(**a**) Schematic diagram of aspiration of cortex. The corpus callosum (indicated as green) appears as a white thick fiber bundle running medial-lateral direction (top view).

(**b**) While proceeding with aspiration of the corpus callosum, the fiber pattern gradually changes to thin fibers running in different directions. These fibers are alveus (indicated as red).

(**c**) Schematic diagram of GRIN lens setting (coronal plane).

(**d**) Photo shows arrangement of GRIN lens and anchor screws fixed by dental cement on mouse skull.

from the edge of the bone wax, the skull is covered with dental cement to fix the GRIN lens to the skull and anchor screws (Fig. 10.3d). The GRIN lens can be protected by the tapered portion of a 0.2 ml PCR tube cut to appropriate size.

More than 2 weeks after the surgery to implant the GRIN lens, mice are anesthetized and placed in a stereotactic apparatus. A miniature microscope, i.e., nVista [19] or UCLA miniscope [20], is attached to the microscope baseplate (Fig. 10.4a) placed above

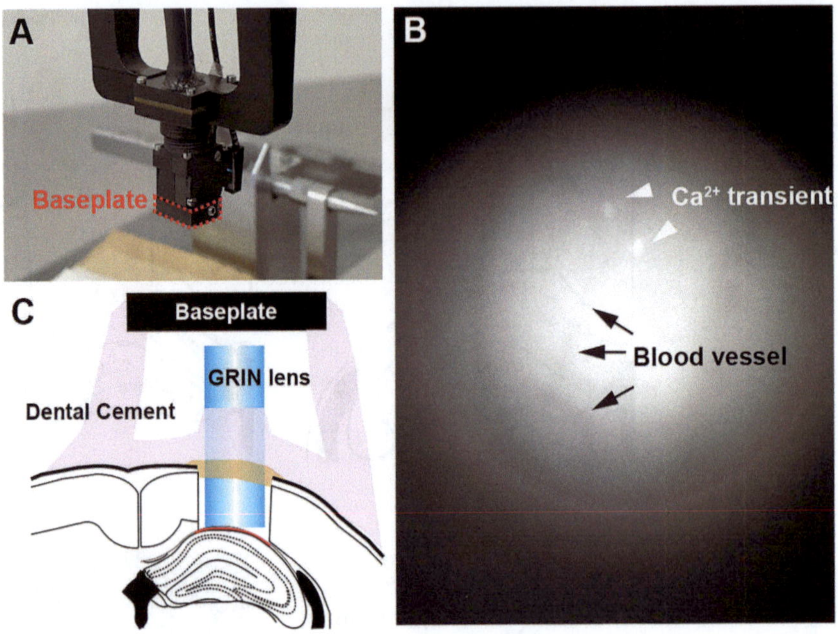

Fig. 10.4 Baseplate positioning
(**a**) Photo shows nVista (the base plate is attached) held by a clamp attached to the manipulator. While imaging, the microscope is slowly lowered to the appropriate position. Notably, the focus setting of the miniature microscope at this time should be set at the middle position to maintain a flexible focus adjustment range (up and down from a fixed position) for actual imaging.
(**b**) View during baseplate fixation position. GRIN lens is centered on the imaging view, and blood vessel can be seen. Ca^{2+} transients may appear under this condition if the anesthesia level is near wakefulness state.
(**c**) Schematic diagram of the baseplate setting (coronal plane). After complete solidification of the dental cement surrounding the GRIN lens, providing a wall and linking the baseplate and skull, the microscope is lifted from the baseplate by a manipulator.

the GRIN lens to observe GCaMP7 fluorescence and blood vessels in CA1 (Fig. 10.4b). The microscope baseplate is linked and fixed to the anchor screw using dental cement which is constructed to shade the GRIN lens; then, the integrated microscope is detached from the baseplate (Fig. 10.4c). The microscope baseplate cover is attached to the baseplate to cover the GRIN lens until it is time for Ca^{2+} imaging.

3.2 Identification and Photoconversion of KikGR Signals from Engram Cells

For photoconversion of KikGR, 365 nm light is passed through the optic fiber with an intensity of 0.6–0.8 mW at tip (Fig. 10.5a). The light is projected into CA1 by placing the tip of the optic fiber close to the outer surface of the GRIN lens beyond the center hole of the baseplate of the miniature microscope. First, a snapshot of green fluorescence of KikGR in the CA1 field is taken by the nVista HD acquisition software (Inscopix Inc.). Then, the 365 nm light is applied using two 15 s pulses separated by a 1.5 min interval to rapidly and irreversibly photoconvert the KikGR signal from green

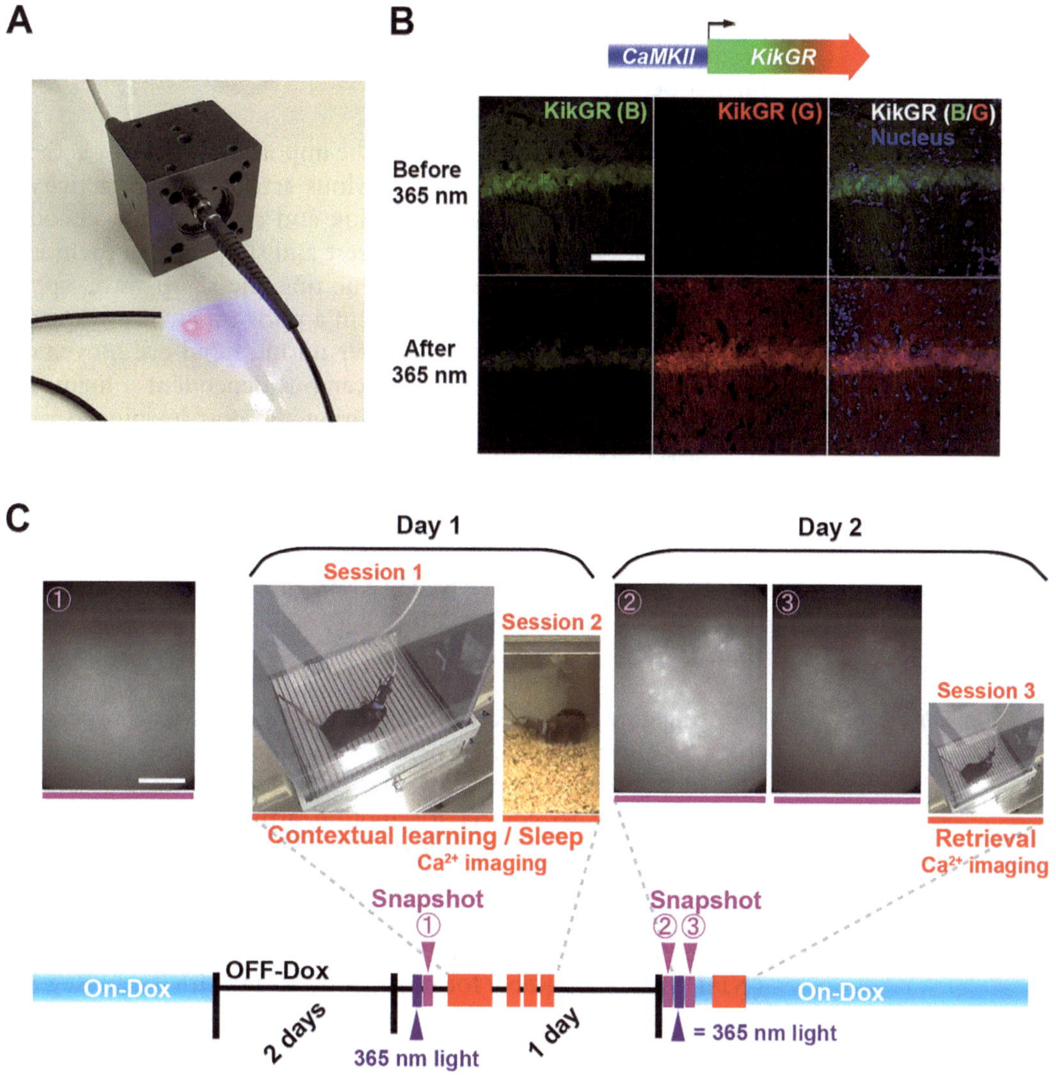

Fig. 10.5 KikGR photoconversion and Ca^{2+} imaging for investigation of memory processing
(**a**) LED light source of 365 nm light. The 365 nm light is projected from the tip of an optical fiber to CA1 through the GRIN lens.
(**b**) Representative image of KikGR expressed under the control of the CaMKII promoter in CA1 before (top) and after (bottom) photoconversion by 365 nm light exposure to a section in vitro. Scale bar, 100 μm.
(**c**) Experimental design for Ca^{2+} imaging during memory processing. Top: snapshots of KikGR before (1) and after (2) contextual learning (novel spatial exploration) and after photo conversion (3) and representative photographs of behavior during contextual learning (session 1), sleep (session 2), and retrieval (session 3). Bottom: Ca^{2+} imaging paradigm. Photoconversion by 365 nm light (dark purple arrowhead) and snapshot (pink arrowhead) before OFF Dox and then Ca^{2+} imaging during contextual learning and sleep with OFF Dox. This is followed by a second snapshot and photoconversion after reinstatement of ON Dox and then a Ca^{2+} imaging session during retrieval. Scale bar, 200 μm. **(b)** and **(c)** Modified from Ghandour et al. (2019) [26]

to red (Fig. 10.5b), rendering it invisible during subsequent Ca^{2+} imaging (Fig. 10.5c). Ca^{2+} imaging is performed 1.5–2 h after this photoconversion, performed during a novel experience (session 1) or memory retrieval (session 3) [26] (Fig. 10.5c) as described further below.

3.3 Ca^{2+} Imaging During Memory Processing

There is growing consensus about the importance of replay activity, i.e., the offline reactivation of previous activity. Replay activity is thought to be critical for conserving and consolidating encoded events [31]. During post-training rest and sleep, spatial memories can be disrupted by selectively inhibiting the sharp wave ripples within which sequential activity from a prior experience reappears [32]. Similarly, reduced theta power during rapid eye movement (REM) sleep weakens hippocampus-dependent memories [33]. These findings imply that during the post-learning period, specific activity in a certain group of cells represents and maintains a given memory trace. Therefore, we performed a series of Ca^{2+} imaging sessions during contextual learning (6 min novel spatial exploration), which corresponds to a novel episode experience, during subsequent non-REM/REM sleep, and then during re-exposure to the context, which corresponds to memory retrieval (Fig. 10.5c).

3.4 Imaging Data Processing

Ca^{2+} imaging movies are processed by Inscopix data processing software (Inscopix Inc.), where the captured images are spatially binned and down sampled by a factor of 2. Then, motion correction is applied to correct for motion artifacts (correction type: translation only, spatial mean [r = 20 pixels] subtracted and spatial mean applied [r = 5 pixels]) (Fig. 10.6a). To ensure that the field of view (FOV) remained unchanged across the recorded frames, a stationary landmark, such as blood vessels, is used. ImageJ software (NIH) is also applicable for movie processing, with TurboReg as a commonly used algorithm for motion correction [34].

Further processing is performed using ImageJ. Each recorded movie is divided (pixel by pixel) from a low-pass (r = 20) filtered version using a fast Fourier transform band-pass filter to remove background florescence. Then, all of the processed movies are concatenated, and the motion correction process is applied again using the previously mentioned parameters except for using a single reference frame for all of the movies to ensure that the FOV was unchanged across the recorded sessions (Fig. 10.6b). This enables us to track the activity of cells of interest across several recorded sessions. Afterward, the relative change in fluorescence (ΔF) is calculated ($\Delta F_{(t)}/F_0 = [F_{(t)} - F_0]/F_0$) for each movie session (Fig. 10.6c), where F_0 is the mean image obtained by averaging the frames for the entire movie from that session. After applying $\Delta F_{(t)}/F_0$ to all recorded sessions, they are concatenated to form one single movie of the entire behavior using Mosaic software (Fig. 10.6d).

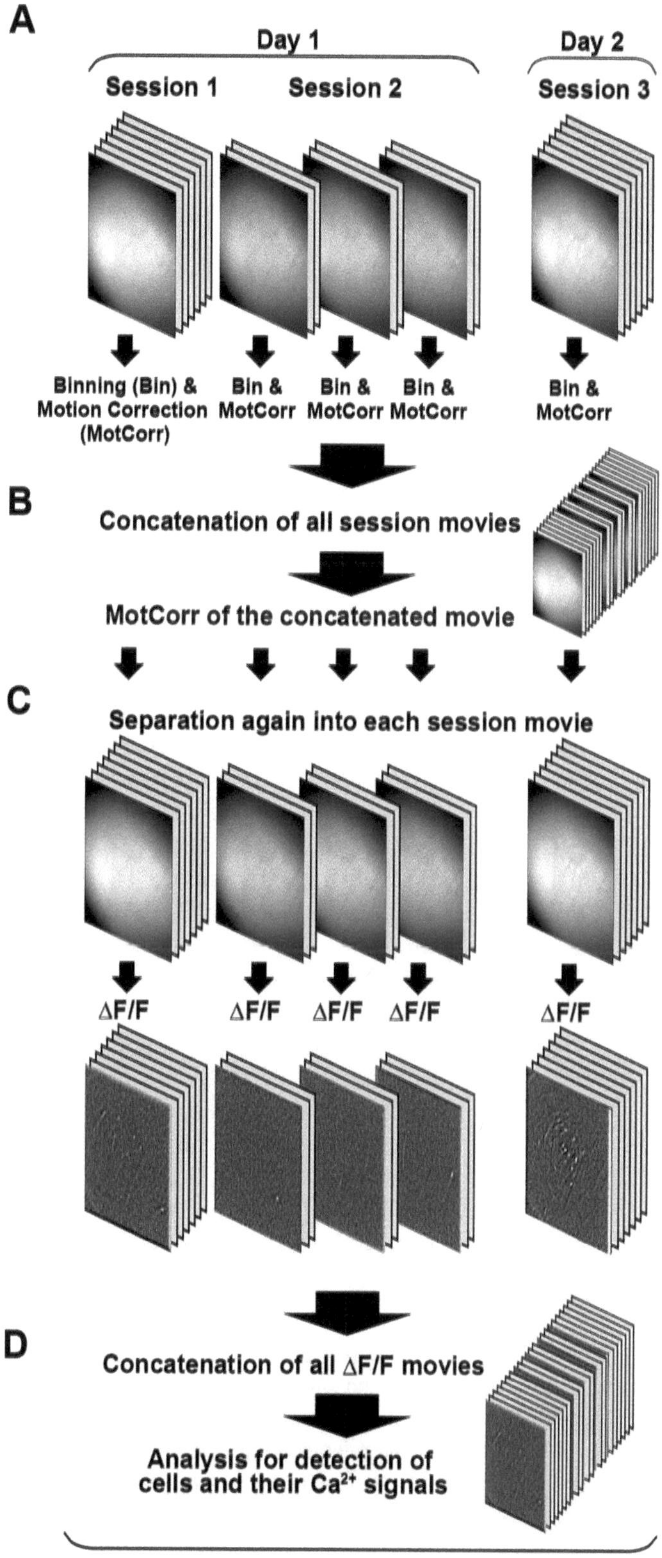

Fig. 10.6 Imaging data processing

(**a**) Representative image after performing the binning and motion correction

3.5 Identification of Cells and Quantification of Their Activities by the Automatic Cell Extraction Program

Ca^{2+} transients are automatically identified by an automatic sorting system, HOTARU (High performance Optimizer to extract spike Timing And cell location from calcium imaging data via lineaR impUlse) developed by Takekawa et al. [26, 35] (https://github.com/tk2lab/HOTARU). Then, cells are categorized into engram and nonengram groups on the basis of the snapshot of KikGR expression taken 24 h after context exposure. With in vivo Ca^{2+} imaging, hundreds or more neurons overlap and fluoresce in the observation field, and the apparent brightness also changes due to the pulsation of blood vessels and uncorrected blurring. Notably, false-positive signals often occur in conjunction with mouse behavior. Therefore, the use of a reliable algorithm to extract the brightness change of each cell with high sensitivity and accuracy while separating and eliminating false positives is directly linked to convincing data acquisition and interpretation. Although we regularly use HOTARU, we also compared it with CNMF-E developed in the Paninski laboratory at Columbia University [36] to obtain accurate data. We concluded that CNMF-E is comparable with HOTARU and can detect cells without any false positives [26] (Fig. 10.7).

3.6 Merging of Ca^{2+} Transients and Engram Cell ROIs

To determine the spatial positions of engram cells, we apply several procedures to the KikGR snapshot taken 24 h after the event, in this case, novel context exposure, using ImageJ. First, the snapshot is spatially binned and smoothed by a factor of 2. Then, using a fast Fourier transform band-pass filter, the snapshot is divided, pixel by pixel, from a low-pass (r = 20) filtered version to remove background florescence (Fig. 10.8a). Second, the snapshot is manually motion corrected (using the [x, y] translation tool) to attain the same FOV as for the recorded Ca^{2+} movie, using blood vessels as landmarks. Third, a heat map is applied to the snapshot, and a mask is drawn on the area containing the visually obvious gene expression positions (Fig. 10.8b); a cutting threshold (maintaining the top ~10% signal intensity) is then applied to further remove background noise. The output image is then binarized (binary options: iteration, 1; count, 2; close [fill small holes between pixels]; and watershed [autoseparation of several combined particles]). Binary options, i.e., the numbers of iterations and counts, can vary slightly

Fig. 10.6 (continued) (Mot Corr) for sessions 1, 2, and 3 (please see Fig. 10.5c), independently.
(**b**) Concatenation of the recorded sessions together to perform Mot Corr using a single reference frame to ensure the same FOV across all sessions.
(**c**) Calculating the relative change in florescence ($\Delta F_{(t)}/F_0$) for all sessions independently after separating the entire movie into single sessions.
(**d**) Concatenation of all movies after calculating the $\Delta F_{(t)}/F_0$ to be ready for automatic Ca^{2+} transient detection. Modified from Ghandour et al. (2019) [26]

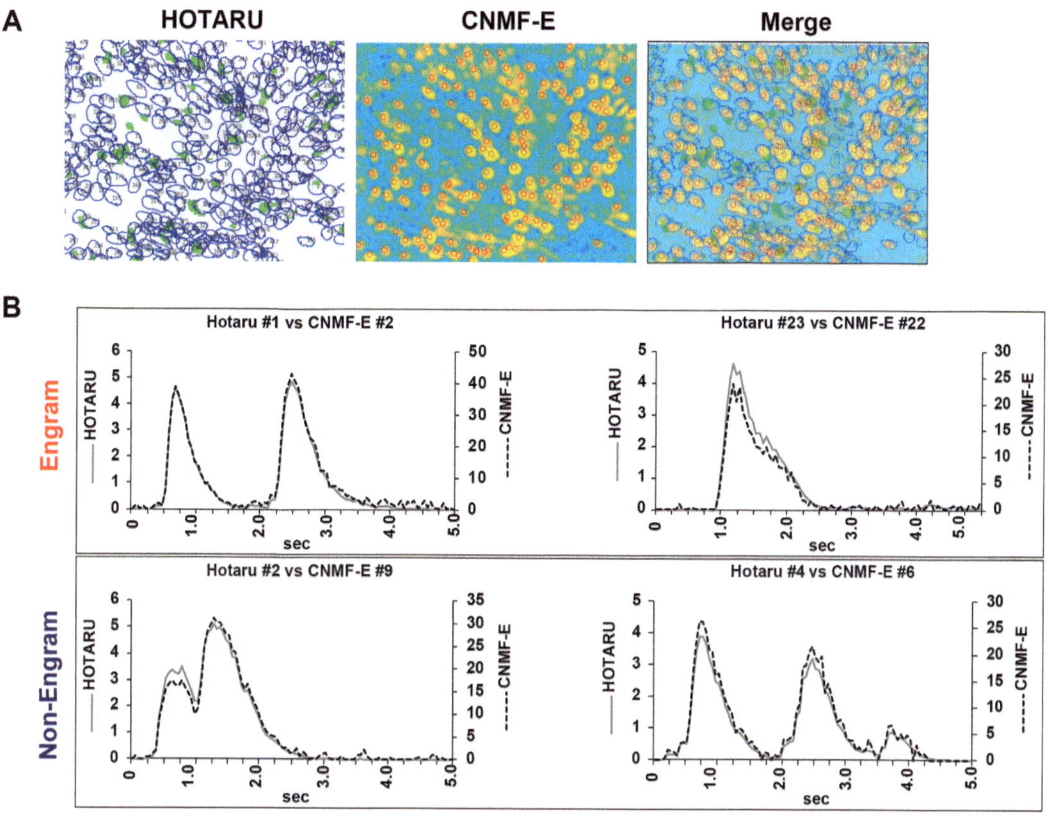

Fig. 10.7 Comparison of automatic sorting system output from the HOTARU and the CNMF-E analyses
(**a**) Representative image of all cells detected from a single animal by an automatic Ca^{2+} activity detection system: HOTARU (left), CNMF-E system (center), and both images overlaid (right). Note: the green-filled circles are the ROIs of engram cells. Blue circles are the cells extracted by HOTARU. Engram cell identification by merging of engram ROI and HOTARU output is explained in Sect. 3.5 and in Fig. 10.8.
(**b**) Representative Ca^{2+} traces for engram (top) and nonengram (bottom) cells, as detected by both systems. Modified from Ghandour et al. (2019) [26]

from one image to another according to gene expression intensity and background in each image. The remaining particles in the resulting image are then automatically detected; these small regions of interest (ROIs) represent engram cell positions (Fig. 10.8c). Finally, the spatial locations of the ImageJ-detected ROIs from the snapshot (representing engram cells) are then overlayed with the spatial contours generated by the automated HOTARU system (representing all cells detected by Ca^{2+} transients) to classify the detected active cells as engram or nonengram transients. Only engram ROIs showing typical neuron morphology with complete, not partial, overlap with contours from the automated system (through visual inspection) are considered engram cell transients (Fig. 10.8d).

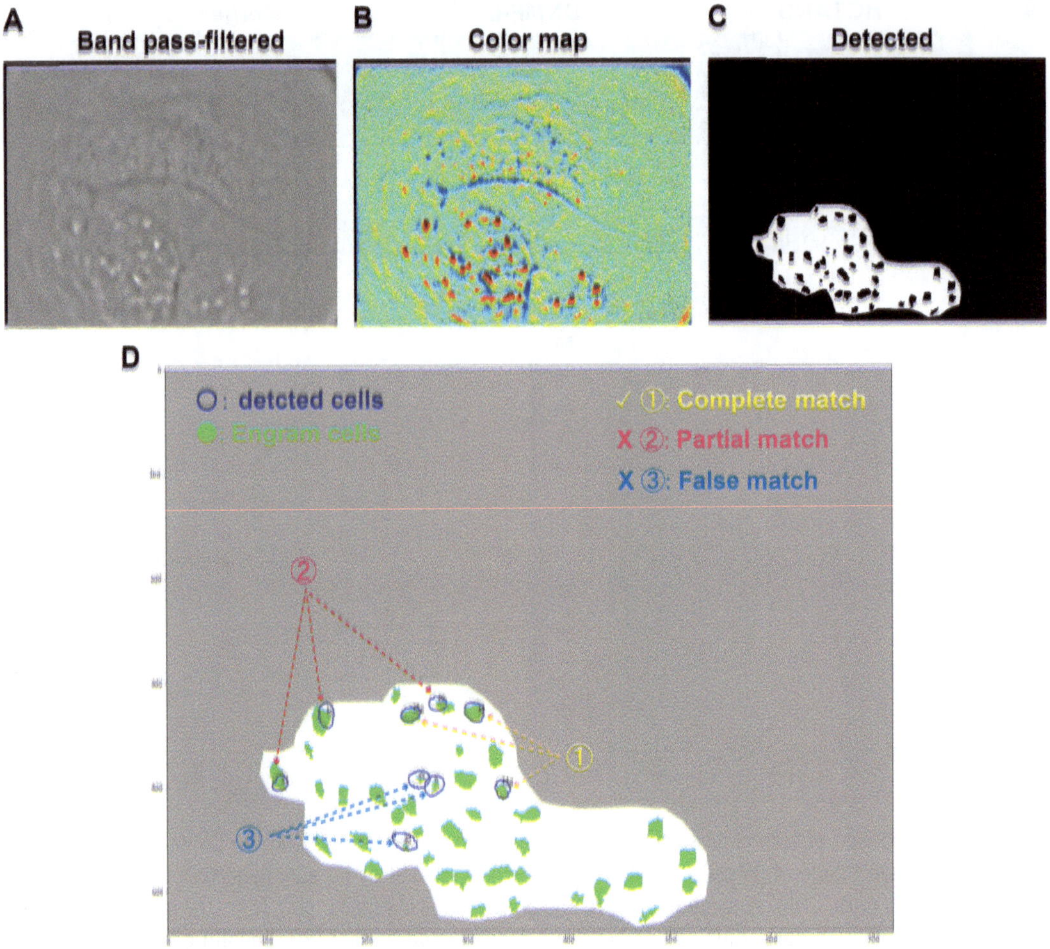

Fig. 10.8 Engram ROI verification and identification

(**a**) Representative snapshot of KikGR$^+$ cells, taken under isoflurane anesthesia after applying a fast Fourier band-pass filter.

(**b**) The same snapshot after applying a heat map to visually identify the KikGR$^+$ cells.

(**c**) Automatically identified engram ROIs.

(**d**) Only engram cells (green-filled contours) that show a complete match with automatically detected Ca^{2+} transients (blue contours) are considered engram cells (category 1). Category 2 cells, in which the Ca^{2+}-detected transients (blue contours) partially overlapped the engram ROI (green-filled contours), and category 3 cells, for which the engram-detected ROIs (green-filled contours) did not show typical neuron morphology, were considered KikGR$^-$ cells. Modified from Ghandour et al. (2019) [26]

4 Conclusions

By using the procedures described above, the reader will be able to collect numerical data of neuronal activities from engram and nonengram cells during memory processing. The data can be structured as a matrix, in which rows and columns correspond to time-lapse data of identified neurons and ensemble activity, respectively

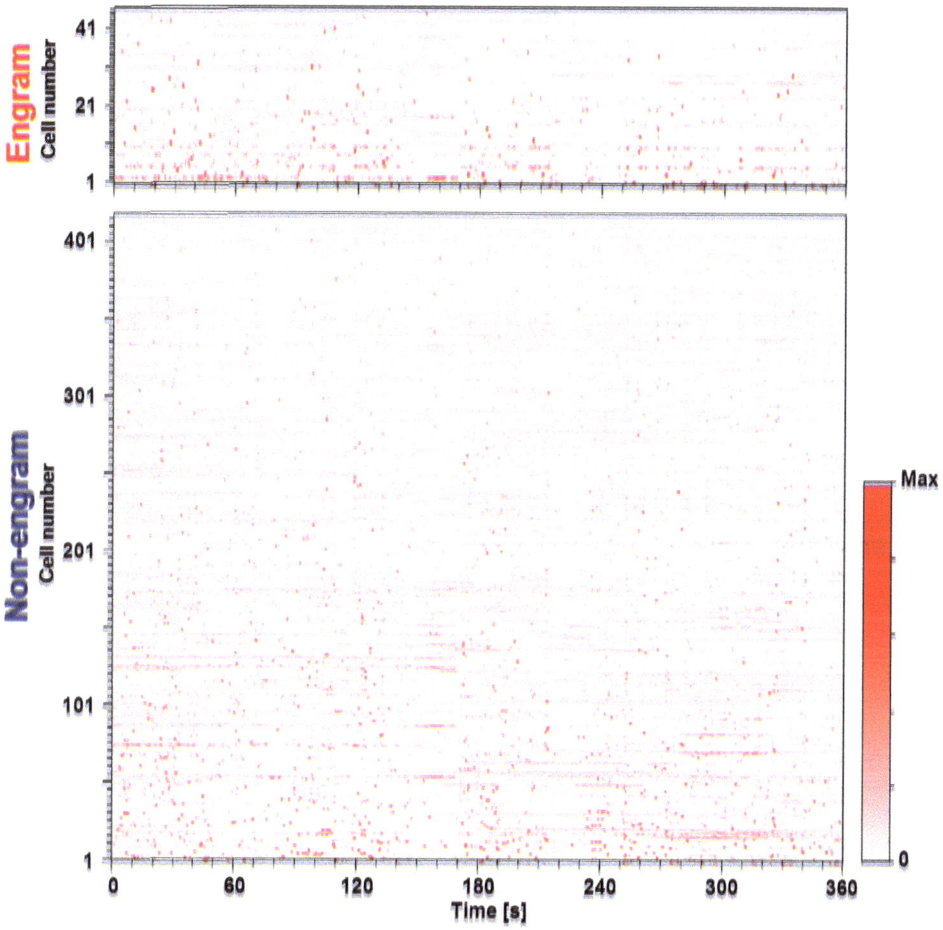

Fig. 10.9 Ca^{2+} transients from engram cells and nonengram cells
Representative output data of Ca^{2+} transients from engram and nonengram cells during the 6 min contextual learning period. Data were normalized by maximum value in each cell and indicated by gradient color scale. Modified from Ghandour et al. (2019) [26]

(Fig. 10.9). The reader can use each row of the matrix data to focus on the features of the activity pattern of each cell. Furthermore, the reader can also carry out analyses of ensemble activities evoked by specific cues, appearing as a columnar pattern in the matrix. Moreover, the matrix data allow the reader to perform multidimensional analyses of neuronal ensembles combined with the timing of cues, specific biological phenomenon, or brain states corresponding to changes in an electroencephalogram or local field potential. Outputs from individual analyses of engram cells and nonengram cells can be compared to extract special features of engram cells. For example, we found that the pattern of repetitive activity of an engram ensemble is significantly larger than that of nonengram cells in a correlation matrix analysis [26].

In summary, we introduced a microscopy strategy to observe two targets, engram cells and Ca^{2+} transients, using a system equipped with optics for only one fluorescent spectrum. This approach should be useful for combining Ca^{2+} imaging and the identification of specific target cells, including cells other than engram cells. Furthermore, it is possible to discriminate two targets, e.g., target X expressing GFP and target Y expressing KikGR can be identified as that with no change in the fluorescence signal and that in which the signal disappears after photoconversion, respectively. In this way, photoconvertible fluorescent proteins combined with miniature microscopy provide opportunities for time-lapse imaging of multiple targets in the brains of behaving animals. We hope this chapter helps to expand the possibilities of the reader's experiments.

Acknowledgments

This work was supported by the Astellas Foundation for Research on Metabolic Disorders (support to N.O.), the Brain Science Foundation (N.O.), the Naito Foundation (N.O.), the Research Foundation for Opto-Science and Technology (N.O.), the Takeda Science Foundation (N.O.), and the Tamura Science and Technology Foundation (N.O.).

References

1. Semon RW (1921) The mneme. G. Allen & Unwin Limited
2. Josselyn SA, Köhler S, Frankland PW (2015) Finding the engram. Nat Rev Neurosci 16: 521. https://doi.org/10.1038/nrn4000
3. Hebb DO (1949) The organization of behavior. Wiley, New York
4. Reijmers LG, Perkins BL, Matsuo N, Mayford M (2007) Localization of a stable neural correlate of associative memory. Science 317(5842): 1230–1233. doi:317/5842/1230 [pii] 10.1126/science.1143839
5. Liu X, Ramirez S, Pang PT, Puryear CB, Govindarajan A, Deisseroth K, Tonegawa S (2012) Optogenetic stimulation of a hippocampal engram activates fear memory recall. Nature 484(7394):381–385. doi:nature11028 [pii] 10.1038/nature11028
6. Garner AR, Rowland DC, Hwang SY, Baumgaertel K, Roth BL, Kentros C, Mayford M (2012) Generation of a synthetic memory trace. Science 335(6075):1513–1516. doi:335/6075/1513 [pii] 10.1126/science.1214985
7. Ramirez S, Liu X, Lin PA, Suh J, Pignatelli M, Redondo RL, Ryan TJ, Tonegawa S (2013) Creating a false memory in the hippocampus. Science 341(6144):387–391. doi:341/6144/387 [pii] 10.1126/science.1239073
8. Ohkawa N, Saitoh Y, Suzuki A, Tsujimura S, Murayama E, Kosugi S, Nishizono H, Matsuo M, Takahashi Y, Nagase M, Sugimura Yae K, Watabe Ayako M, Kato F, Inokuchi K (2015) Artificial Association of pre-stored Information to generate a qualitatively new memory. Cell Rep 11(2):261–269. https://doi.org/10.1016/j.celrep.2015.03.017
9. Ryan TJ, Roy DS, Pignatelli M, Arons A, Tonegawa S (2015) Memory. Engram cells retain memory under retrograde amnesia. Science 348(6238):1007–1013. doi:348/6238/1007 [pii] 10.1126/science.aaa5542
10. Roy DS, Arons A, Mitchell TI, Pignatelli M, Ryan TJ, Tonegawa S (2016) Memory retrieval by activating engram cells in mouse models of early Alzheimer's disease. Nature 531(7595): 508–512. doi:nature17172 [pii] 10.1038/nature17172

11. Nomoto M, Ohkawa N, Nishizono H, Yokose J, Suzuki A, Matsuo M, Tsujimura S, Takahashi Y, Nagase M, Watabe AM, Kato F, Inokuchi K (2016) Cellular tagging as a neural network mechanism for behavioural tagging. Nat Commun 7:12319. https://doi.org/10.1038/ncomms12319
12. Yokose J, Okubo-Suzuki R, Nomoto M, Ohkawa N, Nishizono H, Suzuki A, Matsuo M, Tsujimura S, Takahashi Y, Nagase M, Watabe AM, Sasahara M, Kato F, Inokuchi K (2017) Overlapping memory trace indispensable for linking, but not recalling, individual memories. Science 355(6323): 398–403. https://doi.org/10.1126/science.aal2690
13. Boyden ES, Zhang F, Bamberg E, Nagel G, Deisseroth K (2005) Millisecond-timescale, genetically targeted optical control of neural activity. Nat Neurosci 8(9):1263–1268. doi: nn1525 [pii] 10.1038/nn1525
14. Kim CK, Adhikari A, Deisseroth K (2017) Integration of optogenetics with complementary methodologies in systems neuroscience. Nat Rev Neurosci 18(4):222–235. doi: nrn.2017.15 [pii] 10.1038/nrn.2017.15
15. Alexander GM, Rogan SC, Abbas AI, Armbruster BN, Pei Y, Allen JA, Nonneman RJ, Hartmann J, Moy SS, Nicolelis MA, McNamara JO, Roth BL (2009) Remote control of neuronal activity in transgenic mice expressing evolved G protein-coupled receptors. Neuron 63(1):27–39. doi:S0896-6273(09)00467-X [pii] 10.1016/j.neuron.2009.06.014
16. Roth BL (2016) DREADDs for neuroscientists. Neuron 89(4):683–694. doi:S0896-6273(16)00065-9 [pii] 10.1016/j.neuron.2016.01.040
17. Nakai J, Ohkura M, Imoto K (2001) A high signal-to-noise Ca(2+) probe composed of a single green fluorescent protein. Nat Biotechnol 19(2):137–141. https://doi.org/10.1038/84397
18. Tian L, Hires SA, Mao T, Huber D, Chiappe ME, Chalasani SH, Petreanu L, Akerboom J, McKinney SA, Schreiter ER, Bargmann CI, Jayaraman V, Svoboda K, Looger LL (2009) Imaging neural activity in worms, flies and mice with improved GCaMP calcium indicators. Nat Methods 6(12):875–881. doi:nmeth.1398 [pii] 10.1038/nmeth.1398
19. Ghosh KK, Burns LD, Cocker ED, Nimmerjahn A, Ziv Y, Gamal AE, Schnitzer MJ (2011) Miniaturized integration of a fluorescence microscope. Nat Methods 8(10): 871–878. doi:nmeth.1694 [pii] 10.1038/nmeth.1694
20. Cai DJ, Aharoni D, Shuman T, Shobe J, Biane J, Song W, Wei B, Veshkini M, La-Vu M, Lou J, Flores SE, Kim I, Sano Y, Zhou M, Baumgaertel K, Lavi A, Kamata M, Tuszynski M, Mayford M, Golshani P, Silva AJ (2016) A shared neural ensemble links distinct contextual memories encoded close in time. Nature 534(7605):115–118. doi:nature17955 [pii] 10.1038/nature17955
21. Ziv Y, Burns LD, Cocker ED, Hamel EO, Ghosh KK, Kitch LJ, El Gamal A, Schnitzer MJ (2013) Long-term dynamics of CA1 hippocampal place codes. Nat Neurosci 16(3): 264–266. doi:nn.3329 [pii] 10.1038/nn.3329
22. Cameron CM, Murugan M, Choi JY, Engel EA, Witten IB (2019) Increased cocaine motivation is associated with degraded spatial and temporal representations in IL-NAc neurons. Neuron 103(1):80–91. doi:S0896-6273(19)30350-2 [pii] 10.1016/j.neuron.2019.04.015
23. Daliparthi VK, Tachibana RO, Cooper BG, Hahnloser RH, Kojima S, Sober SJ, Roberts TF (2019) Transitioning between preparatory and precisely sequenced neuronal activity in production of a skilled behavior. Elife 8. https://doi.org/10.7554/eLife.43732. 43732 [pii]
24. Kondo T, Saito R, Otaka M, Yoshino-Saito K, Yamanaka A, Yamamori T, Watakabe A, Mizukami H, Schnitzer MJ, Tanaka KF, Ushiba J, Okano H (2018) Calcium transient dynamics of neural ensembles in the primary motor cortex of naturally behaving monkeys. Cell Rep 24(8):2191–2195. doi:S2211-1247(18)31160-4 [pii] 10.1016/j.celrep.2018.07.057
25. Tsutsui H, Karasawa S, Shimizu H, Nukina N, Miyawaki A (2005) Semi-rational engineering of a coral fluorescent protein into an efficient highlighter. EMBO Rep 6(3): 233–238. doi:7400361 [pii] 10.1038/sj.embor.7400361
26. Ghandour K, Ohkawa N, Fung CCA, Asai H, Saitoh Y, Takekawa T, Okubo-Suzuki R, Soya S, Nishizono H, Matsuo M, Osanai M, Sato M, Ohkura M, Nakai J, Hayashi Y, Sakurai T, Kitamura T, Fukai T, Inokuchi K (2019) Orchestrated ensemble activities constitute a hippocampal memory engram. Nat Commun 10(1):2637. https://doi.org/10.1038/s41467-019-10683-2. 10.1038/s41467-019-10683-2 [pii]
27. Zhang F, Gradinaru V, Adamantidis AR, Durand R, Airan RD, de Lecea L, Deisseroth K (2010) Optogenetic interrogation of neural circuits: technology for probing mammalian brain structures. Nat Protoc 5(3):

439–456. doi:nprot.2009.226 [pii] 10.1038/nprot.2009.226

28. Barretto RP, Ko TH, Jung JC, Wang TJ, Capps G, Waters AC, Ziv Y, Attardo A, Recht L, Schnitzer MJ (2011) Time-lapse imaging of disease progression in deep brain areas using fluorescence microendoscopy. Nat Med 17(2):223–228. doi:nm.2292 [pii] 10.1038/nm.2292
29. Sato M, Mizuta K, Islam T, Kawano M, Sekine Y, Takekawa T, Gomez-Dominguez D, Schmidt A, Wolf F, Kim K, Yamakawa H, Ohkura M, Lee MG, Fukai T, Nakai J, Hayashi Y (2020) Distinct mechanisms of over-representation of landmarks and rewards in the hippocampus. Cell Rep 32(1): 107864. doi:S2211-1247(20)30845-7 [pii] 10.1016/j.celrep.2020.107864
30. Dana H, Chen TW, Hu A, Shields BC, Guo C, Looger LL, Kim DS, Svoboda K (2014) Thy1-GCaMP6 transgenic mice for neuronal population imaging in vivo. PLoS One 9(9): e108697. https://doi.org/10.1371/journal.pone.0108697. PONE-D-14-31023 [pii]
31. Huff ML, Miller RL, Deisseroth K, Moorman DE, LaLumiere RT (2013) Posttraining optogenetic manipulations of basolateral amygdala activity modulate consolidation of inhibitory avoidance memory in rats. Proc Natl Acad Sci U S A 110(9):3597–3602. https://doi.org/10.1073/pnas.1219593110
32. Girardeau G, Benchenane K, Wiener SI, Buzsaki G, Zugaro MB (2009) Selective suppression of hippocampal ripples impairs spatial memory. Nat Neurosci 12(10): 1222–1223. doi:nn.2384 [pii] 10.1038/nn.2384
33. Boyce R, Glasgow SD, Williams S, Adamantidis A (2016) Causal evidence for the role of REM sleep theta rhythm in contextual memory consolidation. Science 352(6287):812–816. doi:352/6287/812 [pii] 10.1126/science.aad5252
34. Thevenaz P, Ruttimann UE, Unser M (1998) A pyramid approach to subpixel registration based on intensity. IEEE Trans Image Process 7(1):27–41. https://doi.org/10.1109/83.650848
35. Takekawa T, Asai H, Ohkawa N, Nomoto M, Okubo-Suzuki R, Ghandour K, Sato M, Hayashi Y, Inokuchi K, Fukai T (2017) Automatic sorting system for large calcium imaging data. bioRxiv. https://doi.org/10.1101/215145
36. Zhou P, Resendez SL, Rodriguez-Romaguera-J, Jimenez JC, Neufeld SQ, Giovannucci A, Friedrich J, Pnevmatikakis EA, Stuber GD, Hen R, Kheirbek MA, Sabatini BL, Kass RE, Paninski L (2018) Efficient and accurate extraction of in vivo calcium signals from microendoscopic video data. elife 7. https://doi.org/10.7554/eLife.28728. 28728 [pii]

Chapter 11

Optical Recording of Brain Neuron Activities from a Male *Drosophila* Behaving on a Treadmill

Soh Kohatsu and Daisuke Yamamoto

Abstract

A wide array of genetic tools is available for monitoring and manipulating the activities of identified neurons in the fruit fly *Drosophila melanogaster* to investigate the mechanisms of neural circuit operation underlying behavior. Although contemporary genetically encoded calcium indicators have attained the high sensitivity that allows detection of a single-action potential, applying them to a behaving adult fly is still challenging because conditions ideal for optical recordings from the brain may constrain movements of the fly. Here we describe a protocol for recording brain neuron activities from a tethered male *Drosophila* in courtship pursuit on a treadmill, with emphasis on reproducible preparation making that takes advantage of a simple custom rig for fly tethering. The method can be applied to other experimental approaches, such as electrophysiology or optogenetic neural activity manipulation with minor modifications.

Key words *Drosophila*, Calcium imaging, GCaMP, Courtship, Vision, Treadmill

1 Introduction

Drosophila melanogaster offers an attractive model for studying how the nervous system processes and integrates sensory signals to generate adaptive behavior. The wide array of genetic techniques available for use in this fly permits the targeting of molecular tools to a defined subset of neurons at cellular resolution, allowing one to induce and inhibit neural activity remotely in a behaving fly by using heat or light [1–4]. These manipulators of neural activity are very powerful in tracking down core components of a neuronal circuit that underlies behavior [5, 6]. In addition, aided by the rapid growth of neuron morphology databases and the progress in connectome projects [7–11], the structure of neural circuits for specific behaviors is now being systemically described. However, to understand how the neural circuit operates, physiological study of the circuit components is indispensable. This is particularly important in view of the finding that neurons involved in sensorimotor transformation dynamically change their response dynamics depending

Daisuke Yamamoto (ed.), *Behavioral Neurogenetics*, Neuromethods, vol. 181, https://doi.org/10.1007/978-1-0716-2321-3_11,

on the behavioral state of the fly [12, 13]. Thus, sophisticated techniques for neural activity recording from an animal during behavior are inevitable for the contemporary study of neurobiology in *Drosophila*. Despite its limitation in temporal resolution, optophysiology with genetically encoded calcium indicators (GECIs) has several advantages over traditional electrophysiology [14, 15]. First, GECIs can be targeted to monitor the activity of genetically identified neurons across individuals. Second, optophysiology with GECIs permits the capturing of both spatial and temporal patterns of neuronal activity. Third, it allows monitoring of the activities of multiple neurons simultaneously, thereby helping to uncover hitherto unknown neurons that show activities relevant to the particular behavior in question [16]. These properties of optophysiology confer great flexibility to in vivo physiological experiments, which generally have many experimental constraints. One of the major challenges in performing physiological recording from a behaving fly is to induce the behavior of interest during recording. To achieve this, most of the prior works in *Drosophila* adopted an approach of tethering the subject fly [12, 13, 17–19], with the exception of recent studies that realized brain imaging in freely walking flies [20, 21]. Treadmill systems, which allow flies to walk under tethered conditions, have been employed in *Drosophila* for the study of both behavior and physiology [13, 17, 22, 23]. In this approach, the tethered fly is allowed to hold an air-suspended lightweight foam ball by its legs. The ball can be easily rotated according to the leg motion of the fly, allowing the fly to walk on it while being tethered (Fig. 11.1). Such treadmill systems are highly useful for both behavioral and physiological studies, but they are not necessarily easy to implement. The critical step in ensuring their success is the angular and positional alignment of the fly relative to the ball. If the fly is misaligned, the ball easily bumps to the ball holder and does not rotate smoothly, disturbing the fly's locomotion, which is highly susceptible to mechanical interference. The use of a treadmill is more difficult in physiological experiments than behavioral experiments, because the integrity of the fly is further jeopardized by the surgery required to expose the brain for recording. Thus, a method for keeping flies healthy while handling them to tether and align in the setup on a routine basis is critical for successful simultaneous monitoring of neural activity and behavior. In this chapter, we describe a protocol for optical recording from brain interneurons in a tethered fly behaving on a treadmill; in this case, the behavior is male courtship. Central control of courtship in the *Drosophila* male involves interneurons that express either or both of the sex determination genes *doublesex* (*dsx*) and *fruitless* (*fru*). When a male's leg touches a female body that is covered by a sex-specific nonvolatile hydrocarbon blend [24], pC1 neurons, a *dsx*-expressing neuron subclass, become activated and trigger the male's vision-dependent courtship pursuit

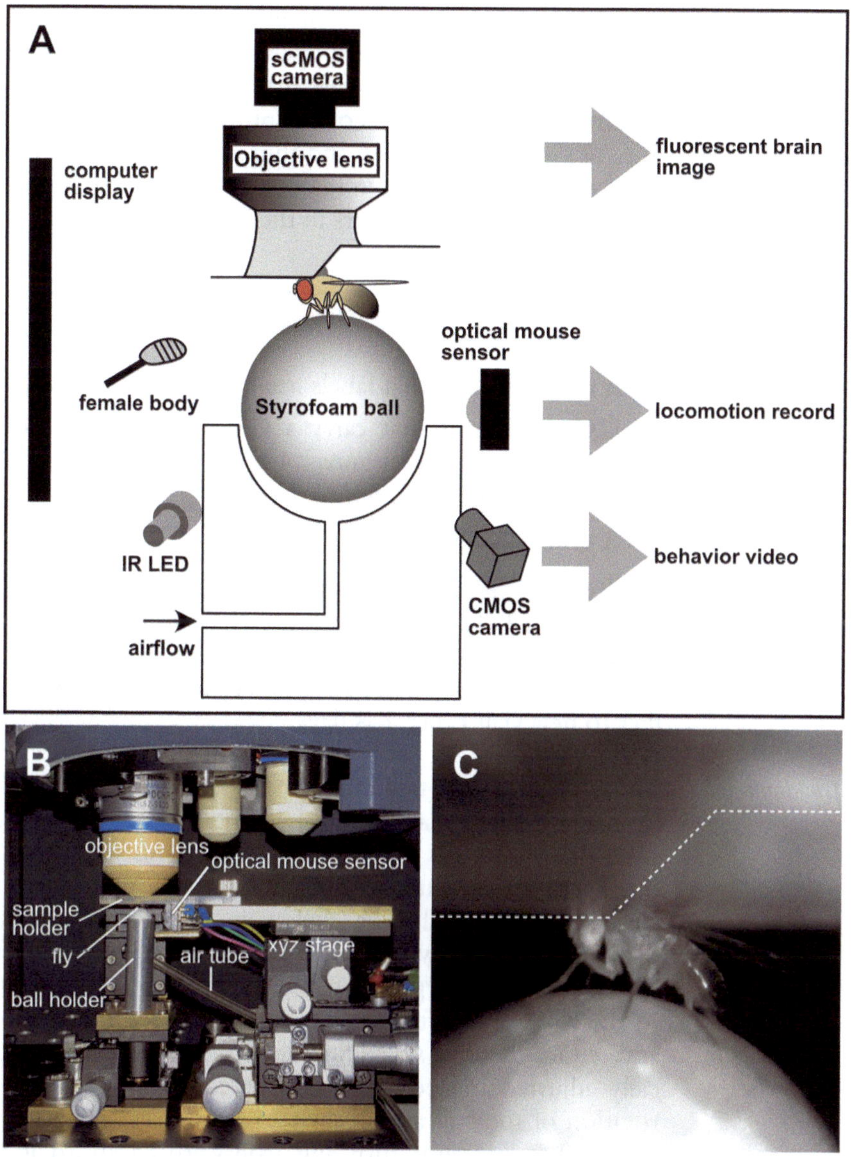

Fig. 11.1 (**a**) Schematic of the experimental setup. (**b**) The arrangement of the treadmill and the components on the microscope stage. Lateral view. The air tube is connected to the air inlet on the sidewall of the ball holder (not seen in this picture). (**c**) A magnified view of a subject male on the treadmill. Lateral view. The dashed lines depict the edges of the metal plate of the sample holder seen through the translucent acrylic frame

[18]. Our protocol features a sample preparation procedure that takes advantage of a simple custom rig and allows, on a routine basis, fly tethering to the sample holder without the use of micro-manipulators. The protocol is also unique in that it uses translucent silicone gel to seal the window opened in the fly's head cuticle to expose the brain. This allows acquisition of fluorescence images of the brain through the overlaid gel, while reducing the motion of

the brain that is a major problem in in vivo imaging experiments. In addition, by sealing the head opening, the fly remains intact up to 1 h without the need to perfuse the brain with aerated saline. Taken together, these advantages of our protocol help simplify the procedure of sample preparation and thus pave the way for physiological studies of unexplored brain regions to decipher circuit mechanisms underlying various behaviors in the fly.

2 Materials

2.1 Flies

- A Gal4 fly strain that labels *dsx*-expressing neurons [25].
- UAS fly strains that carry a GECI transgene (Bloomington Stock Center, stock nos. 42,746 for *UAS-GCaMP6s* and 32,236 for *UAS-GCaMP3.0*).
- A wild-type fly strain (Canton-S).

Rear the flies on a standard cornmeal-yeast medium at 25 °C under the 12/12 h light-dark cycle. Collect newly emerged flies within 4 h after eclosion, and keep them singly in food vials for 4–7 days until use. To keep the flies clean and healthy, transfer the flies to new vials every 2–3 days.

2.2 Fly Treadmill

The core part of the treadmill consists of a polystyrene ball 6–8 mm in diameter and a ball holder, i.e., an aluminum post with a half-spherical pocket on the top (Fig. 11.1). An air-outlet 1 mm in diameter is drilled into the bottom of the pocket to provide airflow to suspend the ball. The airflow is generated by an aquarium diaphragm pump with flow speed control functionality (e-Air 6000WB; GEX Corporation, Osaka, Japan) and then introduced to the ball holder after passing through water in a jar to humidify and reduce the flow speed fluctuation. The airflow speed is empirically adjusted to keep the ball floated stably (*see* **Note 1**). The rotation of the ball along yaw and pitch axes, which corresponds to the fly's left-right turn and frontward-backward locomotion, respectively, was detected as the texture shift of the ball surface with an optical sensor (ADNS-2620; Avago Technologies, San Jose, CA) [18]. The readout of the sensor was sent to the computer via a microcontroller board over serial communication and then stored.

2.3 Preparation Making

2.3.1 Sample Holder

To fix and attach the male to the imaging rig while avoiding undesired interference with the male's appendages, we use a device created by Seelig et al. [17] that consists of a "shampoo-hat" style sample holder made of folded thin stainless plate (0.1 mm in thickness, 518–946; RS components, Corby, UK) with a small window (Fig. 11.2). The fly is glued to the bottom side of the stainless plate by the prothorax so that its head can be observed through the window.

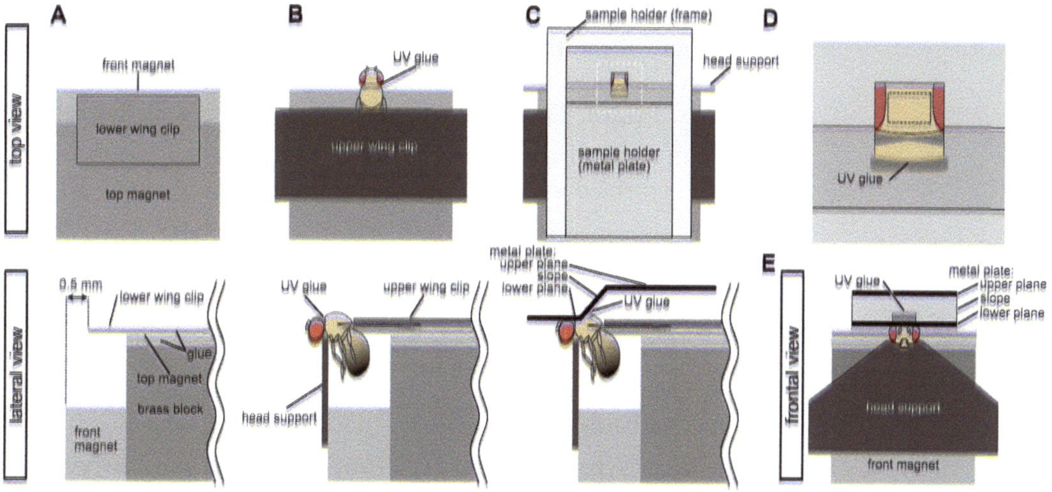

Fig. 11.2 Attaching the fly to the sample holder. (**a**) Preparation block. (**b–e**) Procedure for affixing the fly to the sample holder. Clip the wings of the cold anesthetized male and align the position so that the fly's body axis is perpendicular to the front edge of the wing clip. Align the position of the head support to lift the head and thorax of the fly. Then fix the head by filling the space between the posterior head and prothorax with a tiny drop of UV glue (**b** and **d**). Put the sample holder on the top of the preparation block, and align the position so that the male's head is at the center of the window in the stainless plate. Then slide up the head support to hold the head so that the dorsal head is gently pressed against the edges of the window (**c**). (**d**) A magnified view of the dashed square in (**c**). The black dashed square on the male's head depicts the cuticular area to remove to expose the brain. (**e**) The frontal view of the preparation in (**c**). Note that the preparation rigs and sample holder are not drawn to scale for the sake of clarity. For the same reason, the acrylic sample holder frame in the bottom panels of (**c** and **e**) and the upper wing clip and the male's wings in panel (**e**) are omitted

2.3.2 Preparation Block

We use a custom brass block with magnets and metal plates as the platform for affixing the fly to the sample holder (Fig. 11.2a). By using this "preparation block" and gently clipping the fly's wings with thin metal plates, the fly can be held in position and kept from rotating around the anterior-posterior (roll) axis. The brass portion of the platform serves as an ice block for prolonged cold anesthesia during fly tethering. A magnet sheet and a metal plate that make up the bottom part of the wing clip are glued in layers on the top of the brass block. Another piece of magnet, which is glued on the side-wall of the block, is used to attach the head support. The distance of 0.5 mm between the front end of the protruding portion of the bottom wing clip and the head support is essential to create an appropriately sized gap to accommodate the fly (Fig. 11.2a).

The materials used to make the preparation block are as follows:

- A brass block (width × length × height: 39 mm × 36 mm × 25 mm).
- Razor blades to be used for the metal plates to clip the fly's wings (0.1 mm in thickness, FH-10; Feather Safety Razor, Osaka, Japan) (Fig. 11.2b) and to support the head (0.23 mm

in thickness, FHS-10; Feather Safety Razor) (Fig. 11.2c). The corner of the thicker razor blade should be cut off so that the head support does not interfere with the sample holder (Fig. 11.2e). The blade should be dulled with a file for safety.

- High thermal conductivity glue for fixing the magnet sheet and the razorblades to the brass block (SCV-22; Sunhayato, Tokyo).
- A rubber magnet sheet for fixing the metal plates.

2.3.3 Tools for Surgery

- Syringe needles to be used as micro scalpels (25 Gauge, N-2525R; Terumo, Tokyo).
- Forceps (Taxal No. 5; Fontax Technologies S.A., Bulle, Switzerland).
- Micro scissors (14,124; World Precision Instruments, Sarasota, FL).
- Silicone gel for sealing the cuticular window after exposing the brain (QWIK SIL; World Precision Instruments).
- Tungsten wire for picking up small droplets of UV glue or silicone gel (φ 0.1 mm, W-461167; The Nilaco Co., Tokyo). Bend one end into a hook shape, and attach the opposite end to any appropriate holder.
- UV glue (TB1773E; ThreeBond, Tokyo).
- A UV pen light (SpotCure-B; Kinetic Instruments, Bethel, CT).
- A minutien pin for fixing a stimulant female body (diameter: 0.18 mm, No. 251; Shiga-Konchu Co. Ltd.$_2$ *Tokyo*).

2.4 Recording Setup

- An upright fluorescent microscope with a fixed stage (Axio Examiner; Zeiss, Oberkochen, Germany). An LED light source (HBO 100; Zeiss) and a BP 470/40 nm excitation filter are used to excite GCaMP. The emission light is separated from the excitation light with an FT 495 nm dichroic mirror and BP 525/50 nm emission filter.
- A water immersion Plan Apochromat objective lens (W Plan Apochromat 40×/1.0 NA; Zeiss).
- An sCMOS camera for acquiring fluorescent images (ORCA-Flash4.0; Hamamatsu Photonics, Hamamatsu, Japan).
- A CMOS camera with a 25 mm single focus lens for monitoring the position and behavior of the male (FL3-U3-13S2; Point Grey Research, Richmond, Canada). Remove the IR-cut filter installed in front of the sensor chip of the camera to record the fly illuminated by IR light.
- An IR LED light for illumination of the fly and the foam ball (MCEL-CIR8–2; Moritex, Saitama, Japan).

- A manual xyz stage for alignment of the male's position (TSD-405S; Sigmakoki, Saitama, Japan).
- A manual micromanipulator for handling the female body (MN-4; Narishige, Tokyo).

2.5 Visual Stimulus

To generate and control the visual stimulus, we use a custom python script written using the open-source library Vision Egg [26] or PsychoPy [27]. The visual stimulus is a grayscale square-wave grating pattern drifting in the horizontal direction [18]. The temporal frequency and the lambda value are 8–10 Hz and 30°, respectively. The visual pattern was displayed on a computer monitor with a frame rate of 120 Hz (XL2420T; BenQ, Taipei, Taiwan). The computer monitor is placed in front of the fly at a distance of 15 cm, covering an area that spans approximately 120° in width and 67° in height. The visual pattern is presented on a black background, within a window of 120° in width and 15° in height. The motion of the grating pattern is triggered by a TTL signal.

2.6 Software

- Image acquisition software (Micro-Manager: https://micro-manager.org/wiki/Micro-Manager) [28].
- Image processing software (Fiji: https://fiji.sc) [29].
- Spreadsheet software (LibreOffice: https://www.libreoffice.org/).

3 Methods

3.1 Preparation

1. Cool the preparation block on ice for 3 min.
2. Anesthetize a male in a glass vial on ice for up to 30 s.
3. Put the male on the preparation block on the stereomicroscope stage.
4. Clip the male's wings with the wing clip so that its body axis is perpendicular to the front edge of the wing clip (Fig. 11.2b; *see* **Note 2**).
5. Slide up the head support attached to the front magnet of the preparation block to hold the male's head up (Fig. 11.2b). The head support should be in contact with the fly at the base of the proboscis and the anterior end of the ventral thorax.
6. Immobilize the head by filling the gap between the head capsule and thorax with a tiny drop of UV glue. Use the tungsten wire hook to pick and apply UV glue (Fig. 11.2b; *see* **Note 3**).
7. Put the sample holder on the upper wing clip, and adjust its position so that the male's head is at the center of the window in the stainless plate. Adjust the position of the head support so that the head is gently pressed to the window (Fig. 11.2c–e).

8. Affix the male's dorsal thorax to the posterior edge of the window in the stainless plate with a tiny drop of UV glue (Fig. 11.2c–e).
9. Detach the sample holder with the male from the preparation block. To do this, slide down the head support to release the head, and then slide the sample holder forward while the front edge of the upper wing clip is slightly lifted to release the wings.
10. Attach the sample holder to an appropriate working platform.
11. Apply ~20 μl of fly saline over the fly's head and the window in the sample holder. The saline will stay on the upper side of the metal plate if the gap between the male's head and the window edge is small enough.
12. Cut a window on the male's head capsule by using the edge of the syringe needle as a knife (Fig. 11.2d).
13. Expose the brain by removing fat and other tissues except for the tracheae. Using fine forceps, carefully peel off the tracheae from the brain surface, and push to one side to gain visual access to the brain. Take great care not to damage the brain with the forceps.
14. Seal the head opening with silicone gel. To do this, first remove the saline with a twisted Kimwipe, and then quickly overlay the silicone gel on the head opening with the tungsten hook (*see* **Notes 4–6**).
15. Leave the fly for more than 5 min to recover, and to allow the silicone gel to fully cure.
16. Prepare a female abdomen as a stimulus, which is brought into contact with the male's foreleg. Cut the boundary between the thorax and abdomen of a cold anesthetized wild-type virgin female, and then insert an insect pin into the abdomen through the opening made on the cuticle. Seal the gap between the insect pin and the cut end of the cuticle with UV glue.

3.2 Imaging

The following describes the procedure for making a recording from *dsx*-expressing neurons in the lateral protocerebrum (lpr) upon female contact and during courtship pursuit. By changing the Gal4 driver and/or stimulus procedure, the activity from other neurons can also be monitored in a similar way. As an example of such cases, visual responses recorded during target following are shown in Fig. 11.3.

1. Attach the sample holder with the male to the xyz stage on the microscope stage.
2. Adjust the male position so that the foam ball rotates smoothly as the fly walks (*see* **Note 7**).

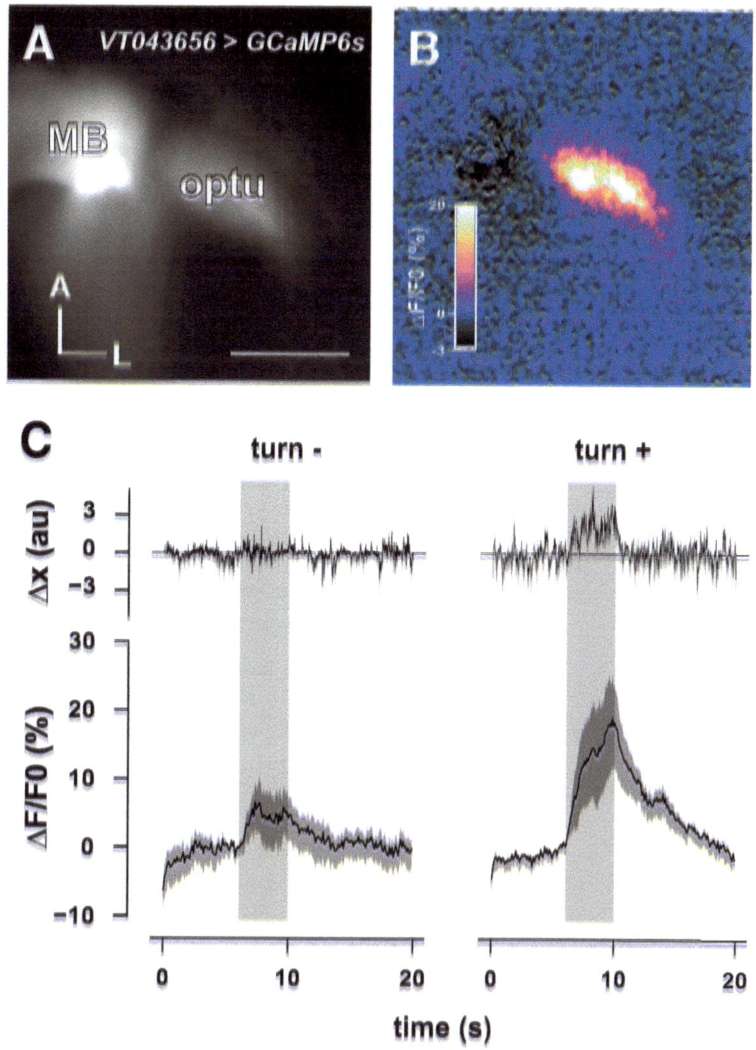

Fig. 11.3 Ca^{2+} activity recording in the visual center during a tethered locomotion turn. (**a**) GCaMP6s fluorescence expressed by *VT043656-Gal4*, which drives expression in the neuronal population including LC10 visual projection neurons. LC10 neurons terminate in the optic tubercle and mediate target following during courtship [31]. *MB* Mushroom body, *optu* optic tubercle. Scale bar, 50 μm. (**b**) Color-coded Ca^{2+} response in the optu to a drifting grating pattern. (**c**) Comparison of the visual responses between trials in which the male showed target following and those in which it did not. The male was given a female contact on the foreleg once and then repeatedly presented with a drifting grating pattern for 4 s, separated by intervals of 10 s, up to 5 times. In this stimulus schedule, the male's following behavior is not fully potentiated and thus is observed only over the limited visual stimulus presentation periods. The simultaneously recorded Ca^{2+} activity and locomotion were paired and pooled and then divided into two groups (turn + or −) depending on whether or not the male showed a locomotion turn to follow the visual target. The means ± sem of each group are shown in the plot. The shaded areas indicate the period when the grating pattern was in motion. The genotype of the flies used was w^{1118}; *UAS-GCaMP6s*/*UAS-GCaMP6s*; *VT043656-Gal4*/+

3. Attach the female abdomen as a stimulant to a holder mounted on the arm of the micromanipulator.
4. Set a water immersion objective lens and introduce the male's brain into the field of the microscope.
5. Turn on the excitation light and focus on the lpr (*see* **Note 8**).
6. Place the computer screen for presenting visual stimuli ~15 cm in front of the male.
7. Trigger fluorescent image acquisition, locomotion recording, and video recording of the lateral view of the fly by sending an electric trigger signal to each of the data acquisition devices. The exposure time and sampling rate for fluorescent image acquisition can be ~100 ms and ~ 10 Hz, respectively, depending on the intensity of the baseline fluorescence signal (*see* **Note 9**). The sampling rate for locomotion and video recording can be 50 Hz and 30–50 Hz, respectively.
8. Touch the body of the female to the male foreleg by manually operating the micromanipulator, and then trigger the visual stimulus presentation immediately (*see* **Notes 10** and **11**).
9. Save the data as 12-bit or 16-bit images.

3.3 Data Analysis

1. Determine the onset of female contact with the male's foreleg by inspecting the video record of the behavior.
2. Open the fluorescence image file in Fiji (*see* **Note 12**).
3. Define regions of interest (ROIs) in the lpr with GCaMP fluorescence for measuring fluorescence signals. For this purpose, select each region using the selection tool, and register it to ROI Manager.
4. Get the time series of the mean signal value of each ROI. Use the "Multi Measure" function in the ROI Manager, and save the result as a text (.txt) file.
5. Open the saved text file in the spreadsheet software.
6. Determine the baseline signal value (F_0) by averaging the signal values of ten frames before the stimulus onset.
7. Determine the signal value change (ΔF) by subtracting the baseline signal value from the signal value of each time point.
8. Determine the relative signal value change ($\Delta F/F_0$) by dividing ΔF by F_0 (*see* **Note 13**).
9. Display the acquired behavior data and traces for calcium activity in a time-locked manner (Fig. 11.4d and e), and then perform further analysis as needed.

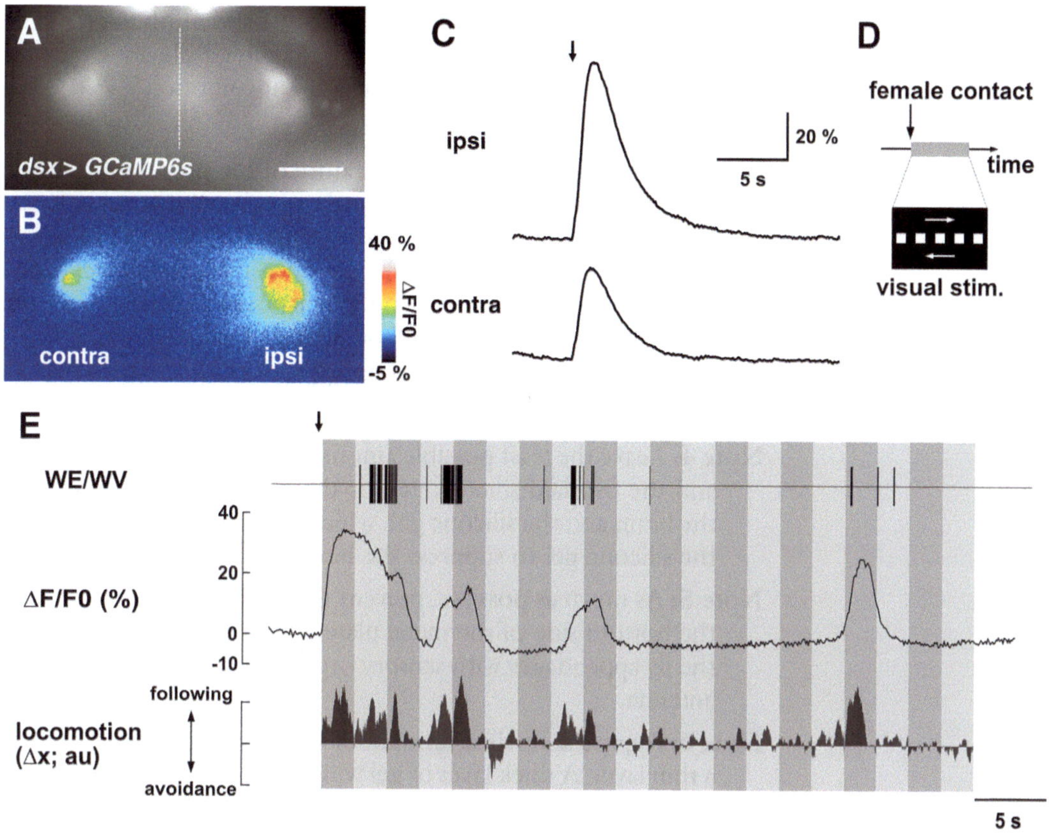

Fig. 11.4 Ca^{2+} activity recording from *dsx*-expressing neurons in the tethered male. (**a**–**c**) Ca^{2+} activity in *dsx*-expressing neurons upon contact of the male foreleg with the female body. The genotype of the flies used was *w*^{1118}; *UAS-GCaMP6s*/+; *dsx-Gal4*/+. (**a**) GCaMP6s fluorescence expressed by the *dsx-Gal4* driver. Horizontal view. Anterior on the top. The dashed white line indicates an approximate midline. Scale bar, 50 μm. (**b**) Color-coded Ca^{2+} response. (**c**) Representative Ca^{2+} response traces. The timing of the female contact on the male's foreleg was determined by inspecting the video record in a frame-by-frame manner and is indicated by the arrow on the plot. (**d** and **e**) Simultaneous recording of behavior and Ca^{2+} activity in a tethered male during courtship. The genotype of the flies used was *w*^{1118}; *dsx-Gal4, UAS-GCaMP3.0*/*UAS-GCaMP3.0*. (**d**) Schematic illustration of the stimulus. (**e**) Representative record of behavior and Ca^{2+} activity in the *dsx*-expressing neurons in the lpr. The arrow shows the timing of the female contact. The region shaded in dark and pale gray indicates when the visual stimulus drifted towards the right and left, respectively. WE/WV denotes the raster plot of wing extension and wing vibration, which are elemental actions in male courtship. The onset and offset of each wing extension/vibration bout were determined by inspecting the video record of the behavior. Δx, shown on the bottom, represents the signals that correspond to the change in the direction of the male's locomotion. To render it more recognizable, the male's activity of following the visual targets that alternately move right and left, plots that correspond to the time window when the target moved to the right (shaded in dark grey) are reflected over the horizontal baseline

4 Notes

4.1 Fly Treadmill

Note 1: Using heavier balls made of material other than polystyrene (e.g., polyurethane foam) yields a more stable locomotion record at the cost of increased physical load for the fly, as it ensures smooth rotation of the ball by preventing the fly from lifting and swinging the ball [17]. Lighter balls are easier for the fly to rotate and yield more active locomotion.

4.2 Preparation

Note 2: Because of the low temperature of the preparation block, the wing clip can have condensation on it, which can moisten and damage the fly's wing. To prevent this, keep the wing clip dry by intermittently wiping it with a piece of tissue paper. Avoid sliding the upper wing clip if it is in contact with the fly's wings, or the wings will be damaged.

Note 3: Residual UV glue and silicone gel on the tungsten hook can be burned off with a lighter.

Note 4: Leave the least possible amount of saline between the brain and the overlaid silicone gel, or there will be a space between the brain and the silicone gel, which will attenuate the ability of the silicone gel to suppress the brain motion.

Note 5: As much as possible, prevent the silicone gel from reaching the bottom side of the metal plate, where it could contaminate the fly appendages with sensory organs, particularly the legs and antenna.

Note 6: Apply the smallest amount of silicone gel necessary to form a thin layer. A thick layer of gel will make the fluorescent images dim and blurry.

4.3 Imaging

Note 7: Check the condition of the male at this point. Damaged flies often exhibit several characteristic behaviors, including periodic proboscis extension separated by intervals of several seconds, active grooming, and low locomotion activity. If these behaviors are observed, try giving gentle air puffs. If any of these behaviors persist for ~3 min, discard the preparation and go to another.

Note 8: Keep the intensity and duration of the excitation light to a minimum to reduce the photobleaching of the GCaMP. The area illuminated by the excitation light should be kept as small as possible by adjusting the iris of the light path to reduce the interference of the excitation light with the male's vision.

Note 9: There is a trade-off between the signal level and temporal resolution. If the sCMOS camera has pixel binning functionality, the signal intensity per image pixel can be increased at the cost of spatial resolution.

Note 10: Stimulating the male with both the female body and visual stimulus can be difficult to carry out on the first attempt. We recommend beginning with an experiment where the male is stimulated only by a female contact. The anticipated result in such a case is shown in Fig. 11.4a–c.

Note 11: The duration of a single recording can be up to 90 s, depending on the duration of the visual stimulus presentation.

4.4 Data Analysis

Note 12: The description of image processing here assumes the use of Fiji software. Any image analysis software with similar functionality can be used.

Note 13: It is critical to check whether the obtained signal change is an artifact derived from brain motion during recording. For this purpose, inspect the time series of the original fluorescent images and the video record as needed. The shift in the position of the sample within a focal plane can be corrected by registering the raw images. The moco, a plugin of Fiji, can be a good starting point to try this approach [30].

Acknowledgments

This work was supported by MEXT grants (nos. 18H02476 to S.K. and 16H06371 to D.Y.).

References

1. Kitamoto T (2001) Conditional modification of behavior in *Drosophila* by targeted expression of a temperature-sensitive *shibire* allele in defined neurons. J Neurobiol 47:81–92
2. Lima SQ, Miesenböck G (2005) Remote control of behavior through genetically targeted photostimulation of neurons. Cell 121:141–152
3. Lin JY, Knutsen PM, Muller A et al (2013) ReaChR: a red-shifted variant of channelrhodopsin enables deep transcranial optogenetic excitation. Nat Neurosci 16:1499–1508
4. Klapoetke NC, Murata Y, Kim SS et al (2014) Independent optical excitation of distinct neural populations. Nat Methods 11:338–346
5. Flood TF, Iguchi S, Gorczyca M et al (2013) A single pair of interneurons commands the *Drosophila* feeding motor program. Nature 499:83–87
6. Kimura K-I, Ote M, Tazawa T et al (2005) Fruitless specifies sexually dimorphic neural circuitry in the *Drosophila* brain. Nature 438:229–233
7. Chiang AS, Lin CY, Chuang CC et al (2011) Three-dimensional reconstruction of brain-wide wiring networks in *Drosophila* at single-cell resolution. Curr Biol 21:1–11
8. Jenett A, Rubin GM, Ngo T-TB et al (2012) A GAL4-driver line resource for *Drosophila* neurobiology. Cell Rep 2:991–1001
9. Milyaev N, Osumi-Sutherland D, Reeve S et al (2012) The virtual fly brain browser and query interface. Bioinformatics 28:411–415
10. Zheng Z, Lauritzen JS, Perlman E et al (2018) A complete electron microscopy volume of the brain of adult *Drosophila melanogaster*. Cell 174:730–743.e22
11. Scheffer LK, Xu CS, Januszewski M et al (2020) A connectome and analysis of the adult *Drosophila* central brain. elife 9:e57443
12. Maimon G, Straw AD, Dickinson MH (2010) Active flight increases the gain of visual motion processing in *Drosophila*. Nat Neurosci 13:393–399
13. Chiappe ME, Seelig JD, Reiser MB et al (2010) Walking modulates speed sensitivity in *Drosophila* motion vision. Curr Biol 20:1470–1475
14. Chen T-W, Wardill TJ, Sun Y et al (2013) Ultrasensitive fluorescent proteins for imaging neuronal activity. Nature 499:295–300
15. Akerboom J, Carreras Calderón N, Tian L et al (2013) Genetically encoded calcium indicators for multi-color neural activity imaging and combination with optogenetics. Front Mol Neurosci 6:2
16. Jung Y, Kennedy A, Chiu H et al (2020) Neurons that function within an integrator to promote a persistent behavioral state in *Drosophila*. Neuron 105:322–333.e5
17. Seelig JD, Chiappe ME, Lott GK et al (2010) Two-photon calcium imaging from head-fixed *Drosophila* during optomotor walking behavior. Nat Methods 7:535–540
18. Kohatsu S, Yamamoto D (2015) Visually induced initiation of *Drosophila* innate courtship-like following pursuit is mediated

by central excitatory state. Nat Commun 6: 6457

19. Seelig JD, Jayaraman V (2015) Neural dynamics for landmark orientation and angular path integration. Nature 521:186–191
20. Grover D, Katsuki T, Greenspan RJ (2016) Flyception: imaging brain activity in freely walking fruit flies. Nat Methods 13:569–572
21. Grover D, Katsuki T, Li J et al (2020) Imaging brain activity during complex social behaviors in *Drosophila* with Flyception2. Nat Commun 11:623
22. Buchner E (1976) Elementary movement detectors in an insect visual system. Biol Cybern 24:85–101
23. Schroll C, Riemensperger T, Bucher D et al (2006) Light-induced activation of distinct modulatory neurons triggers appetitive or aversive learning in *Drosophila* larvae. Curr Biol 16: 1741–1747
24. Antony C, Jallon J-M (1982) The chemical basis for sex recognition in *Drosophila melanogaster*. J Insect Physiol 28:873–880
25. Rideout EJ, Dornan AJ, Neville MC et al (2010) Control of sexual differentiation and behavior by the doublesex gene in Drosophila melanogaster. Nat Neurosci 13:458–466
26. Straw AD (2008) Vision egg: an open-source library for real time visual stimulus generation. Front Neuroinform 2:4
27. Peirce JW (2007) PsychoPy–psychophysics software in python. J Neurosci Methods 162: 8–13
28. Edelstein AD, Tsuchida MA, Amodaj N et al (2014) Advanced methods of microscope control using μManager software. J Biol Methods 1:e10
29. Schindelin J, Arganda-Carreras I, Frise E et al (2012) Fiji: an open-source platform for biological-image analysis. Nat Methods 9: 676–682
30. Dubbs A, Guevara J, Yuste R (2016) moco: Fast motion correction for calcium imaging. Front Neuroinform 10:6
31. Ribeiro IMA, Drews M, Bahl A et al (2018) Visual projection neurons mediating directed courtship in *Drosophila*. Cell 174:607–621.e18

Chapter 12

In Vivo Brain Imaging in Freely Moving and Socially Interacting *Drosophila*

Dhruv Grover, Takeo Katsuki, and Ralph J. Greenspan

Abstract

The fruit fly *Drosophila melanogaster* offers an array of powerful genetic tools useful for recording and manipulating neural activity in a relatively small brain that can display highly complex behaviors. One outstanding challenge in *Drosophila* neuroscience is to monitor neural activity in untethered, freely behaving animals. Here, we describe a method named Flyception2, which allows for real-time brain activity imaging from a freely walking and socially interacting fly. We provide detailed practical procedures subdivided into four main parts: (1) preparing the fly sample with a chronic imaging window and a retroreflective bead marker on the head, (2) assembling and configuring the imaging apparatus, (3) operating the system, and (4) analyzing the data.

Key words *Drosophila*, Fluorescence imaging, Real-time tracking, Social behaviors

1 Introduction

A major goal of modern neuroscience is to be able to see and understand the myriad signals that coarse through the brain during cognitive activities [1]. While brain imaging without physical restraint has been achieved in mice [2], zebrafish [3], and *Caenorhabditis elegans* [4, 5], conventional imaging paradigms in insects such as *Drosophila* have required immobilizing the animals [6, 7]. As a result, the only means of recording brain activity from a behaving fly is to simulate the world around it, for instance, using virtual reality paradigms in which a tethered fly walks on a trackball or is immersed in a flight simulator [8, 9]. Although such methods have allowed for the study of certain neural functions [10, 11], including some social interactions [12–14], studying complex social behaviors remains technically challenging with restrained flies. This restraint introduces a major impediment to observing behavioral activity under naturalistic conditions. Hence, we developed a novel technology for long-term chronic brain imaging in naturally behaving fruit flies [15, 16].

Daisuke Yamamoto (ed.), *Behavioral Neurogenetics*, Neuromethods, vol. 181, https://doi.org/10.1007/978-1-0716-2321-3_12,

The system, initially introduced as Flyception and later improved as Flyception2, has begun to shed light on the neural dynamics of complex social and cognitive behaviors that have otherwise remained inaccessible under conventional imaging paradigms. The method involved first creating a chronic transparent window on the fly's head to gain optical access to the brain while allowing the animal to behave without restraints. Brain activity recordings were then performed by stabilizing the image of the freely walking fly's head under tracking and florescence camera sensors (1x magnification) using two-axis galvanometer mirrors (at 1000 Hz). A retroreflective marker-based tracking, backed by the fast-synchronous electronic focus control and ratiometric dual-color brain imaging system, permits brain activity imaging in flies during behaviors in which their heads may move in and out of the plane of focus or tilt from the imaging axis. With these uniquely developed methods of sample preparation, optomechanical hardware and software, as well as an analysis pipeline, Flyception2 provides a state-of-the-art means of recording brain activity from freely behaving flies.

2 Materials

Fly stocks: In our Flyception-based experiments, we have utilized a variety of Gal4 drivers of different specificity and spatial distribution. Unless otherwise stated, stocks with initials BDSC were obtained from the Bloomington *Drosophila* Stock Center. Our olfactory stimulation experiments made use of the olfactory projection neuron driver *GH146-Gal4* (BDSC #30026). For mating behavior experiments, we utilized male flies from two driver lines: first, *P1a-Gal4* (gift from K. Asahina at the Salk Institute), a split-Gal4 that labels the P1 interneurons, a cluster of ~20 male-specific dorsal posterior *fruitless* gene-expressing neurons, and second, *R25E04-Gal4* (BDSC #49125) that labels the mAL neurons, a set of sexually dimorphic GABAergic interneurons of ~30 cells/hemisphere. Gal4 drivers for all ratiometric in vivo calcium imaging experiments were crossed to *20XUAS-GCaMP6s, UAS-myr-tdTomato*, a gift from A. Calhoun at Princeton. We utilized *UAS-CD8-GFP* and *UAS-myr-tdTomato* as calcium-independent ratiometric imaging controls. These flies were gifts from W. Joiner at UCSD.

Surgical station: A stereomicroscope with total magnification of 10X to 40X (Stemi, Zeiss) and an LED gooseneck illuminator (AmScope).

Retroreflective bead marker: Barium titanate solid glass microspheres hemispherically coated with aluminum (BTGMS-4.15 ϕ5–22 μm, Cospheric), UV glue (KOA-300, Poly-Lite), UV spot-curing system (LED-200, Electro-Lite).

Micro-sized coverslip: A 0.05-mm-thick coverslip (No. 000, Matsunami) diced to a 400 × 600 μm size (Opto-Line International).

Tape stage: Double-sided tape (Scotch), transparent single-sided tape (Scotch), microscope glass slide, single-edge razor blade, and a ruler.

Reusable surgical platform: Drill a 5 mm hole at the center of the lid of a 35 mm plastic Petri dish.

Temperature-controlled surgical platform: Circulate iced water from a flask in the ice bucket through an aluminum water liquid block (95 × 40 ×12 mm in size) using an aquarium pump (Model 1046, EHEIM). Alternatively, if available, a Peltier-based temperature-controlled stage can be used.

Silicone adhesive: We use a semi-transparent two-part bio-compatible silicone elastomer adhesive (KWIK-SIL, World Precision Instruments) for covering the imaging window on the fly head. The two-part silicone adhesive typically comes in a syringe with two plunges linked. Unlink the plunges by cutting the plastic connection with scissors, and extrude each component into a separate 1.5 ml microcentrifuge tube by slowly pressing one of the plunges at a time. Use the head of the insect pins (Size 00, BioQuip) to take small amount of each component, and mix them quickly on a microscope glass slide.

Forceps: Three different tip sizes are used. Coarse forceps (tip size 100 μm, Dumont #5 Forceps) are used for handling tapes and flies. Fine forceps (tip size 25 μm, Dumont #55 Forceps) are used for cutting out a window on the tape and for assisting surgery. Ultrafine forceps (tip size 5 μm, Dumont #5SF Forceps resharpened by Corte Instrument to tip size A) are exclusively used for cutting cuticles. Clean forceps with 70% ethanol before each use. Use a water-soaked swab (Absorption Spears – Mounted/Sterile, #18105-01, Fine Science Tools) for removing tissue debris from the ultrafine forceps during and after surgery.

Fly arena: The fly behavioral arena consists of a transparent acrylic chamber with a concave elliptical arena of dimensions 44 mm (major axis) × 40 m (minor axis) and surrounded by a 10-mm-wide water-filled moat. A custom AR-coated (500–850 nm) glass window (76 mm diameter, 0.7 mm thickness, Edmund Optics) was placed above the arena to contain the flies in the chamber during an experiment.

Tracking illumination sources: The arena was backlit with a custom-made diffuse 4 in × 4 in, 850 nm infrared LED light (~1 mW cm^{-2}). We used an 850 nm collimated fiber (SMA905 Fiber Collimation Package, f = 18.45 mm, Thorlabs) as an incident light source for fly-view tracking.

Dual-color fluorescence illumination sources and excitation light path optics: We used 50 mW blue (473 nm) and yellow (561 nm) lasers (Gem, Laser Quantum) to activate GCaMP6s and tdTomato signals, respectively, in the brain. The lasers were set to average power densities of 4.85 mW cm^{-2} (blue) and 3.8–4.77 mW cm^{-2} (yellow) when measured at the arena surface. The two laser beams were merged with a combination of kinematic mirrors (M1, Newport) and a green longpass dichroic beamsplitter (LM01-503-25, Semrock). Additionally, we placed a 2x beam expander (BE02M/A, Thorlabs) in the laser beam path.

Tracking machine vision cameras: The arena and fly-view tracking cameras used were the Flea3 USB3.0 (FL3-U3-13Y3M-C, Point Grey) and the Gazelle Camera Link (GZL-CL-22C5M-C, Point Grey), set to resolutions of 512 × 512 px at 50 Hz and 240 × 240 px at 1000 Hz, respectively. We used a full configuration camera link frame grabber (XCelera PX4, OR-X4C0-XPF00, Teledyne Dalsa) for image acquisition from the fly-view camera.

Fluorescence sCMOS imaging camera: The camera used for fluorescence imaging was a Prime sCMOS (Photometrics), set to a resolution of 2048 × 256 px at 50 Hz and controlled by μManager.

Optical lenses: The arena-view camera was fitted with a fixed focal 8 mm c-mount lens (M0814-MP2 2/3, Computar). The fly-view and fluo-view cameras used Canon EF 180 mm f/3.5 L Macro USM lenses. Apertures of both macro lenses were set at f/4.0. The macro lenses were attached to the two cameras via EOS EF-mount to C-mount adapters (Fotodiox).

Camera filters: A light-tight tube system was created between the lens mount and the fly-view camera sensor (SM1A9, SM1A10, SM1L03, Thorlabs) to include a red longpass (625 nm, 84746, Edmund Optics). Image splitting optics (W-VIEW GEMINI, Hamamatsu) were placed in the light path of the fluo-view camera (between the lens and camera) for simultaneous image acquisition of dual-wavelength images to allow for ratiometric imaging. These included a 580 nm dichroic beamsplitter (FF580-FDi01-25x36, Semrock) and two bandpass filters for GCaMP (520/35 nm, FF01-520/35-25, Semrock) and tdTomato (607/36 nm, FF01-607/36-25, Semrock) emission wavelengths.

Galvanometer mirror system: A two-axis galvanometer mirror system was mounted above the fly arena (GVS012, Thorlabs). The galvanometer analog position range was set from −10 V to +10 V with an output scaling factor of 0.5 V deg^{-1}. Voltages for each output channel were relayed to the galvanometer servo driver board using a 16-bit digital-to-analog converter (PCIE-6351, National Instruments).

Imaging beamsplitters: The fly-view camera was placed behind a far-red super-resolution longpass dichroic beamsplitter (660 nm, Di03-R660-T1-25X36, Semrock). Also placed in the light path (between the far-red dichroic beamsplitter and camera) was a 10: 90 (reflectance/transmission ratio) plate beamsplitter (BSN11R, Thorlabs). A second dual-band dichroic beamsplitter (488/561 nm, Di01-R488/561-25x36, Semrock) was placed in the fluo-view camera's light path, permitting only the transmission of green GCaMP6s fluorescence and red tdTomato fluorescence to the fluo-view camera.

Camera synchronization electronics: The exposure timing and frame rate of the arena-view, fly-view, and fluo-view image acquisitions were controlled by external triggers generated using FPGA hardware-based clock timers on a MyRIO device (National Instruments). A flash (TT560, Neewer) was used to generate a visually detectable cue for alignment of the frames of the three cameras. The MyRIO timer also triggered the firing of the flash, set to coincide with the trigger onset of the arena-view, fly-view, and fluo-view cameras.

3 Methods

3.1 Sample Preparation

1. Preparing Retroreflective Bead Marker Coverslips

   To allow for robust tracking of the fly head, we use a retroreflective marker [17] consisting of three barium titanate solid glass microspheres hemispherically coated with aluminum. The beads are placed on a micro-sized coverslip. The beads are placed to resemble the vertices of an isosceles triangle with its apex facing the forward direction of the fly. The first bead is positioned centrally along one long edge of the coverslip to coincide with the intersection of the anterior and medial axes of the fly body. The second and third beads are positioned along the opposite long edge of the coverslip, coinciding with the posterior lateral edges of the fly's head (Fig. 12.3n). This simple marker arrangement enables robust tracking of the fly's head at high frame rates while simultaneously allowing access to the brain, without obstruction, for in vivo optical recording.

   a. Secure a piece of micro-sized coverslip on a double-sided tape (Fig. 12.1D).

   b. Place small amounts of UV glue at the three locations on the coverslip with an insect pin (Fig. 12.1E arrows).

   c. Spread out retroreflective beads on a glass slide and pick up individually with an insect pin using capillary action and then place on the spots of glue on the coverslip (Fig. 12.1F).

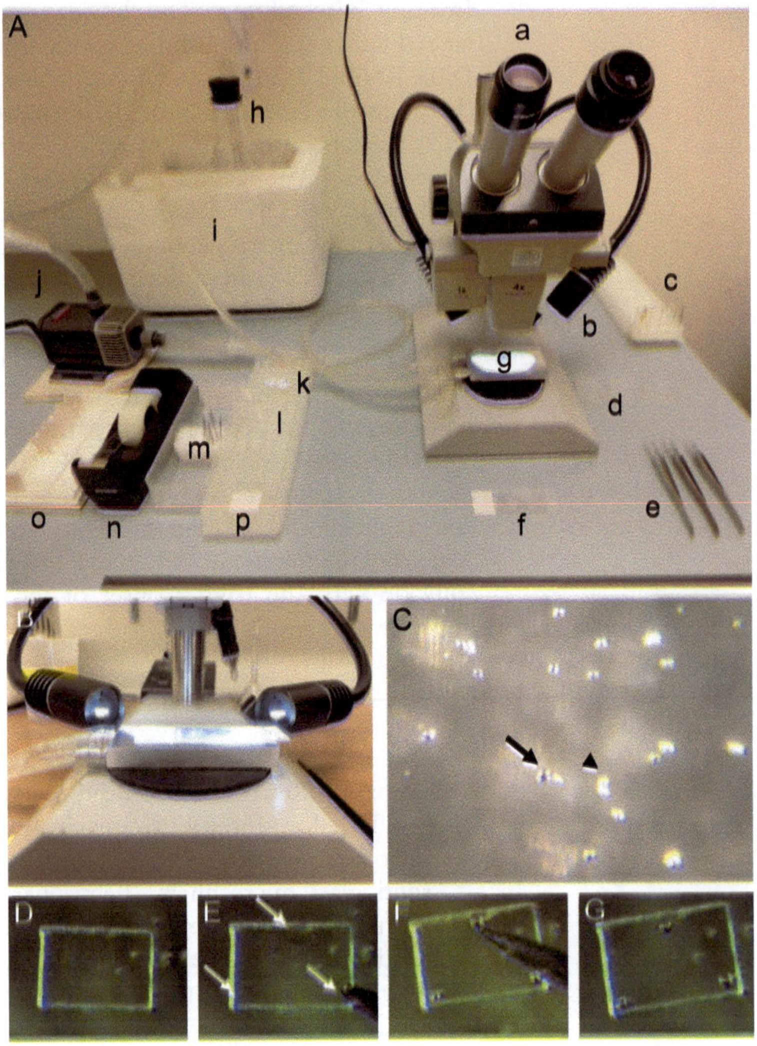

Fig. 12.1 Surgical station and preparation of retroreflective bead marker coverslips. (**A**) (a) A stereomicroscope with total magnification of up to 40X. (b) LED illumination. (c) Food vials. (d) Surgery platform. (e) Forceps. (f) A taped glass slide. (g) Cold surgical stage. (h) A filtering flask. (i) Ice bucket. (j) An aquarium pump circulating iced water between the flask and the surgical stages. (k) Tubes containing KWIK-SIL. (l) Water-soaked swab for cleaning forceps. (m) Insect pins. (n) Adhesive tape. (o) Slides with tapes. (p) A slide with retroreflective beads. (**B**) LED illumination lighting the glass slide from the sides. Retroreflective beads are spread on the slide. (**C**) When side-lit, beads in the correct orientation appear dark when viewed through the eyepiece (arrow). Tilted beads appear brighter (arrowhead). (**D**) A micro-sized coverslip secured on a double-sided tape. (**E**) UV glue placed at the three locations using the insect pin. (**F**) Retroreflective beads placed on the glue with the tip of the insect pin. (**G**) Retroreflective beads after correcting the orientation for proper retroreflection

d. The orientation of the beads on the coverslip is critical for achieving good retroreflective properties during tracking. Correct the orientation of the beads by touching the beads with the insect pin under the stereomicroscope with the side-lit fiber LED illumination (Fig. 12.1B). Because the beads are hemispherically coated with aluminum, when side-lit, beads in the correct orientation appear dark when viewed through the eyepiece (Fig. 12.1C arrow and G). On beads that are tilted, however, varying degrees of light reflect off the aluminum coating up to the eyepiece and appear brighter (Fig. 12.1C arrowhead).

e. Use a UV spot-curing system to cure the UV glue for 60 s.

2. Fly Brain Window Surgery

Successful imaging experiments with Flyception require the preparation of the imaging window on the fly head while maintaining the fly's health. The surgical procedure starts by immobilizing the fly on the surgical platform with the adhesive tape.

a. Prepare 0.5-mm-wide, 15-mm-long strips of tapes by attaching a tape on a glass slide and cutting it with a razor blade guided with a ruler (Fig. 12.2a).

b. Prepare the tape stage for immobilizing a fly. Place a double-sided adhesive tape on a glass slide, and attach the non-sticky side of a 20-mm-long one-sided clear adhesive tape to the double-sided tape (Fig. 12.2b).

c. Start cooling the surgical stage to ~4 °C.

d. Anesthetize adult flies by placing a glass vial containing the flies on ice for a few minutes until the flies fall on the bottom of the vial. Use flies older than 4 days. Younger flies can be used, but surgery will be more difficult due to the softness of the cuticle.

e. Handle flies and tapes with the coarse forceps.

f. Observing through a stereomicroscope under 10x magnification, place a fly ventral side up onto the tape stage (the fly's dorsal side is attached to the tape).

g. Secure the fly head using a strip of tape (Fig. 12.2c). Be sure that the head is not tilted.

h. Immobilize the abdomen using a strip of tape (Fig. 12.2c).

i. Using hands, remove the tape with the fly from the glass slide, and attach it to the surgical platform positioning the fly in the center of the hole (Fig. 12.2f and g).

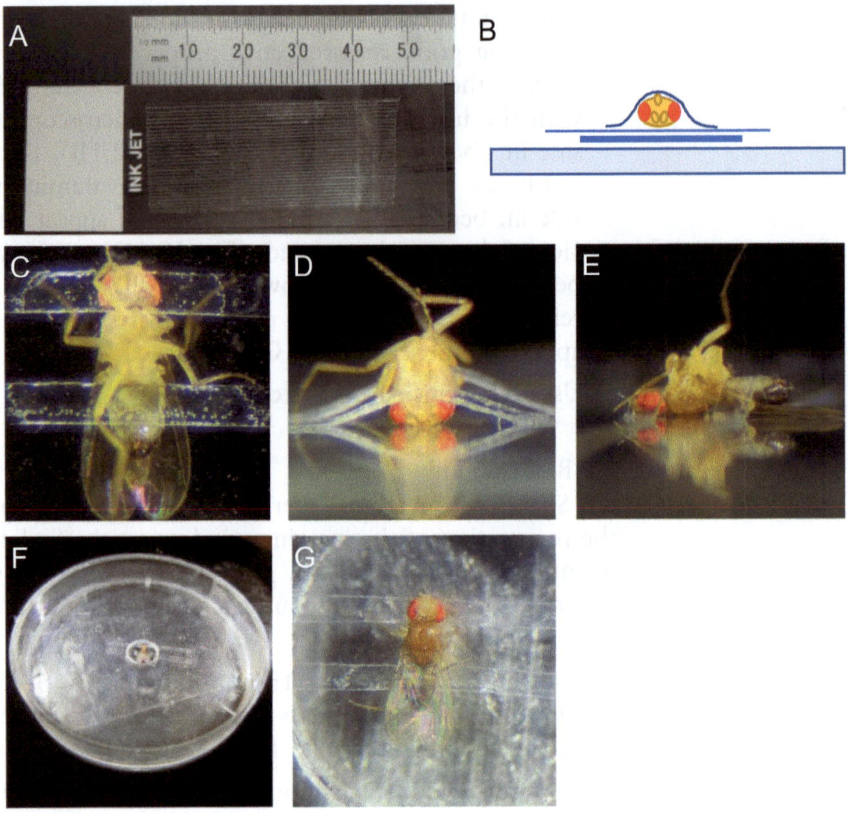

Fig. 12.2 Securing the fly to the surgical platform. (**a**) 0.5-mm-wide, 15-mm-long strips of tape prepared on a glass slide. (**b**) Schematic drawing of the tape stage. From bottom to top: glass slide, double-sided tape, one-sided tape (sticky side up), a fly, and strips of tape holding the head and the abdomen of the fly. The double-sided tape provides a reusable adhesive substrate for the single-sided tape that holds a fly. (**c**–**e**) A fly secured on the tape slide, viewed from the bottom, front, and side. (**f** and **g**) A fly mounted on the surgery platform

j. Ensure that the dorsal side of the head is in contact with the tape (Fig. 12.2d and e). If there is a space between the head and the tape, the tape is too loose, and the head may move during surgery. In this case, flip the surgical platform, and press the strips of tape around the head so that the tape applies more pressure on the face of the fly.

k. Repeat steps (b)–(k) for the number of flies being operated.

3. Surgical Procedures

Once the fly is secured to the surgical platform, we create a window on the tape and then a window on the fly head using ultrafine point forceps. The window on the head is sealed with a silicone elastomer and covered with the micro-sized coverslip prepared in the previous steps. Successful surgery requires practice, but an experienced fly surgeon can achieve >80% success rate (for some reason, the rate also depends on the genotype).

a. Set the magnification of the stereomicroscope to 40x.

b. Using the fine forceps, cut out a rectangular opening in the tape around the head. First, poke holes on the four corners, and then move the forceps from one corner to the next to cut the side (Fig. 12.3b). Work on the sides lateral and in front of the fly first (Fig. 12.3b arrow 1–3). Finally, cut the back side around the neck by lifting the flap and peeling it sideways (Fig. 12.3b arrow 4). Do not expose more than half of the thorax, because it will make the head and the thorax unstable during the subsequent surgical procedures.

c. For opening the cuticle on the dorsal side of the head, use the ultrafine forceps. Use a water-soaked swab to clean the ultrafine forceps as necessary.

d. Make the first incision near the ocelli at the dorsal posterior edge of the head. Extend the cut along the ridge towards one of the eyes (Fig. 12.3d arrow 1 and e).

e. Grab the cuticle with the forceps, and peel anteriorly along the dorsal edge of the eye (Fig. 12.3d arrow 2 and f).

f. Rotate the fly by 180 degrees, and repeat the same procedure on the other side (Fig. 12.3d arrow 3–4 and g).

g. Grab the flap near the front end, and peel the flap from one side to the other. If the head is not stable, use a second pair of forceps (fine tip) for holding the base of the flap while peeling (Fig. 12.3d arrow 5 and h).

h. Remove the white sheaths (air sacs) that cover the brain (Fig. 12.3i).

i. Seal the opening with a two-part translucent silicone elastomer adhesive. Take a small amount of each component of the silicone adhesive with the head of an insect pin, and immediately mix them on a glass slide for 2–3 s (also using the insect pin). Try to use roughly a 1:1 ratio of components to achieve an equi-volume mixture.

j. Using the tip of the fine forceps, take a small amount of the mixed silicone adhesive, and apply it on the opening of the head (Fig. 12.3k–m). If desired, add the silicone adhesive to the neck to immobilize the head relative to the thorax.

k. Immediately after applying the silicone adhesive, pick up the coverslip with the retroreflective marker prepared in step 1 from the adhesive tape, and place it on the head (Fig. 12.3n). Handle coverslips with the fine forceps. Complete this step within 2 min of mixing the two silicone components.

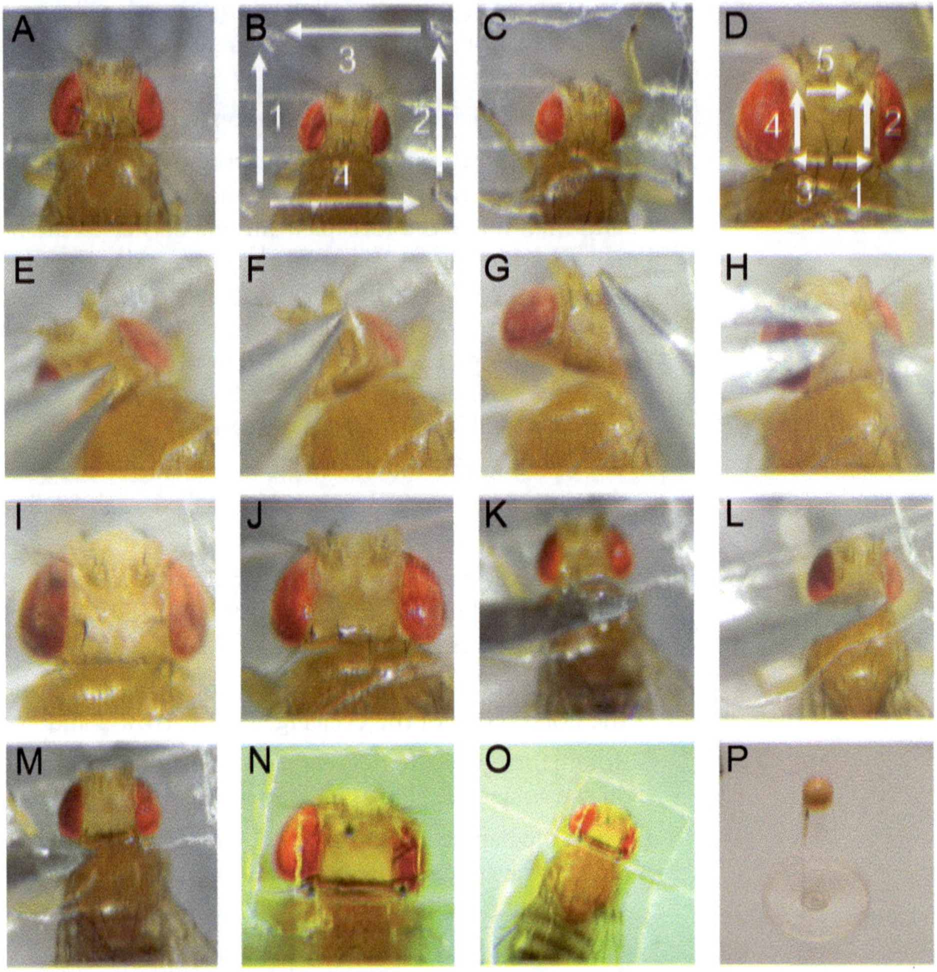

Fig. 12.3 Step-by-step surgical procedures. (**a**) A fly mounted on the surgery platform. (**b**) Perforations at the four corners around the fly head. Arrows indicate the order of cutting the tape. (**c**) After creating a window on the tape. (**d**) Five major steps of surgery. Each numbered step is further described in (**e–h**). (**e**) Creating the first incision near the ocelli along the ridge between the dorsal and posterior head. (**f**) Cutting the cuticle along the right eye by peeling the cuticle toward the front of the head. (**g**) Cutting the cuticle along the left eye by peeling the cuticle anteriorly. (**h**) Removing the cuticle by peeling sideways while holding the base of the cuticle near the antennae with the second forceps. (**i**) After removing the cuticle from the dorsal side of the head. (**j**) After removing the air sacs. (**k**) Right before applying the silicone adhesive with one of the tips of the forceps. (**l**) Applying the silicone adhesive to the opening and the neck. (**m**) After applying the silicone adhesive to the opening and the neck. (**n**) After placing a coverslip on the opening. (**o**) After removing the tape to expose the thorax. (**p**) A food vial placed on the surgical platform to contain the fly after surgery

l. Remove the tape from the fly using the fine forceps to peel the tape covering the thorax towards the back of the fly (Fig. 12.3o).

m. Set the Petri dish holding the fly aside, and place a food vial/tube upside down over the fly (Fig. 12.3p). If the surgery is successful, the fly will naturally remove itself from the tape and start walking. The fly will then walk up the tube via negative geotaxis. If necessary, however, help release the fly from the tape using the coarse forceps, and transfer it to the food vial.

n. Once the fly is in the vial, seal the vial with a piece of tape. Poke an air hole in the tape with the fine forceps. Keep the flies on the food overnight to let them recover from the surgery. Poorly operated flies usually die within a day of surgery.

o. Assess behavior and image the brain with Flyception. Flies that recover typically show walking speeds >10 mm s^{-1} and negative geotaxis.

3.2 System Hardware Assembly

Assembling the system hardware (Fig. 12.4) can be split into the following major components – galvanometer mirror assembly, dual-laser setup and beam alignment, fly arena structure, fly-view and fluo-view camera alignment, Z-focus control mechanism, referencing camera images with the lasers, and finally arena- and fly-view image-world coordinate calibration. We suggest that these components be assembled in the order presented below.

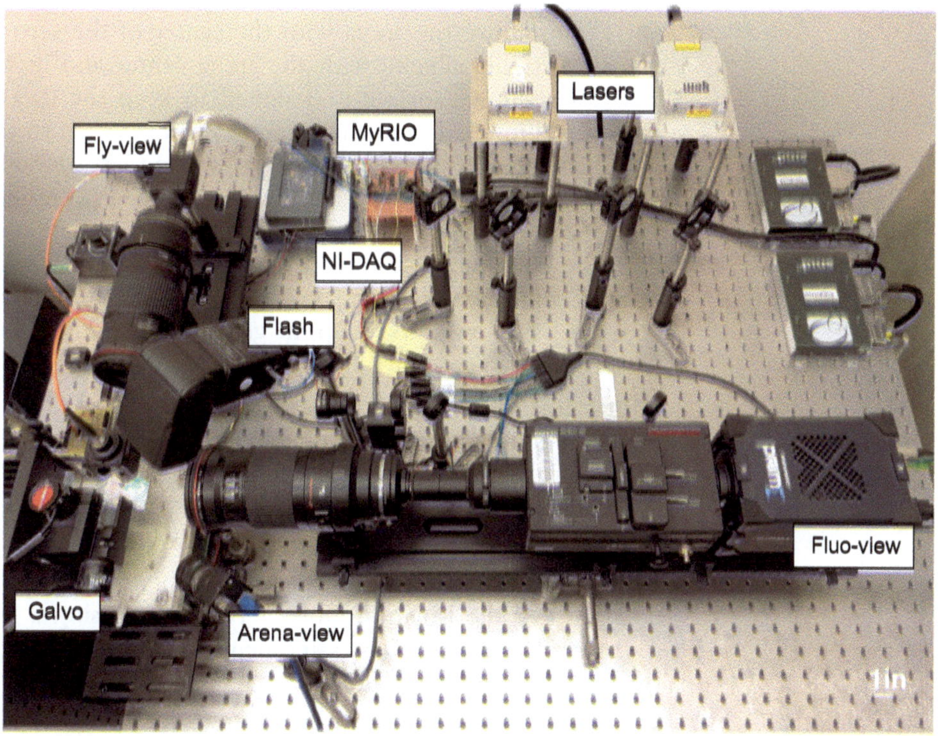

Fig. 12.4 Flyception system hardware overview. Outlined are the major components: (1) Galvo and fly arena assembly; (2) DPSS lasers; (3) fly-view, fluo-view, and arena-view cameras; and (4) MyRIO, DAQ, and Flash

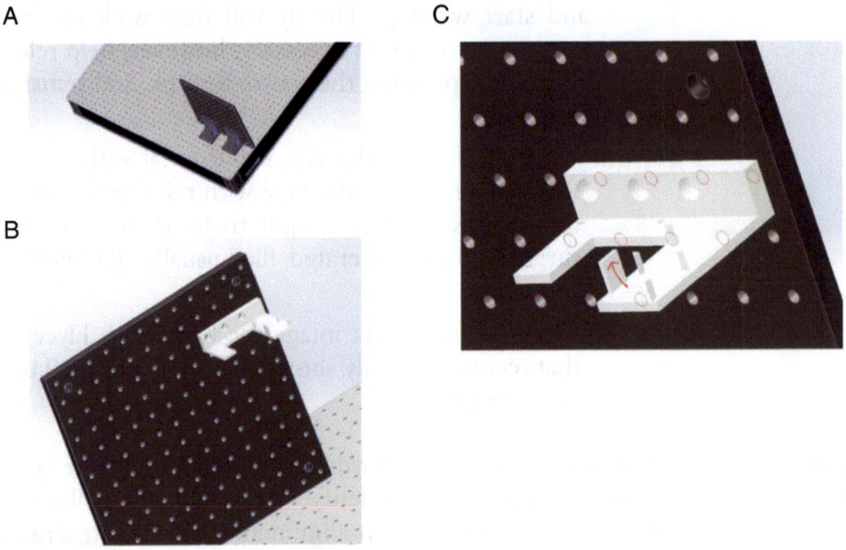

Fig. 12.5 Galvanometer mirror assembly. (**a**) Placement of galvanometer mirror assembly vertical breadboard mount on the isolation table. (**b**) Galvo holder placement on vertical breadboard mount. (**c**) Alignment hole on the galvo holder (red arrow) at a height of 242.5 mm from the surface of the isolation table

1. Galvanometer Mirror Assembly

   Briefly, a two-axis galvanometer mirror system is mounted 68.17 mm above the arena with the center of the secondary mirror vertically aligned with the arena center and its major axis. The concave surface is designed to keep the fly's head orthogonal in relation to the rotating mirror system throughout the arena. Because the primary and secondary galvanometer mirror centers are separated from each other by 15.174 mm, rotations of the two mirrors produce arcs of different curvatures, leading to the concave arena contour with an elliptical perimeter. The following step-by-step procedure can be followed to set up the galvanometer mirror assembly as described above.

   a. Attach AP90 brackets to the MB12 breadboard (Thorlabs) and roughly position the breadboard near the bottom left corner of the isolation table and flush to the bottom edge of the table (Fig. 12.5a).

   b. Attach the galvo holder to the breadboard as shown (Fig. 12.5b). Measure the height of the holder from the surface of the isolation table at two edges using a height gauge, and set it at the same height so that the holder is parallel to the isolation table.

   c. Measure the height of the small hole on the galvo holder (Fig. 12.5c, red arrow). It should be around 242.5 mm from the surface of the isolation table. The hole is at the height equivalent to that of the center of the primary galvo mirror (x-axis mirror).

d. Attach the two-axis galvo to the galvo holder with screws. Make sure that the back side of the galvo is parallel to the back of the holder by pushing the galvo toward one side of the holder while screwing it.

2. Dual-Laser Light Path Alignment

The two Gem lasers (473 nm blue and 561 nm yellow) are placed in an orthogonal relation to the galvanometer assembly near the top right edge of the isolation table (see Fig. 12.4). Their beams are merged with a combination of kinematic mirrors (M1, Newport) and a green longpass dichroic beamsplitter (LM01-503-25, Semrock) before being directed toward the custom-machined galvanometer assembly holder setup in step 1 (Fig. 12.6a). Additionally, a 2x beam expander (BE02M/A, Thorlabs) is placed in the laser beam path to increase the surface area of the laser spot and provide consistent illumination over the fly.

a. Place the two lasers on custom laser mounting plates, and mount them with 9 in posts (using 8 in and 1 in posts) and 4 in post holders.

b. Roughly position the height of the laser aperture at 241 mm from the table surface.

c. Align beams along the holes of the breadboard using two mirrors on kinematics mounts and two targets set at the height of the small hole measured at step 1c above (e.g., 241.52 mm). For basic laser alignment techniques, see the following link: https://www.edmundoptics.com/resources/application-notes/lasers/simplifying-laser-alignment/.

d. Fold the blue laser beam 90° to the left along the holes using a 520 nm dichroic mirror post-mounted in an LMR1 mirror holder (Thorlabs). This ensures that the green beam hits the dichroic at 45° at the next step (note that a transmission curve of dichroic mirrors is angle dependent).

e. Fold the green beam 90° to the left with a mirror post-mounted on an LMR1. Adjust the position of the mirror such that the green laser through the dichroic mirror completely overlaps with the blue laser. Now the two lasers are combined. Do not place a beam expander yet.

f. Fold the combined beam by 90° toward the bottom with a mirror mounted on an LMR1 and a clamp (Fig. 12.6b). Use two targets to align the beam along the holes. Move the galvo assembly if it is in the way of the laser. This will determine the position of the rest of the components including the galvo, fly arena, and cameras.

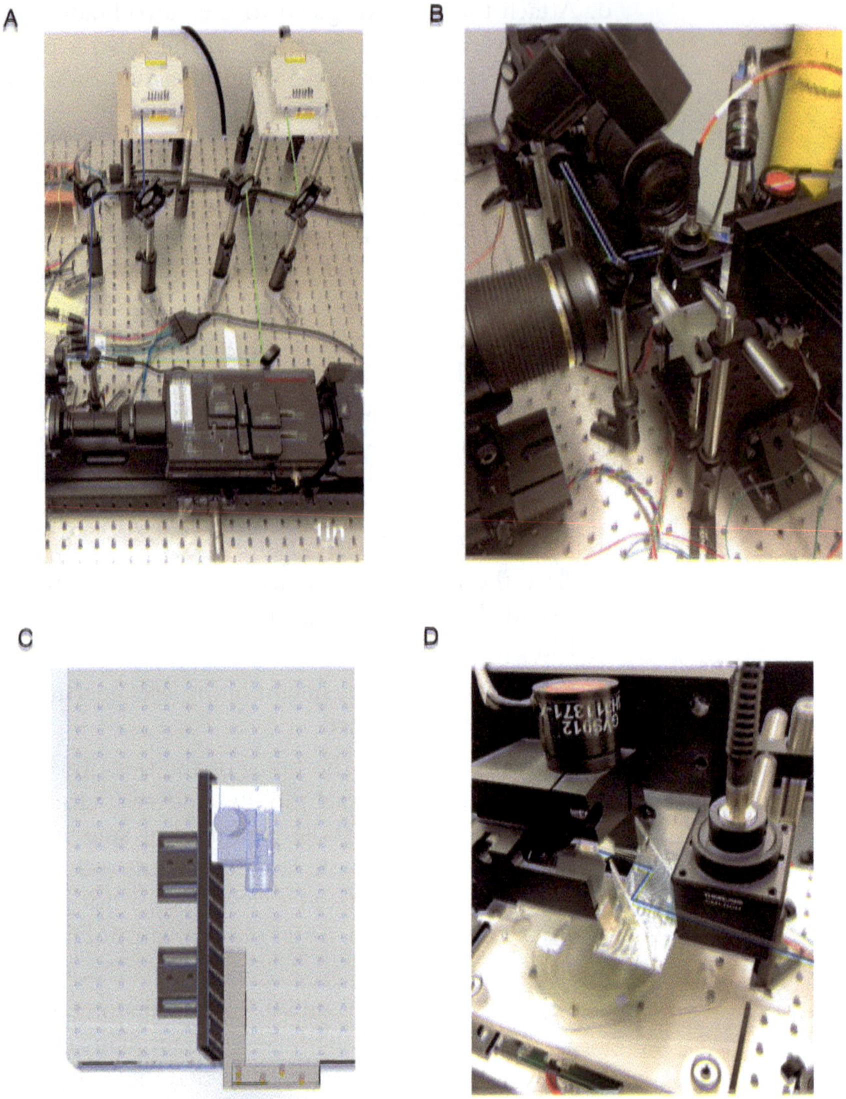

Fig. 12.6 Dual-laser light path configuration and alignment. (**a**) Two free-space Gem laser (473 nm blue and 561 nm yellow) beams merged with a combination of kinematic mirrors and a green longpass dichroic beamsplitter. (**b**) Merged laser beams being directed toward the custom-machined galvanometer assembly holder. (**c**) Using a speed square, align the galvo assembly to ensure it is orthogonal to the incoming merged laser beam and that the beam is aligned with the small hole on the galvo holder. (**d**) Attachment of dual-band (488/561 nm) reflective and 660 nm longpass dichroic mirrors to the galvo holder

g. Move the galvo assembly back into the path of the lasers, and adjust the position of the assembly such that the lasers pass through the alignment hole of the galvo holder. Make sure that the breadboard of the galvo assembly is parallel to the laser using a square. Screw down the brackets (Fig. 12.6c).

h. Attach a Di01-R488/561-25x36 dichroic mirror (Semrock) to the galvo holder. Secure the mirror to the holder with a small amount of silicone rubber (KE45W, Shin-Etsu silicone) on the edges of the dichroic mirror. This minimizes the strain to the mirror (Fig. 12.6d).

i. Check that the laser beam redirected by the dichroic mirror hits the alignment hole on the back of the galvo holder.

j. Attach a Di03-R660-T1-25X36 dichroic mirror (Semrock) to the galvo holder. Minimizing strain is particularly critical for this second dichroic mirror as the reflected light is imaged with the camera for fluorescence imaging (Fig. 12.6d).

3. Fly Arena Structure Assembly

a. Assemble the fly arena holder with the 562-XYZ stage (Newport), 38-baseplate (Newport), an MB4 breadboard (Thorlabs), and 4x 3 in posts. Add 4x 1.75 in posts (using 1 in and 0.75 in posts) in between the 3 in posts along the short edges of the breadboard (Fig. 12.7).

Fig. 12.7 Fly arena assembly. Fly chamber and Delrin diffuser inserted in outer corner posts and resting on inner smaller posts. This structure is then mounted on an XYZ stage. Height of the top surface of the fly chamber is set to 184.28 mm from the isolation table

b. Place the fly arena holder under the galvo assembly.
c. Insert the custom Delrin diffuser into the outer posts of the fly arena holder such that it rests on the inner posts.
d. Insert the fly chamber into the outer posts on top of the diffuser.
e. Measure the height of the top surface of the arena, and adjust it to 184.28 mm by moving the XYZ stage micromanipulator along the z-axis. The height of the top surface of the acrylic chamber should be 241.52 mm + sin(15 deg) * 15.174 mm − 68.167 mm + 7 mm = 184.28 mm, where 241.52 mm is the height of the x mirror center (or the height of the laser measured in step 1c), 15.174 mm is the distance between the x and y mirror centers (as per two-axis galvo technical specifications, Thorlabs), 68.167 mm is the distance between the y mirror center and the arena center, and 7 mm is the distance between the arena plate surface and the center of the arena.
f. Replace the arena with a custom resolution target holder.
g. Move the arena assembly such that the laser hits the center of the target. A rough alignment is sufficient at this step.

4. Fly-View Camera Setup

The fly-view camera is to be aligned with the center of the primary galvanometer mirror (Fig. 12.8a) and placed behind the far-red super-resolution longpass dichroic beamsplitter (Fig. 12.6d, step 2j). Also placed in the light path (between the far-red dichroic beamsplitter and camera) is a 10:90 (reflectance/transmission ratio) plate beamsplitter (BSN11R, Thorlabs). This plate beamsplitter is used to direct infrared collimated fiber light down to the arena through the galvanometer mirrors as well as to allow retroreflected infrared light back up from the fly to the fly-view camera (Fig. 12.8c).

a. Assemble the fly-view camera on a PRL-12 rail (Newport) and a 562-XYZ stage. Level the camera and the lens with a digital angle gauge. Note that a longpass filter is installed inside a SM1L03 lens tube.

5. Fluo-View Camera Setup

The fluorescence imaging camera is placed in an orthogonal position in relation to the fly-view camera and aligned with the reflected light path of the far-red longpass dichroic filter (Fig. 12.8b). The camera is to also be in line with the second dual-band dichroic beamsplitter (step 2h), permitting only the transmission of green GCaMP6s fluorescence and red tdTomato fluorescence to the fluo-view camera (Fig. 12.8d).

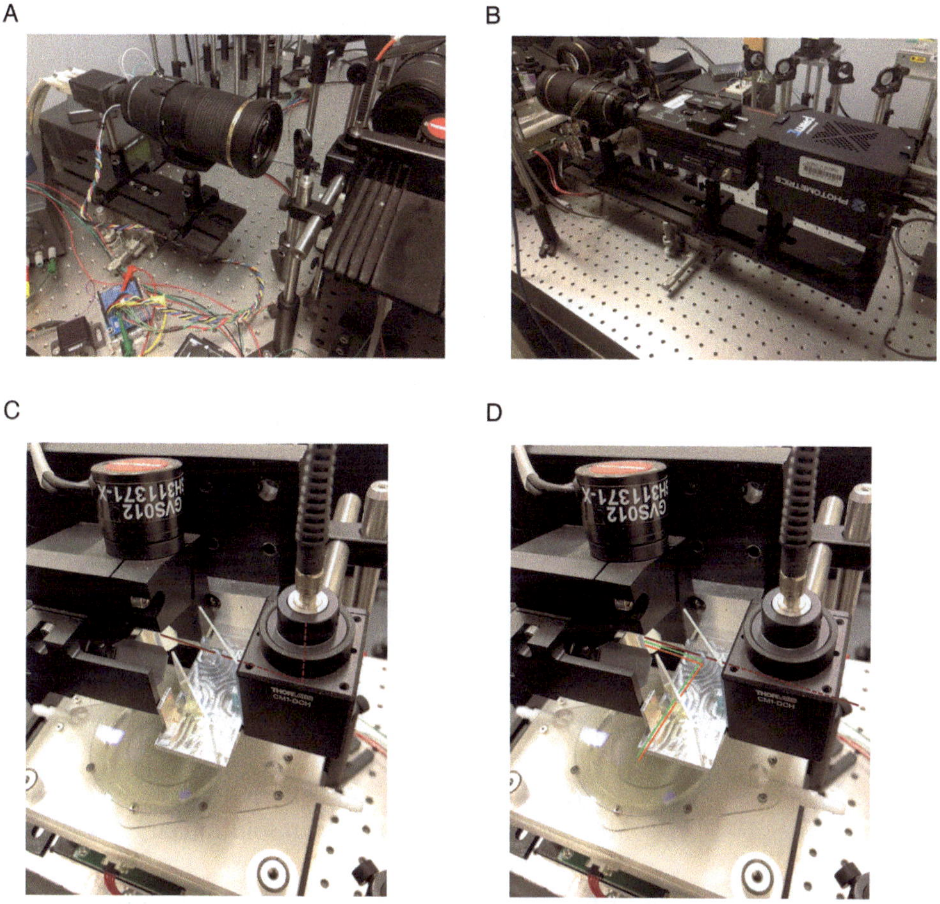

Fig. 12.8 Fly-view and fluo-view camera assemblies. (**a**) Fly-view tracking camera mounted on a rail and XYZ stage and aligned with the center of the primary galvanometer mirror. (**b**) Fluo-view camera placed in an orthogonal position in relation to the fly-view camera and aligned with the reflected light path of the far-red longpass dichroic filter. The camera is to also be in line with the second dual-band dichroic beamsplitter. (**c**) In red is the IR fiber LED beam path passing through the far-red longpass dichroic mirror to the galvanometer mirrors. The dotted red line represents the IR fiber LED beam path internal to the filter cube, emanating from the source fiber patch cable, passing through a collimating lens, and reflecting off a 10:90 reflectance/transmission beamsplitter mirror. (**d**) Fluorescence emission and retroreflection light paths from the fly arena through the two-axis galvanometer mirror assembly being split via dichroic and beamsplitter mirrors to fly-view and fluo-view imaging cameras. In green and orange are the emission wavelength light paths for GCaMP and tdTomato channels reflecting off the far-red longpass dichroic mirror and transmitting through the multi-bandpass dichroic mirror to the fluo-view camera. In red is the IR retroreflected light path passing through the far-red longpass dichroic mirror and internally through the filter cube that houses the beamsplitter mirror (dotted line segment) to the fly-view camera

a. Assemble the fluo-view camera on a PRL-24 rail (Newport) and a 562-XYZ stage. The Gemini is supported with a BA2 (Thorlabs) and a post from underneath.

b. Install a dichroic mirror (FF580-FDi01-25x36) and filters (FF01-520/35-25 and FF01-607/36-25) in the

HAMAMATSU Gemini following the manufacturer's instructions. Do not overtighten the screws that hold the dichroic mirror.

c. Place the overhead IR illumination assembly next to the galvo holder and flush to the edge of the galvo holder.

6. Centering the Laser Beam in the Camera Images

   The following procedure helps align the fly-view and fluo-view camera image centers with the dual-laser beam spot on the arena surface.

   a. Run the fly-view camera. Follow the manufacturer's instructions for initializing the camera.
   b. Place the custom 850 nm LED array under the arena.
   c. Connect the LED array to a DC power supply, and provide 14 V, 0.1 mA constant current.
   d. Adjust the position of the fly-view camera using the 562-XYZ stage such that the resolution target is at the center of view and in focus.
   e. Run the fluo-view camera with overhead white light (flashlight). Follow the manufacturer's instruction for how to start up the camera.
   f. Adjust the position of the fly-view camera using the 562-XYZ stage such that the resolution target is at the center of view and in focus.
   g. Adjust Gemini according to the manufacturer's instruction. If the image is severely distorted or blurry, check that the dichroic mirrors are not strained.
   h. Now both fly-view and fluo-view are roughly in the right positions. Fine adjustment will be done in the next steps.
   i. Turn off the white LED and turn on the lasers. Set exposure to 100 ms, and adjust the position of the arena such that the laser points the center of the target in the fluo-view camera.
   j. Once the laser is at the center, add the beam expander to the beam path between the 520 nm dichroic mirror and the mirror near the fly-view lens. Looking at the fluo-view image, adjust the height and angle of the expander such that the expanded beam hits the center of the target.
   k. Adjust the position of the fly-view camera again using the 562-XYZ stage such that the resolution target is at the center of view and in focus.
   l. Now the arena, fly-view camera, and fluo-view cameras are all aligned.

m. Place the arena-view camera on a post looking down the arena at an angle. Adjust the position of the arena-view camera while monitoring the image of the arena on Fly-Capture software such that the whole arena fits in a centered image size of 512 × 512 pixels.

7. Z-Focus Control Mechanism

For integrating electronic focal plane adjustment into the system, we built the interface used for communication between the camera body and lens body of a standard DSLR camera (Canon EOS). This allowed electronic control of the lens motor system of both the fly-view and fluo-view camera lenses for simultaneous electronic focus control.

For the physical connection, we modified the EOS EF-mount to C-mount adapters (Fotodiox) used to connect the lenses to the fly- and fluo-view cameras. Holes were drilled in the adapter, giving access to the interface pins on the mounting flanges of the lenses, and wires were attached, creating a custom harness for each lens (Fig. 12.9a, b). Both harnesses were connected to the Serial Peripheral Interface (SPI) module of a standard Arduino Uno microcontroller, via a custom PCB shield, that was programmed to mimic the behavior of the camera body (Fig. 12.9c). Command signals were then sent simultaneously to both lenses controlled by the tracking software via a custom interface between the NI DAQ and the Arduino's input pins. Ultrasonic motor lenses (USM) were chosen over standard lenses for reasons of focus repeatability and image stability.

8. Arena-View Camera Calibration

The system is initialized with tracked positions of the focal fly in the arena camera view. A custom nearest-neighbor-based algorithm is used to track the identity of multiple flies in the arena-view at 50 Hz. A mapping of 2D image coordinates in the arena-view to real-world coordinates in the arena is pre-computed offline by detecting the position of a laser beam spot on the arena surface for varied galvanometer mirror positions. This mapping is used in real time to back-project 2D pixel centroids of the detected flies to real-world coordinates. The galvanometer mirrors are then directed to the 2D tracked location of a user-defined fly in the arena. The arena-view tracker is used to provide a coarse approximation of the focal fly's position in the arena and initially guide the galvanometer mirrors to the approximate location of the fly's head.

a. Turn on the 473 nm blue laser (with beam expander removed from light path) at low power (~1 mW). When the calibration raster code is executed, a live stream of the arena-view camera is displayed, and the intensity/

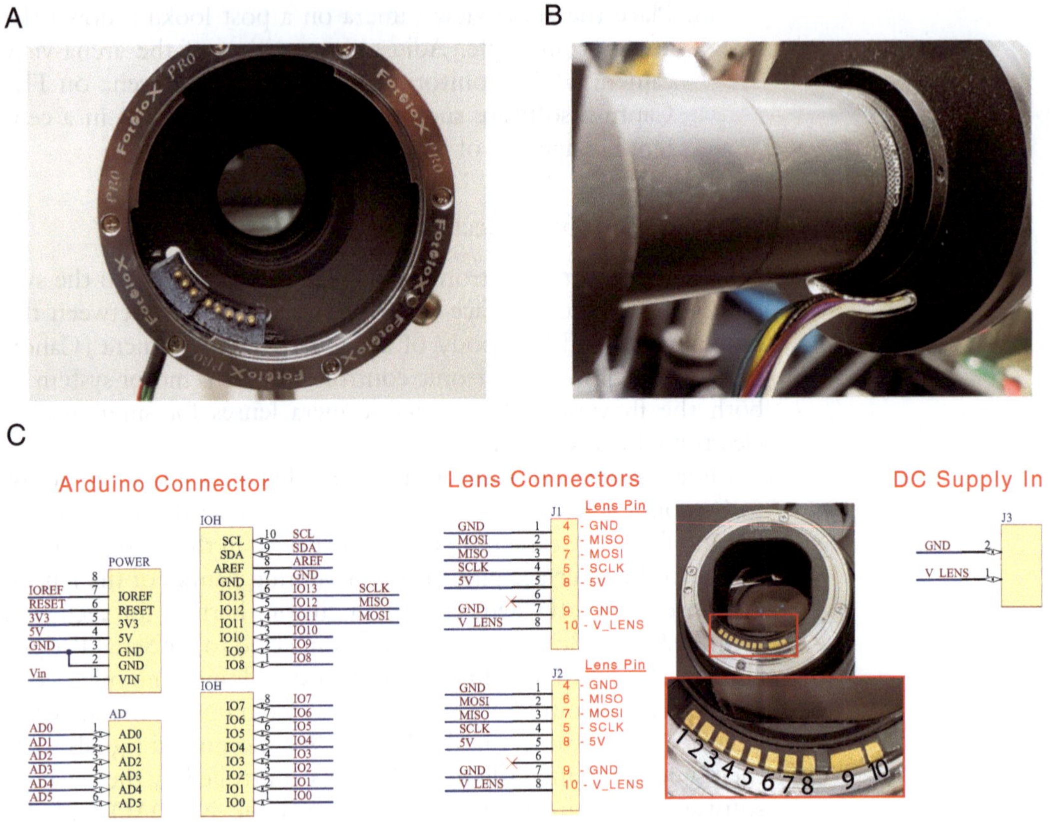

Fig. 12.9 Schematic of electronic lens control mechanism. (**a**) Front side of EF to C-mount lens adapter (to which fly- and fluo-view camera lenses are attached) with custom interface connector for driving lens via Arduino. (**b**) Backside of lens adapter shown in (**a**). (**c**) Left, labeled Arduino Uno shield pin connections. Middle, identical lens connector ports for fly-view and fluo-view camera lens control. Top image shows the backside of lens and pin interface used for electronic lens control. Bottom inset view shows labeled pin interface corresponding to the lens pin labels shown in the lens connector schematic. Right, input port for DC power supply

thresholding of the laser dot can be adjusted to be as tight a spot as possible.

b. F2 initializes the raster scan from min to max X and Y mirror angles for a defined step size. In our experiments, we set the angle range to be [−9.0 9.0] degrees for a step size of 0.1°.

c. The output is saved in a space delimited text file of the following format per row [x-angle y-angle x-image-center y-image-center]. This text file is read into the main tracking program for arena-view tracking.

9. Converting Fly-View Pixel Coordinates to Real-World Coordinates

Once the focal fly's head is in view of the fly-view camera running in parallel, control of the galvanometer mirror system switches to receive positional updates from the fly-view camera tracker at 1000 Hz. The positional deviation of the fly's head from the center of the fly-view image is calculated in each frame, and the galvanometer mirrors are then rotated to compensate for the deviation. We use a resolution test target (R1L3S10P, Thorlabs) to determine the scaling coefficient to convert fly-view pixels to rotation angles for the galvanometer mirrors.

3.3 System Operation

1. Hardware Configuration

Our hardware configuration includes a Dell Precision T7810 Workstation (Dual 10-core Intel Xeon v3 at 2.8 Ghz), 128 GB RAM, 2x512 GB SSD drives (first for OS, second for experimental video recordings), Teledyne DALSA XCelera PX4 full frame-grabber PCIe card, NI PCIe-6351 DAQ, Point Grey 2-port USB3 PCIe card, Photometrics Prime PCIe card, Arduino Uno Rev3, and a NI MyRIO.

2. Prerequisite Software

The Flyception tracking code is designed to run on a PC running Microsoft Windows 7/8 or 10 64-bit. The code is freely available for download from the public repository https://github.com/dgrover/flyception2. Microsoft Visual Studio 2013–2019 (at minimum the community edition with Desktop development with C++ enabled) is required to compile and run the code. Additionally, CMake, OpenCV (compiled with WITH_OPENMP flag enabled), Point Grey Flycapture SDK (64-bit), Teledyne DALSA Sapera LT SDK, PuTTY, NI-DAQmx, NI-LabVIEW, and Arduino are also required. The Laser Quantum Remote App is required for excitation laser control. Finally, the Photometrics Prime Camera SDK and μManager are required for fluorescence camera control and recordings.

The Flyception2 source code has the following folder hierarchy:

- Flyception2 – root folder
- Flyception2\myRIO – source code to be installed on the NI myRIO FPGA for camera and flash triggering
- Flyception2\canon_lens_control – source code to be installed on the Arduino Uno for lens control of fly-view and fluo-view cameras
- Flyception2\flybam – source code for the tracking program
- Flyception2\calibration – scripts to calibrate the arena-view camera to convert pixel coordinates to real-world coordinates
- Flyception2\ccf – camera configuration files for Point Grey Gazelle camera (fly-view) and Photometrics Prime (fluo-view)
- Flyception2\offline – arena-view k-means offline tracking and behavior labeling program

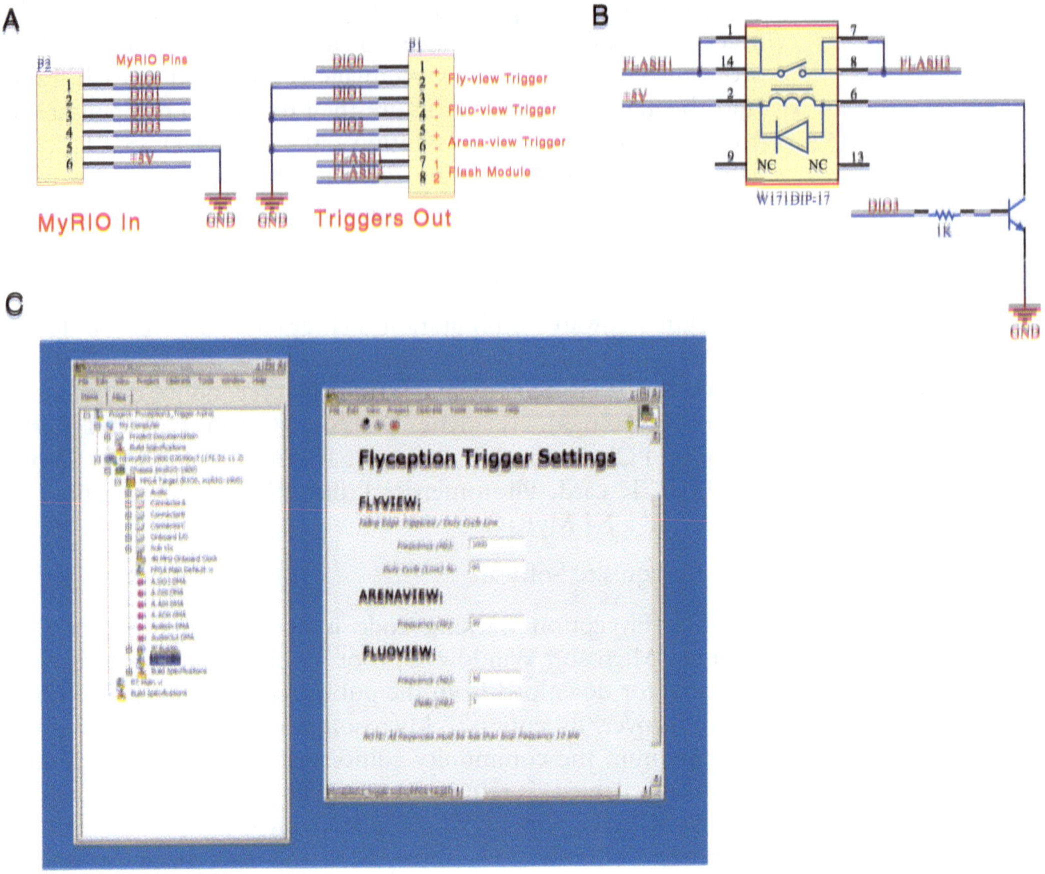

Fig. 12.10 Schematic of camera triggering and flash circuitry. (**a**) Left, input connection port from National Instruments MyRIO embedded device labeled corresponding to its output lines. Right, output port to camera GPIO trigger pins and flash module with corresponding labels. (**b**) Schematic of flash module that takes as input a digital signal from the MyRIO and switches a read relay to fire the flash speedlite. (**c**) Flyception trigger LabVIEW application control panel to set parameters for the various cameras

3. Triggering Cameras with LabVIEW and MyRIO

Recording of the three camera images (arena-view, fly-view, and fluo-view) is manually triggered to start at roughly the same time. The exposure timing and frame rate of the arena-view, fly-view, and fluo-view image acquisition are controlled by external triggers generated using FPGA hardware-based clock timers on a MyRIO device (National Instruments) (Fig. 12.10a). Furthermore, to generate a visually detectable cue to align the frames of the three cameras temporally, a flash (TT560, Neewer) is used (Fig. 12.10b). The MyRIO timer also triggers the firing of the flash, set to coincide with the trigger onset of the arena-view, fly--view, and fluo-view cameras. The arena-view camera is set to capture 512 × 512 px images at 50 Hz with a 3 ms exposure; the fly-view camera captures 240 × 240 px images at 1000 Hz with a 0.9 ms exposure; and the fluo-view camera captures 2048 × 256 px images at 50 Hz with an 18 ms exposure.

The Flyception2_trigger.lvproj file (located in the Flyception2 \myRIO\Flyception2_trigger folder) opens the project solution in LabVIEW. Navigate to MyRIO device -> Chassis -> FPGA Target and open the Trigger.vi file. Compiling (to FPGA module) and executing this vi file will start sending triggers to the various devices. The project also contains a control panel to set parameters for the various cameras (Fig. 12.10c), for instance, fly-view frequency and duty cycle for falling edge triggering and exposure in bulb shutter mode, arena-view frequency (exposure set in Flyception tracking system code), and fluo-view frequency (exposure set in μManager; see below) and delay (as we have determined there to be a time difference between the Photometrics Prime camera receiving the trigger and opening the shutter) to align fluo-view camera exposure with those of arena-view and fly-view cameras.

4. Configuring Fly-View Camera Mode

   The settings below initialize the fly-view camera (using serial communication) with an ROI of 240 × 240 pixels centered on the sensor, with falling edge external trigger and bulb shutter (simultaneous exposure/readback) mode. The MyRIO LabVIEW code will then trigger the camera at the preset frequency (default: 1000 Hz) and duty cycle (default: 90% low).

   1. Open PuTTY and connect to the camera via Serial COM port assigned in Sapera Configuration utility.
   2. Type dump and return (Device Information); this might need to be done twice if the first dump command returns nothing.
   3. Type setroi 0 424 240 and return.
   4. Type w 240 and return.
   5. Type ox 904 and return.
   6. Type trsrc GPIO0 and return (trigger source).
   7. Type trm bulb and return (bulb mode).
   8. Type tra fe and return (falling edge).

5. Compiling Flyception2 Code

   a. Navigate to the Flyception2\flybam folder and run the flybam.sln solution file.
   b. Verify that the x64\RELEASE solution platform is selected.
   c. The provided project file includes the path to all prerequisite libraries; however, if custom folders were used, the library paths need to be changed here.
   d. Navigate to Project -> flybam properties -> configuration properties -> vc++ directories.
   e. Verify the following directories are included:

Include Directories

C:\Program Files\Point Grey Research\FlyCapture2\include

[OpenCV root folder]\opencv-3.x.x\build\install\include (replace 3.x.x with installed version)

C:\Program Files x86\National Instruments\Shared\ExternalCompilerSupport\C\include

C:\Program Files\Teledyne DALSA\Sapera\Examples\Classes\Common

C:\Program Files\Teledyne DALSA\Sapera\Classes\Basic

C:\Program Files\Teledyne DALSA\Sapera\Include

Library Directories

C:\Program Files\Point Grey Research\FlyCapture2\lib64

[OpenCV root folder]\opencv-3.x.x\build\install\x64\vcXX\lib (replace 3.x.x with installed version)

C:\Program Files x86\NationalInstruments\Shared\ExternalCompilerSupport\C\lib64\msvc

C:\Program Files\Teledyne DALSA\Sapera\Lib\Win64

f. Under Linker -> Input, verify the following .lib files are included. Note, for opencv libraries, replace 3.x.x with specific compiled version.

nidaqmx.lib

FlyCapture2.lib

FlyCapture2GUI.lib

opencv_core3xx.lib

opencv_highgui3xx.lib

opencv_video3xx.lib

opencv_imgproc3xx.lib

opencv_imgcodecs3xx.lib

opencv_calib3d3xx.lib

SapClassBasic.lib

g. With these folder paths included, the Flyception2 software is ready to be compiled and executed.

6. Flyception Tracking

   When the Flyception2 code is executed, the arena-view and fly-view cameras are initialized (Fig. 12.11). The software automatically starts tracking flies in the arena-view video and turns

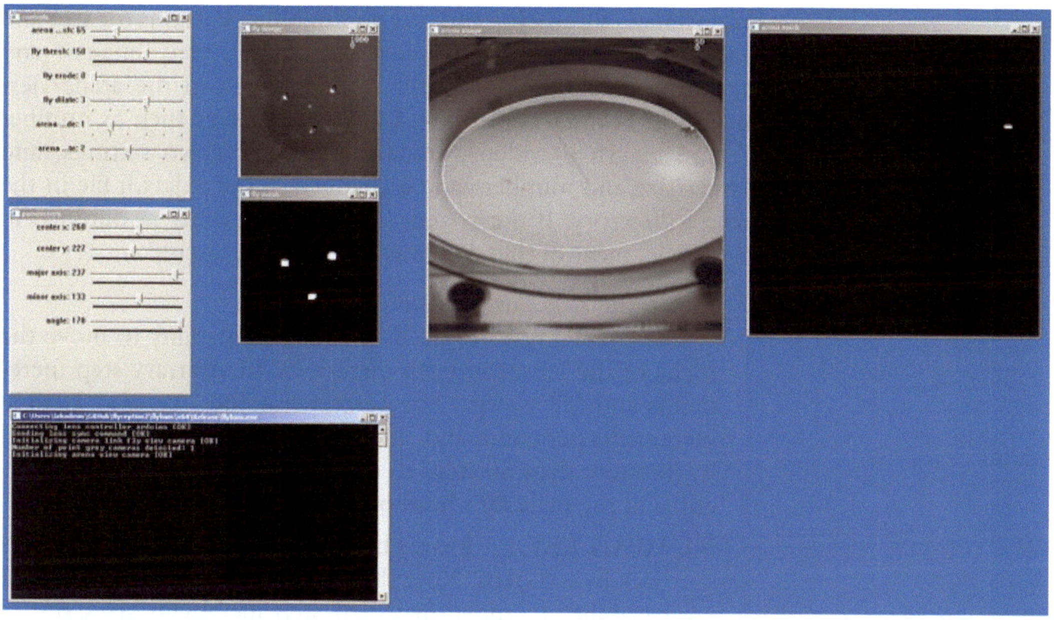

Fig. 12.11 Flyception tracking software. Default configuration includes a live view of arena-view camera with tracked flies and background subtracted masked image of fly in arena-view at 50 Hz. Fly-view (at 1 kHz) with masked image of the retroreflective marker being segmented out from the image. Control panels for customization of variety of image processing parameters for both camera views

the galvanometer mirrors to follow the first fly tracked. The following keyboard shortcuts are used for controlling the software:

F1 – initializes fly-view tracking overriding arena-view tracking updates.

F2 – starts/stops recording of arena-view and fly-view images and tracking coordinates. If the fluo-view camera is initialized via μManager (see next section), recording of fluo-view videos will also begin.

F3 – recomputes arena-view background image.

F4 – fires flash.

UP, DOWN, LEFT, RIGHT keys – manual control of galvanometer mirror assembly overriding fly-view or arena-view tracking updates.

HOME – resets galvo mirror position to the center of arena.

TAB – if more than one fly is present in the arena, tab cycles through each fly as the focal one for fly-view tracking.

ESC – exits the program.

In addition to the keyboard shortcuts, the software also allows the user to tweak various tracking parameters via a control panel.

For instance, thresholds and morphological operations for size and shape of detected flies in arena-view and fly-view images, and the position of the edge outline of the arena in the arena-view image can be user controlled. Other parameters such as number of flies, length of videos, arena dimensions, galvo settings, and marker size and number can be changed in the stdafx.h file in the project directory. If the system is replicated exactly as outlined in this article, most of these settings need not be altered.

7. Electronic Z-Focus Adjustment

   The tracking software included the capability to move the focus of the lenses in both directions, in arbitrary step increments and full sweeps. For our experiments, we used 10 step increments per command, and lens focus position was adjusted by no more than 50 steps to compensate for the height differential in the male fly's head position due to copulation.

   NUMPAD KEY 1 – Moves both fly-view and fluo-view lenses back by ten steps. Note that the Arduino code provides the capability to move in varied step increment sizes of 10^x where $[x = 0, 1, 2, 3]$ and the key bindings can be changed accordingly.

   NUMPAD KEY 2 – Moves both fly-view and fluo-view lenses forward by ten steps. Note that the Arduino code provides the capability to move in varied step increment sizes of 10^x where $[x = 0, 1, 2, 3]$ and the key bindings can be changed accordingly.

   NUMPAD KEY 4 – Sweeps both fly-view and fluo-view lenses to their shortest focal distance.

   NUMPAD KEY 5 – Sweeps both fly-view and fluo-view lenses to their longest focal distance.

The autofocus capability is achieved by continuously monitoring the variance of Laplacian of the fly-view image and compensating for the drop of variance by electronically moving the focal plane of the lens. Because the variance of Laplacian does not provide the direction of out of focus, we move the lens focus in a positive or negative direction randomly. If a lens motion decreases the variance of Laplacian, that means that the lenses have moved in the wrong direction; in this case, we move the lenses two steps in the opposite direction. If, on the other hand, a lens motion increases the variance of Laplacian, that means that the lenses have moved in the correct direction; thus, we keep the lenses move in the same direction until the variance of Laplacian surpasses the previous baseline. We have tested different step sizes of the lens motion, sampling rate of the variance of Laplacian, and thresholds for detecting the drop of the variance of Laplacian and found that a step size of 1, sampling rate of 50 Hz, and variance threshold of 5 resulted in a robust tracking of motion along the z-axis.

F6 – records current variance of Laplacian score as base focus measure.

F7 – starts/stops lens autofocus to maintain base focus measure.

8. Excitation Laser Control

We run two instances of the Laser Quantum Remote App Laser Control software, one for each of the two Gem lasers (473 nm and 561 nm) used in our system. Each laser controller (Smd12) is connected to the PC via serial cables, and the COM port assignments are set in the Remote App software settings for each instance. In our experiments, we use the lasers in power mode, and the power levels are set in mW. We use a laser power meter (Thorlabs, PM100D) to determine the appropriate power levels of each laser on the arena surface (set to average power densities of 4.85 mW cm^{-2} for the 473 nm laser and 3.8–4.77 mW cm^{-2} for the 561 nm laser).

9. Fluorescence Recording

Fluorescence images are captured with a Photometrics Prime CMOS camera using μManager [18]. We recommend enabling the camera's metadata option, to support accurate timestamp recording for analysis purposes. This can be accomplished by accessing the Device Property Browser under the Tools submenu (Fig. 12.12a). Set the Prime-MetadataEnabled flag to YES. If desired, this option can be added as a Preset to Configuration settings for easy access. This flag will need to be set each time μManager is started.

Additional camera settings are controlled in the Configuration settings shown in the screen capture below with the following parameters:

- Trigger type: Edge trigger (default: 50 Hz)
- Fan speed: Low (The camera can be liquid cooled, in which case, the fan can be turned off)
- Exposure: 18 ms
- Binning: 1 × 1
- Image size: 2048 × 256 × 2

F2 key needs to be bound to multi-D acquisition function so that the image acquisition starts roughly at the same time as fly-view and arena-view. Number of images to capture is set to 5000 in Multi-D Acquisition setting.

Start the camera by hitting Live button in the main console (Fig. 12.12b), and then start recording with F2 shortcut key. Images will be saved in TIFF format, and metadata will be saved in TXT format, with a user-specified filename prefix.

A

B

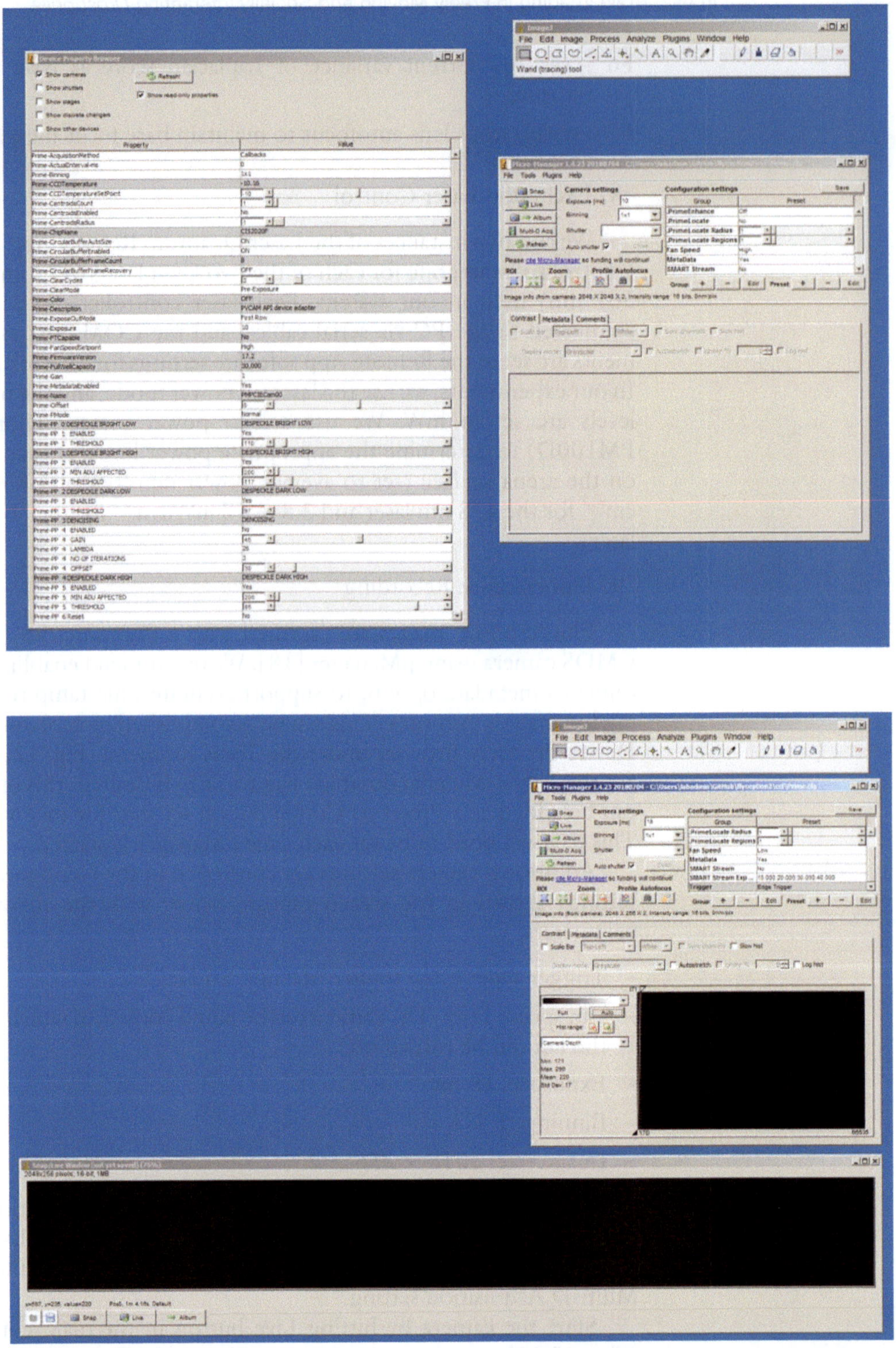

Fig. 12.12 Photometrics Prime CMOS camera fluorescence recordings with μManager. (**a**) μManager device property manager view to set metadata logging, trigger type, fan speed, exposure, binning, and image size parameters. These parameters need to be initialized each time μManager is started. (**b**) Live view of the fluo-view camera after initialization. The F2 key will then start recording videos from both Flyception tracking software (arena-view and fly-view cameras) and μManager (fluo-view camera)

10. Experiment Recordings

Each recording typically lasts 100 s, though this is customizable in the source code. Arena-view and fly-view sequences are saved in an uncompressed video format known as fly movie format (fmf) [19], and fluo-view data are saved in a TIFF format. Included in each recording are camera timestamp log files, an arena-view background image, and arena- and fly-view output trajectory files.

The output of the system after a successful recording session is as follows:

- Arena-view video (default length 100 s at 50 Hz) in fmf format (essentially an uncompressed monochrome avi format)
- Fly-view video (default length 100 s at 1000 Hz) in fmf format (essentially an uncompressed monochrome avi format)
- Arena-view and fly-view video log files to ensure cameras are functioning at preset frame rates
- Trajectory file, space delimited, with the following parameters per frame at 1000 Hz [*Frame Number, Arena x-coordinate, Arena y-coordinate, Fly-view head center x-coordinate, Fly-view head center y-coordinate, Galvo x-angle, Galvo y-angle*]

3.4 Analyzing the Data

1. Ratiometric Fluorescence Quantification

We measure fluorescence intensity by creating a mask of the labeled neurons for each fluo-view video frame in the calcium-independent (tdTomato) channel and ratiometrically comparing the intensity values of the corresponding pixels from both GCaMP and tdTomato channels in that masked region. We performed post-acquisition image analyses using R (CRAN), with its image processing software packages EBImage [20], dipr (https://github.com/tkatsuki/dipr), and Flyception2R (https://github.com/tkatsuki/Flyception2R). The function *Flyception2R()* included in the Flyception2R package can process and analyze the data acquired with Flyception2 semi-automatically. The minimal input needed to run the *Flyception2R()* function is the location of the directory that contains the Flyception2 video files, although you would normally want to specify which frames of the video to be analyzed; otherwise, it will analyze the entire video and likely run out of memory. It is recommended to use a computer with more than 64 GB of RAM for analysis. The sample dataset from a *P1a-Gal4, UAS-GCaMP6s,* and *UAS-myr-tdTomato* courtship experiment is available for download from the Flyception2R website above. Once the R environment and necessary packages are installed, run the script below.

```
> library(Flyception2R)
> dir <- "CHANGE_THIS/P1-Gal4_UAS-GCaMP6s_tdTomato_5/P1-Ga-
14_UAS-GCaMP6s_tdTomato_5/"
> Flyception2R(dir=dir, FOI=c(2000, 3446), window_size=list(c
(70, 20)), window_offset=list(c(5, -5)), flash=1)
```

Because the marker position, and consequently the center of the image, relative to the fly brain varies from fly to fly, the position of the window (ROI) for segmenting neurons needs to be manually adjusted. Flyception2R will ask the user if the window_size and window_offset are acceptable. Check the file _redwindow.tif, and adjust the window size or offset so that the neurons of your interest are in the window. To move the window up, decrease the y-value. To move the window left, decrease the x-value. Multiple ROIs can also be created. Check _frgcombined_goodfr20_normalized.tif that the neurons are correctly segmented.

Figure 12.13a shows representative results from ratiometric calcium imaging of P1 neurons [21] during courtship and mating. The analysis pipeline generates several plots and image files besides the video file such as fly trajectories, dF/F curves, and fly-fly angle and distance if applicable (Fig. 12.13c–e). One can observe in pseudo-color that the activity of P1 neurons increases as the male exhibited courtship behavior, approached, and contacted the female (Fig. 12.13a, first and second rows). Further, Z-focus electronic control allowed for maintaining focus on the male fly as he mounted the female during copulation (Fig. 12.13a, third row).

Note that even though tracking is robust, not all video frames recorded are analyzable. *Flyception2R()* omits those data points that do not meet a set of criteria (e.g., motion and angle change per frame, blurriness) from analysis. Therefore, if the angle of the head is too steep (Fig. 12.13b), or if the retroreflective beads are obstructed by the wings of the other flies, or if the fly jumps, then those frames will not be quantified and marked as such.

2. Behavioral Metrics Acquisition

The arena-view camera provides a full view of the behavioral arena with a resolution of 512 × 512 px at 50 Hz. The system tracks each fly in real time, concurrently saving the uncompressed raw arena-view camera video and a trajectory file containing the real-time xy-positional data and pixel size of each fly in the arena. Proximity or contact between flies in arena-view could lead to merging of their detected contours and consequently an interchanging of identity. It is important to note that the effect of identity switching in arena-view has no effect on tracking the head of the focal fly in fly-view. Nevertheless, to aid studies of social behavior that require accurate tracking of the behavior of interacting flies (and to better correlate those behaviors with imaging of brain activity), we provide a k-means clustering-based offline arena-view

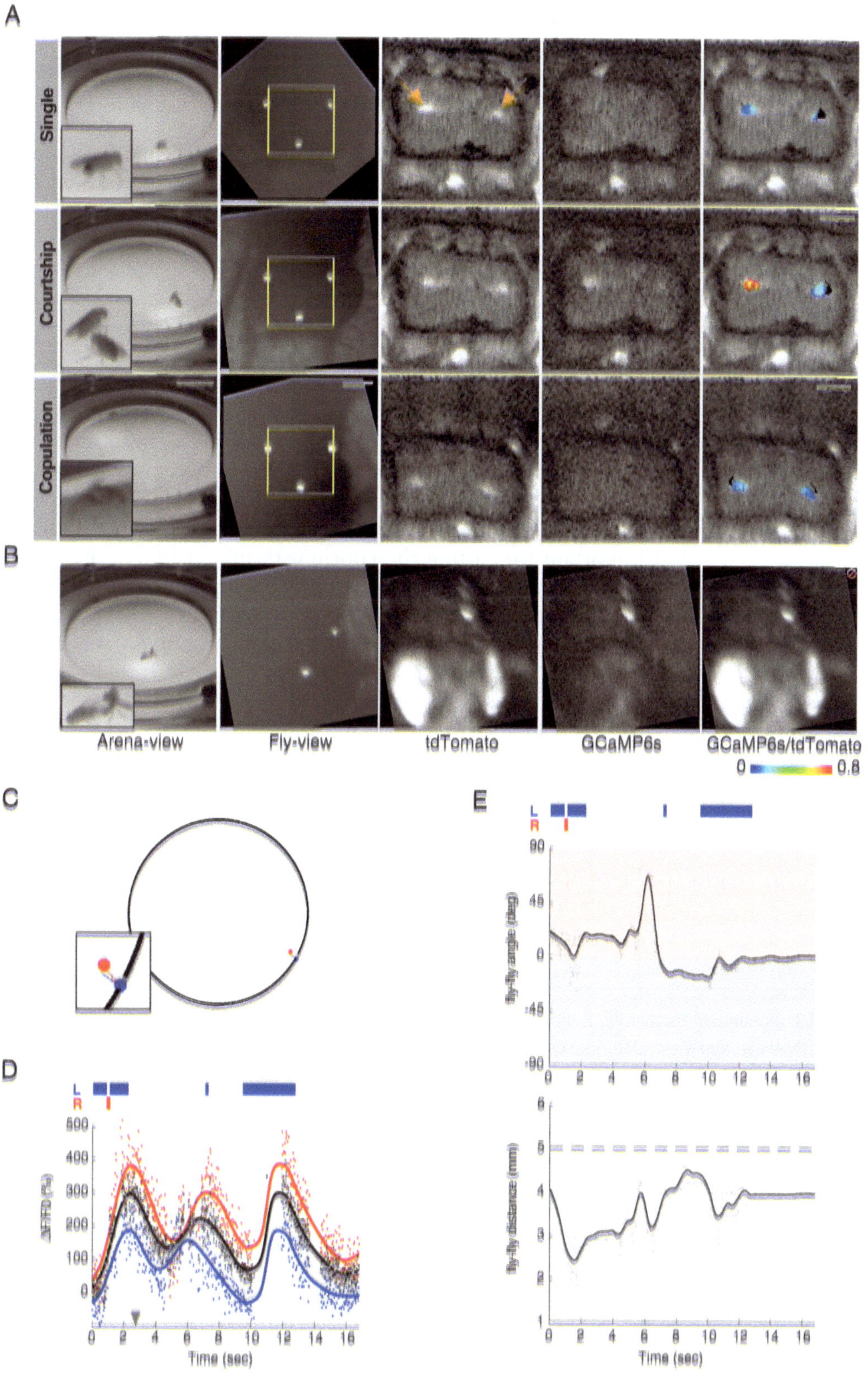

Fig. 12.13 Data analysis outputs with R. (**a**) A virgin male fly with *P1a-Gal4*, *UAS-GCaMP6s*, and *UAS-myr-tdTomato* transgenes courting and mating with a female fly. (**a**) Top, frame without activity in P1 neurons in a single fly in the arena. Middle, the same male fly courting a female fly. Bottom, the male fly copulating with the

video tracker [22] that can separate merged contours and maintain accurate tracking, even during copulation events (Fig. 12.13c). As part of this offline tracking software, we also provide the user with the ability to manually label, on a frame-by-frame basis, specific behavioral events of interest. For example, in our study, we hand-labeled left- and right-wing extension events using this strategy (Fig. 12.13d, e).

4 Notes

4.1 Fly Brain Window Surgery

One of the critical factors for successful surgery is the stability of the fly head during the procedure and the sharpness of the forceps. As described in the methods, one should tape the head first, even if tempted to tape the abdomen first. This may seem trivial, but if the abdomen is immobilized first, then securing the head in the correct angle will be harder. Sharpness and straightness of the forceps are important for cutting the cuticle with minimal force. Because the finest tip commercially available is 25 μm wide, we use forceps resharpened to 5 μm wide. Caution should be taken when handling and cleaning the ultrafine forceps as the 5 μm tip is easily bent.

4.2 Customizing Flyception Light Paths for Fluorescence Sensors of Other Wavelengths

While our use case involved a dual-excitation laser light path (473 nm and 561 nm) and emission light splitting optics to enable GCaMP/tdTomato ratiometric imaging in the fly brain, this configuration can be easily modified to enable imaging of fluorescence sensors of other wavelengths. Further, imaging biosensors (e.g., FRET-based) which necessitate pulsed dual-wavelength excitation lights and a single emission wavelength channel can also be enabled by placing out-of-phase optical choppers in the laser light paths, synchronized with the fluo-view camera exposure.

Fig. 12.13 (continued) female fly. From left to right: arena-view with an inset showing a zoomed-in view of the flies, fly-view, fluo-view tdTomato channel, fluo-view GCaMP6s channel, and fluo-view tdTomato channel with a pseudo-color $F_{ratio} = F_{GCaMP6s}/F_{tdTomato}$ representation of the GCaMP6s signal in P1 neurons. Orange arrows indicate P1 neurons. (**b**) A frame in which the male fly attempted copulation and the head of fly tilted so steeply that one of the retroreflective beads was undetectable. This frame did not qualify for further analysis and was excluded from quantification, which is indicated as the red stop icon on the top right corner of the image. (**c**) Arena-view tracking of flies in the arena. Blue dot represents tracked centroid of the male fly, and red dot represents tracked centroid of the female fly. The blue line indicates the head direction of the male fly (blue), and the red line indicates the direction of the female fly (red) relative to the male fly (visible in inset). (**d**) $dF_{\mathrm{ratio}}/F_{\mathrm{ratio}}$ raw values for regions of interest encompassing both hemispheres of P1 neurons (black scatter), left hemisphere (blue scatter), and right hemisphere (red scatter), with corresponding LOESS (local polynomial regression) curve fits. Blue and red boxes indicate an extension of the left-wing and right-wing, respectively. The gray arrowhead indicates the time point at which the images in (**a**) middle row were captured. (**e**) Top, angular difference of male fly's heading relative to female fly. A positive (red region) or a negative (blue region) angular difference indicates the female fly is on the right-hand or left-hand side of the male, respectively. Bottom, Euclidean distance between the flies. Dashed line at 5 mm indicates the upper distance threshold for interaction and contact events

5 Conclusion

We believe that this technology is the first step toward a much-needed shift in methodological paradigms of brain imaging and behavioral neuroscience, to make possible the study of neural activity in naturally behaving animals. Such methods will radically change the way we study brain functions in the context of behavior, providing researchers with the ability to address previously unexplored questions.

To demonstrate the power of the system, we chose to study mating behavior that has remained an experimentally intractable (under in vivo imaging) yet ethologically critical paradigm and to allow new aspects of behavioral physiology, not accessible previously, to be revealed by this technical innovation. However, this technology should be compatible with a wide range of behavioral assays except free flight, involving multiple sensory modalities including touch, hearing, vision, taste, smell, and temperature in studies of complex behaviors, both social (e.g., aggression) and cognitive (e.g., sleep) in nature. Moreover, this method can extend the imaging duration in a single individual anywhere from seconds to weeks, much longer than what is currently possible, thereby enabling studies of long-term plasticity and aging of the nervous system. In sum, the continued development of such methods is vital to shining a spotlight onto the neural instantiation of naturally occurring complex behavior.

Acknowledgments

This work was supported by awards from the Air Force Office of Scientific Research (FA9550-14-1-0211, FA9550-16-1-0229, and FA9550-19-1-0280), Mathers Foundation (20154167), National Science Foundation (1212778), and Kavli Institute for Brain and Mind (Innovative Research Grant 2016 and 2019).

References

1. Alivisatos AP, Chun M, Church GM, Greenspan RJ, Roukes ML, Yuste R (2012) The brain activity map project and the challenge of functional connectomics. Neuron. https://doi.org/10.1016/j.neuron.2012.06.006
2. Flusberg BA et al (2008) High-speed, miniaturized fluorescence microscopy in freely moving mice. Nat Methods. https://doi.org/10.1038/nmeth.1256
3. Symvoulidis P et al (2017) NeuBtracker – imaging neurobehavioral dynamics in freely behaving fish. Nat Methods. https://doi.org/10.1038/nmeth.4459
4. Zheng M, Cao P, Yang J, Xu SXZ, Feng Z (2012) Calcium imaging of multiple neurons in freely behaving C. elegans. J Neurosci Methods. https://doi.org/10.1016/j.jneumeth.2012.01.002
5. Nguyen JP et al (2016) Whole-brain calcium imaging with cellular resolution in freely behaving Caenorhabditis elegans. Proc Natl

Acad Sci USA. https://doi.org/10.1073/pnas.1507110112
6. Reiff DF et al (2005) In vivo performance of genetically encoded indicators of neural activity in flies. J Neurosci. https://doi.org/10.1523/JNEUROSCI.4900-04.2005
7. Fiala A et al (2002) Genetically expressed cameleon in Drosophila melanogaster is used to visualize olfactory information in projection neurons. Curr Biol. https://doi.org/10.1016/S0960-9822(02)01239-3
8. Seelig JD, Jayaraman V (2015) Neural dynamics for landmark orientation and angular path integration. Nature. https://doi.org/10.1038/nature14446
9. Aimon S et al (2019) Fast near-whole-brain imaging in adult drosophila during responses to stimuli and behavior. PLoS Biol. https://doi.org/10.1371/journal.pbio.2006732
10. Seelig JD et al (2010) Two-photon calcium imaging from head-fixed Drosophila during optomotor walking behavior. Nat Methods. https://doi.org/10.1038/nmeth.1468
11. Van Swinderen B, Greenspan RJ (2003) Salience modulates 20-30 Hz brain activity in Drosophila. Nat Neurosci. https://doi.org/10.1038/nn1054
12. Kohatsu S, Yamamoto D (2015) Visually induced initiation of Drosophila innate courtship-like following pursuit is mediated by central excitatory state. Nat Commun. https://doi.org/10.1038/ncomms7457
13. Kohatsu S, Koganezawa M, Yamamoto D (2011) Female contact activates male-specific interneurons that trigger stereotypic courtship behavior in Drosophila. Neuron. https://doi.org/10.1016/j.neuron.2010.12.017
14. Clowney EJ, Iguchi S, Bussell JJ, Scheer E, Ruta V (2015) Multimodal chemosensory circuits controlling male courtship in Drosophila. Neuron. https://doi.org/10.1016/j.neuron.2015.07.025
15. Grover D, Katsuki T, Greenspan RJ (2016) Flyception: imaging brain activity in freely walking fruit flies. Nat Methods. https://doi.org/10.1038/nmeth.3866
16. Grover D, Katsuki T, Li J, Dawkins TJ, Greenspan RJ (2020) Imaging brain activity during complex social behaviors in Drosophila with Flyception2. Nat Commun. https://doi.org/10.1038/s41467-020-14487-7
17. Mischiati M, Lin HT, Herold P, Imler E, Olberg R, Leonardo A (2015) Internal models direct dragonfly interception steering. Nature. https://doi.org/10.1038/nature14045
18. Edelstein AD, Tsuchida MA, Amodaj N, Pinkard H, Vale RD, Stuurman N (2014) Advanced methods of microscope control using μManager software. J Biol Methods. https://doi.org/10.14440/jbm.2014.36
19. Straw AD, Dickinson MH (2009) Motmot, an open-source toolkit for realtime video acquisition and analysis. Source Code Biol Med. https://doi.org/10.1186/1751-0473-4-5
20. Pau G, Fuchs F, Sklyar O, Boutros M, Huber W (2010) EBImage-an R package for image processing with applications to cellular phenotypes. Bioinformatics. https://doi.org/10.1093/bioinformatics/btq046
21. Hoopfer ED, Jung Y, Inagaki HK, Rubin GM, Anderson DJ (2015) P1 interneurons promote a persistent internal state that enhances intermale aggression in Drosophila. Elife. https://doi.org/10.7554/eLife.11346
22. Arthur D, Vassilvitskii S (2007) K-means++: the advantages of careful seeding. In: Proceedings of the annual ACM-SIAM symposium on discrete algorithms

Chapter 13

Drosophila as a Model to Explore Individuality

Kuo-Ting Tsai and Ya-Hui Chou

Abstract

Drosophila is an attractive animal model for exploring the fundamental relationships between genetics, neural circuitry, and behavior. Like mammals, *Drosophila* exhibit individuality that is reflected by the varied responses of single animals to the same environmental cues. Up to now, only a few behavioral paradigms have been developed to test the mechanisms underlying individuality in single flies behaving alone. Here, we describe the application of a *Drosophila* olfactory treadmill system to assess the behavioral kinetics of individual flies upon exposure to a well-controlled odor plume. We provide step-by-step instructions for building a camera-mode treadmill system, and we also present methods to extract four key features of the fly's ambulatory response: orientation, distance, straightness, and curl. Furthermore, potential factors that may influence the behavioral readouts of individual flies are discussed.

Key words *Drosophila*, Treadmill system, Individuality, Internal status, Olfaction, Virtual reality, Neural circuit, Walking trajectories

1 Introduction: The Individuality of Flies

Social animals may behave differently when exposed to the same environmental conditions, and this inter-individual variation in behavioral response can be called individuality. The causes of such differential behavioral outcomes may include multiple factors, including variations in circuit wiring, brain structure, experiences, sex, and age [1–5]. However, the influence of these factors on individuality is difficult to study because of its complex nature and the requirement that experiments be performed on intact behaving animals. Fortunately, the rich genetic tools [6–11], fully mapped hemibrain connectome [12, 13], and relatively uncomplicated brain circuits [14–18] of *Drosophila* make it an excellent animal model for exploring the neural mechanisms of individuality at molecular, cellular, circuit, and behavioral levels. In addition, the recent finding that *Drosophila* can

Daisuke Yamamoto (ed.), *Behavioral Neurogenetics*, Neuromethods, vol. 181, https://doi.org/10.1007/978-1-0716-2321-3_13,

transmit mating preferences culturally across generations [19] and their well-characterized group behaviors [20–22] may provide new avenues for studies on *Drosophila* individuality.

1.1 The Possible Neural Basis of Individuality

Factors such as developmental stochasticity, neural variability, internal status, neuromodulation, and previous experience may collaboratively influence how an individual animal deciphers and responds to environmental cues. For example, the numbers and connections of *Drosophila* dorsal cluster neurons (DCN) are stochastically varied across individuals on the same isogenic background [23]. Such developmental stochasticity causes different degrees of left/right asymmetry in the visual circuit, which then produces differential objective orientation responses. In addition to circuit connections, neural variations can be observed at the level of single neurons [24–31]. The morphologies, numbers, electrophysiological properties, and ion channel compositions can vary in the same identifiable neurons across individuals, and these properties will clearly affect circuit outputs. On top of these structural differences, it is well established that internal status [32–35], neuromodulation [36–38], and previous experiences [39–41] will also greatly influence the behavioral outputs of individual animals.

1.2 Behavior Paradigms Used to Study Drosophila Individuality

Many behavioral assessment tools have been developed to monitor and quantify the behavioral outputs associated with different fly sensory modalities, and several of these assays are described in other chapters of this book. Two methods have recently proven to be particularly useful to address individuality, and in this chapter, we will provide a detailed introduction to another promising method, the treadmill system. First, Buridan's paradigm comprises a round arena contoured by water moat, which is further surrounded by a high LED wall (Fig. 13.1a). Two black vertical strips are located opposite each other on the wall as visual landmarks. Since *Drosophila* has a strong phototaxis response, flies in the arena tend to walk toward one of the strips. However, the water-filled moat hinders the fly from moving forward, and the fly turns back to another strip. All the way along, the fly walks back and forth between the two black strips [23, 42]. This behavior was used to identify individual flies with distinct walking patterns, which then were shown to depend on the degree of asymmetry in DCN circuits (also see Sect. 1.1) [23]. The second behavioral paradigm is a custom-designed odor-preference experimental apparatus [36]. This apparatus is composed of multiple parallel long arenas and a system to precisely deliver two different odor mixtures at the two ends of each arena (Fig. 13.1b); the odors are evacuated at the middle of each arena. In addition, the design allows to create a series of two odor streams along the arena [36]. A cohort of flies (one fly per arena) can be monitored simultaneously to assess walking patterns and tendencies. This assay has been useful for measuring the

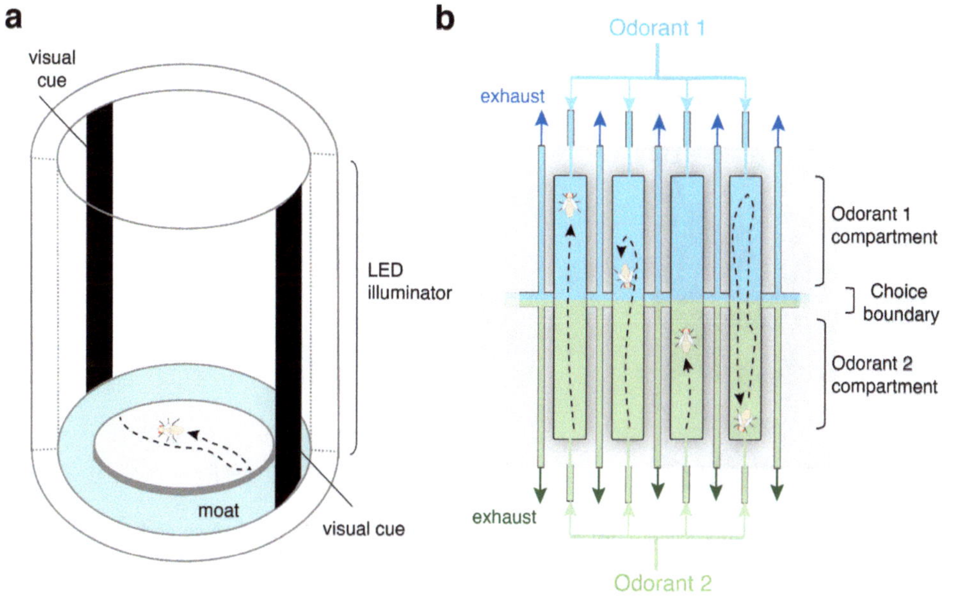

Fig. 13.1 The behavior paradigms for assessing *Drosophila* individuality. Schematic diagrams show Buridan's paradigm (**a**) and the odor-preference experimental apparatus describe in [36] (**b**)

contributions of neuromodulation to individual walking patterns [36]. The *Drosophila* treadmill system that was originally designed by Erich Buchner [43] will be the main focus of this method chapter (see below).

1.3 The Drosophila *Olfactory System*

In this chapter, we will discuss a method to monitor and distinguish the differential walking strategies of individual flies toward a food odor. The outcomes of this method rely heavily on the individual fly's olfactory system [44]. Each *Drosophila* has two pairs of sensory organs, antennae and maxillary palps, which house the cell bodies of olfactory receptor neurons (ORNs; ~ 1400 per brain hemisphere) [45] (Fig. 13.2a, c). ORNs project their axons to the first olfactory information processing center, the antennal lobe (Fig. 13.2b, c), which is equivalent to the olfactory bulb in mammals. The axons of ORNs expressing the same odorant receptor converge in a specific glomerulus in the antennal lobe, and most cross the brain midline to innervate a paired glomerulus in the contralateral antennal lobe [46, 47]. Projection neurons (PNs, ~ 135 per brain hemisphere [48], Fig. 13.2c) cognate to each class of ORNs also target their dendrites to the same glomerulus and form synapses with the ORN axons [49]. After receiving olfactory inputs, PNs relay the information to higher brain centers, including the mushroom body and lateral horn [50]. In this way, the ORN-PN-glomerulus structure creates a one-to-one olfactory information channel. Yet, the targeting and synapse of PN axons to calyx (dendrites of MB neurons) are random [51]. In addition, the same

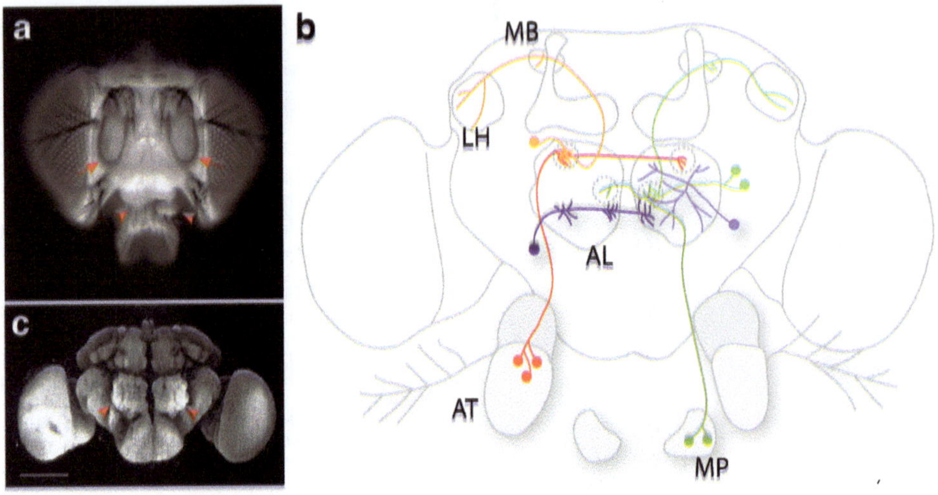

Fig. 13.2 *Drosophila* olfactory system. (**a**) Anterior view of an adult male fly. Arrows and arrowheads indicate antennae and maxillary palps, respectively. (**b**) The projected confocal image shows the anterior view of an adult brain stained with neuropil marker Bruchpilot. Arrowheads indicate antennal lobes. Scale bar, 100 μm. (**c**) Schematic diagram of the adult olfactory system. Olfactory receptor neurons have their cell bodies in either the antennae (AT) or maxillary palps (MPs). The same class of olfactory receptor neurons (dark orange or dark green) converge and project axons to a specific glomerulus (dashed circle) in the ipsilateral antennal lobe (AL) and then may cross the brain midline and innervate the same glomerulus in the contralateral antennal lobe. In the glomerulus, cognate projection neurons (light orange or light green) form synapses with the corresponding olfactory receptor neurons to receive olfactory information; projection neurons project their axons to mushroom body (MB) and lateral horn (LH) to relay information. In the antennal lobe, unilateral local interneurons (light magenta) that only innervate ipsilateral antennal lobe or bilateral local interneurons (dark magenta) that innervate both antennal lobes have processes covering multiple glomeruli and form horizontal connections in the antennal lobe

odor mixture at different concentrations may active different populations of glomeruli [52, 53]. A second large population of neurons around the antennal lobe is the olfactory local interneurons (LNs, ~ 200 in one brain hemisphere, Fig. 13.2c) [24]. These neurons can be either unilateral LNs innervating only the ipsilateral antennal lobe or bilateral LNs that innervate both antennal lobes. Therefore, LNs form multiple horizontal connections of glomeruli within and between two antennal lobes [54–61]. The fact that these neurons are highly diverse and variable [24, 25, 62–65] further adds to the complexity of coding and decoding olfactory information. Another important aspect of the olfactory circuit, albeit a less studied topic, is that multi-glomerular PNs (~200 per brain hemisphere) also innervate glomeruli and relay information to the MB, LH, and other brain centers [66–68]. Therefore, the *Drosophila* olfactory circuit and its corresponding behavioral outputs are a practicable system with which to explore the neural and circuitry-based mechanisms of individuality.

2 Materials

2.1 Choices of Fly Genetic Backgrounds

Several *Drosophila melanogaster* strains with different genetic backgrounds are available from stock centers and individual laboratories (Table 13.1). Because many factors, such as developmental stochasticity [23] and circuit plasticity [63, 65, 69, 70], contribute to the behavioral outputs of individual flies, a single fly strain must be selected as a general genetic background for all tested flies [71, 72]. In general, at least six backcrosses are required to establish flies with a required component (e.g., transgene, mutation, etc.) on a particular genetic background [72–75]. If the component is on the X chromosome, establishment of the line will require around ten backcrosses [76]. Whether a single genetic background or an isogenic background should be used depends on the main questions to be answered in the research project (also see Sect. 2.7) [76–78]. One interesting point that may be worth consideration is that humans are not isogenic and exhibit many polymorphisms. Therefore, an isogenic background may not always be the best choice for projects studying some human diseases.

2.2 Culture Conditions for Flies

2.2.1 The Environment

Drosophila has circadian day-night cycle of 24 h [38, 79], during which it exhibits two feeding peaks [80]. In addition, environmental temperature, humidity, and light can all strongly influence the behaviors of *Drosophila* [3, 81]. Therefore, a controlled environment is essential when monitoring fly behaviors, especially for deciphering mechanisms of individuality. The controlled environment applies to both the incubator and the behavior room. It is ideal to rear flies in an incubator with 12 h–12 h light-dark cycle, at 25 °C and 70–75% humidity; the same well-controlled environment is suitable for flies during development and for adults. A behavior room with temperature, humidity, and light-dark control set to the same levels as the incubator is also ideal. In some studies, behavior rooms with constant room temperature, humidity, and light control have also been described [78, 82].

2.2.2 Fly Food Ingredients

With the exception of commercially available instant fly food (e.g., Formula 4-24, Carolina), the ingredients of most commonly used cornmeal fly foods are not standardized between research groups. The major difference between formulations is the sugar source; dextrose, molasses, corn syrup, or saccharose may all be used. Importantly, flies fed with dextrose- or molasses-based cornmeal diets were shown to have different molecular activities in the olfactory circuit [83]. In addition, food with milk whey has been shown to reduce the severity of seizure-like phenotypes [84]. These pioneering studies describing the effects of fly food on behavioral readouts suggest that consistent diet formulation is a key to producing unambiguous data.

Table 13.1
Commonly used *Drosophila* strains

Strain	Isogenic	Stock no.	Reference[a]
w^{1118}		BDSC 3605, BDSC 5905	[73, 120–122]
w^{1118}	Yes	BDSC 6326	[74]
Canton-S		BDSC 64349 (original BDSC 1), Kyoto 105666	[19, 74, 76, 77, 84, 120]
Oregon-R	Yes		[74, 123]
Oregon-R	Yes	Thomas Clandinin lab	[78]
Oregon-R-C	Yes	BDSC 5, Kyoto 105669	
Oregon-R-P2	Yes	BDSC 2376, Kyoto 107658	
DL		Michael Dickinson lab	[78, 124, 125]
HCS (or PR)		Michael Dickinson lab[b]	[78, 93]
TB		Michael Dickinson lab[c]	[78]
Dahomey			[77]
w(isoCJ1)	Yes		[36, 82, 120, 126]
iso^{KH11}	Yes		[36]
DGRP-639	Yes	BDSC 25199	[23]
DGRP-859	Yes	BDSC 25210	[23]

[a]Only a subset of papers with clear fly strain information are listed as examples
[b]Heisenberg Canton-S (HCS) is used in more current publications (communication with Dr. Michael Dickinson)
[c]Collected from Top Banana fruit stand in Seattle on Sep. 2013, Dickinson lab's most current wild type (communication with Dr. Michael Dickinson)

2.2.3 Preparation of Flies: Single versus Group Housing

Since sex, age, and culture conditions are widely known to influence behavioral readouts, siblings derived from the same group of parents are most commonly used as control populations. Yet, group-housed flies of the same sex may exhibit aggressive behaviors [85–87], and group-housing flies of both sexes may allow for mating experiences [39, 88–92]. In addition, the density of flies is an important consideration, as overcrowded rearing conditions may also introduce confounding factors [76, 93, 94].

2.3 Current Treadmill Systems for Drosophila

The current insect treadmill systems can be categorized into two types based on their detection system, either mouse-mode or camera-mode (Table 13.2). Mouse-mode treadmill systems detect the motion of a ball using the sensor chip from a computer mouse (usually a CMOS sensor); of note, sometimes this sensor has been referred to as the camera of a computer mouse [43, 95–104]. In general, the sensor chip detects speckles or coarse textures on the ball surface as a flurry of images. Then, it compares the images back to back to analyze the position change of the ball in real time, finally

Table 13.2
Comparisons of mouse-mode and camera-mode treadmill systems

	Mouse-mode	Camera-mode
Detector	Computer mouse chip	High-speed CCD camera
Temporal resolution	500–4000 signals/second[a]	50–500 frames/second[b]
Signal	Pulse signal	Image (frame)
Real-time tracking	Yes	Possible
Verification of focusing spot	No	Yes
Properties of focusing spot	Single spot[c]	Frame[d]
Detection of rotation center drift	No	Yes
System establishment	Easier	–
Additional tracking software	No	Yes[e]

[a]Depending on the type of computer mouse [98, 102]. Logitech mouse (G9) and Avago Technologies laser senor for optical mouse (ADNS-6090) were used as examples here
[b]Depending on the maximal resolution of camera, frame rates of camera, the size of region of interest, camera-data interface, the interface of image processing, etc. [78, 107, 109]
[c]Single spot focus does not allow for the identification of the rotation axis; this limitation can be overcome by using two orthogonally aligned mouse chips
[d]Additional software is required to translate frames of the video to the ball motion
[e]The software FicTrac [108–113] and optic flow [78, 107] are commonly used

sending a signal indicating the position change to the computer. The signal pulses can be sent at speeds as fast as 1000 Hz, making this detection mode ideal for real-time tracking in virtual reality systems [105, 106]. However, because a single light beam, typically from an infrared light-emitting diode (LED), is used to detect ball motion, the motion is only detected along two axes: either the pitch and yaw (from a camera facing the posterior of the ball) or the roll and yaw (from a camera facing the lateral surface of the ball). This limitation has been overcome by using two orthogonally aligned mouse sensor chips (e.g., two cameras aligned at 45° left and right to the posterior of the ball, which represents 0°) to provide three-axis ball motion that defines its rotation [102]. Yet such a design does not offer information about the shift of rotation center during recoding (see Sect. 2.4). The second limitation of sensor chip detectors is that the distance between the detected ball surface and the sensor must be within the working distance of the sensor. A compensatory optic design may be necessary to acquire the motion of the ball in such cases, but no reports have described such a compensatory strategy. The third limitation of the sensor chip detector is that it lacks a method to validate the accuracy of signal. Therefore, determining whether there is a difference between signal and real ball motion in real time remains impractical [78, 107].

In another approach, high-speed CCD cameras have recently been used to detect ball motion in treadmill systems [78, 107–113]. Accordingly, these systems are called camera-mode treadmill systems [78] or camera-based optical flow tracking systems [107]. The moving ball was captured in a series of photos, which were directly transferred to the computer. Next, ball motion was then calculated according to a frame-by-frame comparison. In an earlier camera-mode treadmill system, each frame was compared with a predefined three-dimensional map of the patterns on the ball surface, a method called FicTrac [108]. The limitation of this method is that each ball needs a pre-built map. This requirement was later bypassed by using optic flow (https://docs.opencv.org/3.4/d4/dee/tutorial_optical_flow.html) [114] for back-to-back frame comparisons [78, 107]. Because high-resolution images are required to make accurate comparisons, the frame rate is highly dependent on the performance of the CCD camera and data transfer strategies, which may be USB or camera-link. Both factors limit the frame rate, and since the image comparison is performed post hoc, the speed of data analysis makes the method difficult to apply in a virtual reality system. This limitation has recently been overcome by applying pixel-binning to images acquired with small region of interest (ROI) size and by using CUDA 9.1 to accelerate image comparison. Although the frame rate is slower than that of computer mouse chip, a 500 Hz frame rate is achievable in a camera-based optical flow virtual reality system; one iteration of image acquisition and processing can be performed in less than 2 ms [107].

2.4 Design of the Camera-Mode Fly Treadmill System

The camera-mode treadmill system is composed of three parts: a ball-floating system, a video system, and a ball motion identification system (bottom, Fig. 13.3 and Table 13.3). For the ball-floating system, room air is pumped into an air tank using an air compressor (bottom left, Fig. 13.3). After the air tank, air flow is split by a five-way joint and injected at four orthogonal sides of the flow mount to create mutually counteracting air streams that support a ball suspended in the center of the flow mount (bottom right, Fig. 13.3). The shape of the center of the float mount is curved such that the ball will not contact the surface of the float mount when supported by the four air streams (Fig. 13.4). Therefore, the friction between the ball and flow mount is negligible, and the motion of the ball is mainly driven by movement of the fly on the ball. Accordingly, the ball motion will reflect the walking and movement of the fly.

The design of this flow mount is different from other reported flow mounts, as it has four air inlets (Fig. 13.4). We noticed turbulence occasionally occurs when using a flow mount with one air inlet. Hence, we designed a flow mount with four inlets and four outlets to support the ball. This design greatly reduced ball rotation resistance and occasional turbulence. To build four air streams with

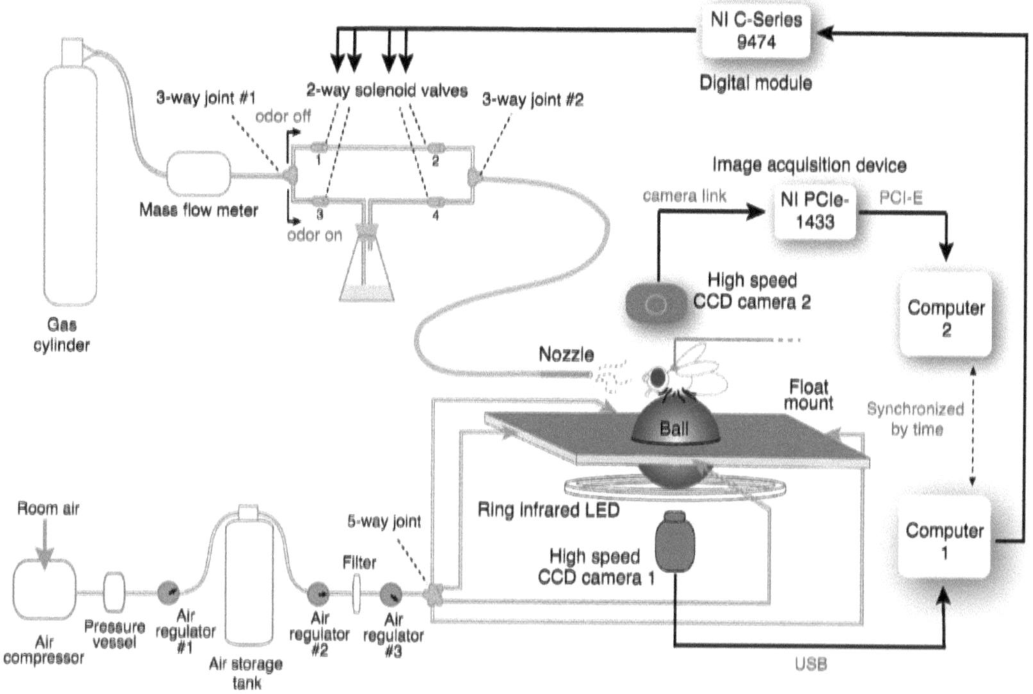

Fig. 13.3 Design of the camera-mode treadmill system and odor delivery system. Schematic diagram of the design and connections of the camera-mode treadmill system (right), with an air pipeline (left bottom) to support ball suspension, and an odor delivery system (left top). The air was 21% O_2 and 79% N_2. See Sects. 2.4 and 2.5 of the text for details. (The figure was adapted from [78] under CC BY-NC-ND License)

stable and steady flow rates to support the ball, two design features were implemented. First, to stabilize the air pressure throughout the tubing, we included a pressure vessel and an air storage tank between the air compressor and the flow mount (bottom left, Fig. 13.3). This feature offers buffering space for the inlet air. Second, three air regulators were inserted between the pressure vessel, air storage tank, and float mount. The pressure of pumped room air was gradually decreased until 0.12 kg/cm^2 air pressure was achieved at the end point of the third regulator (bottom left, Fig. 13.3). In addition, a set of oil mist filters was added to the air-line in order to remove possible oil vapor from the air compressor and small particles from the environmental air.

The second main component of the treadmill system is the video system, which is composed of a high-speed CCD camera and an infrared light source. We aligned the tethered fly on the right-top of the floating ball. The best camera alignment would be along the same axis of the fly, either on the right-top of the fly or below the right-bottom of the ball. To capture full images of the ball center, we aligned the high-speed CCD camera (camera #1 in Fig. 13.3) with the bottom of the ball and created a 192 × 200 pixel ROI that covers the bottom-center of the ball surface. The

Table 13.3
List of materials and parts

Section	Parts	Company/manufactory	Catalog number	Comments
2.4. Treadmill	High-speed CCD camera	Point Grey	GS3-U3-23S6M-C	(1) Equipped with a zoom lens (AF 18–35 mm f/3.5–4.5D ED, NIKON) (2) 100 frames/sec; camera #1 in Fig. 13.3
2.4. Treadmill	High-speed CCD camera	BASLER	acA2040-180kmNIR	(1) Equipped with a lens (1"SWIR 8 mm C-mount lens (LM8HC-SW), Kowa American Corp.) (2) 30 frames/sec; camera #2 in Fig. 13.3
2.4. Treadmill	Flow mount	High precision machine shop, Institute of Physics, Academia Sinica	Custom-design	See Fig. 13.4
2.4. Treadmill	Ball	Holo pack	6 mm-PP ball	Weight, 0.11 g; diameter, 5.95 mm
2.4. Treadmill	Ring infrared LED illuminator	X-loupe	AgileLite system	Equipped with X-Core control unit
2.4. Air compressor	Air compressor	JUN-AIR	3–4 lubricated quiet air compressor	Equipped with air regulator #1 (Fig. 13.3)
2.4. Air compressor	Pressure vessel	Dana-tank A/S	3050005	4 L
2.4. Air compressor	Air storage tank	PUMA	AST-22	Equipped with air regulator #2 (Fig. 13.3)
2.4. Air compressor	Oil mist filter	CKD	M4000–15-W	
2.4. Air compressor	Controller for air pump	Master Pneumatic-Detroit	R55M-2L15	Air regulator #3 (Fig. 13.3)
2.5. Odor delivery	Gas cylinder	PGC Corp.		(1) 21% O_2 and 79% N_2 (2) Equipped with pressure regulator (BFR5K, Parker Veriflo)
2.5. Odor delivery	Mass flow meter	Brooks	5850E	Coupled with a power supply (PC-510 MFC readout)

2.5. Odor delivery	Flow meter power supplier	PROTEC instruments	PC-510 MFC Readout	
2.5. Odor delivery	Two-way solenoid valves	SUNWELL	SC03-2Z	The electric control of solenoid valves: Nass magnet (108-030-0264)
2.5. Odor delivery	24-V power supply	Exlite	LP-24V-50-U	To control two-way solenoid valves
2.5. Odor delivery	Erlenmeyer flask	Pyrex	4980, stopper no. 5	125 ml
2.5. Odor delivery	Teflon tubes	Saint-Gobain performance plastics	Tygon R-3603, AAC00002	1/8” OD
2.5. Odor delivery	Stainless steel tube		1/8” OD	The terminal ends of tubing to lead air in and out of the flask
2.5. Odor delivery	Nozzle/needle	Fairy polymer industry	23G disposable needle	23G, 1”
2.5. Odor delivery	Nozzle/tube	Becton Dickinson	BD Intramedic PE tubing, 427411	Polyethylene tubing (PE50); 1/32” OD
3.3.1. Fly starvation	Agar	Gene Teks bioscience	GT-PA001	
3.3.1. Fly starvation	Food vial	Sun min medical supplies		PS, 30 mm (D) X 100 mm (H)
3.3.2. Odor preparation	Ethanol	Sigma-Aldrich	3221–2.5L	Absolute, ≥99.8% (GC)
3.3.2. Odor preparation	Pentanoic acid (valeric acid)	Sigma-Aldrich	240370-100ML	≥99%; working solution was diluted with distilled water
3.3.2. Odor preparation	Lubricant	Klüber lubrication	UNISILKON TK M 1012; article no. 022090; 75 g	
3.3.2. Odor preparation	Air purifier	Honeywell	HPA5150WTW	
3.3.4. Fly tethering	Pin holder	FINE SCIENCE TOOLS	26018–17	
3.3.4. Fly tethering	Insect pin	Entomoravia - Austerlitz insect pins®	Minutiens ∅ 0.10 mm	
3.3.4. Fly tethering	UV gun	DENTAMERICA	LITEX 696 PLUS	470 ± 5 nm

(continued)

Table 13.3
(continued)

Section	Parts	Company/manufactory	Catalog number	Comments
3.3.4. Fly tethering	UV glue	DENTAMERICA	LIPIT, 12 gm	
3.3.4. Fly tethering	Forceps	FINE SCIENCE TOOLS	Dumont #5/45, 11251-35	Can be replaced by Dumont #5 (11252-20) forceps with tips manually bent to 45°
3.3.4. Fly tethering	Cold plate	Pyrex	Pyrex 9 Depression Spot Plate, 7220-85	The plate was cut into three 3-spot plates
3.3.4. Fly tethering	3 M tape	Scotch	Magic tape, 810	
3.3.6. Monitoring room condition	Thermohygrometer and barometer	Testo	Testo 622	
3.3.7. Climbing test	Setting	Min hung industry CO.	Customized	
3.3.7. Climbing test	Tube	FALCON	PN. 352057	
3.4. PID	Photoionization detector	Hal Technology	HAL-HVX501	For detecting EtOH; sensitivity: 0.01 ppm (0–20/0–200 ppm) and 0.1 PPM (0–2000 ppm)
3.4. PID	Photoionization detector	Ion science	PhoCheck TIGER	For detecting pentanoic acid; sensitivity: 1 ppb (0–20,000 ppm)

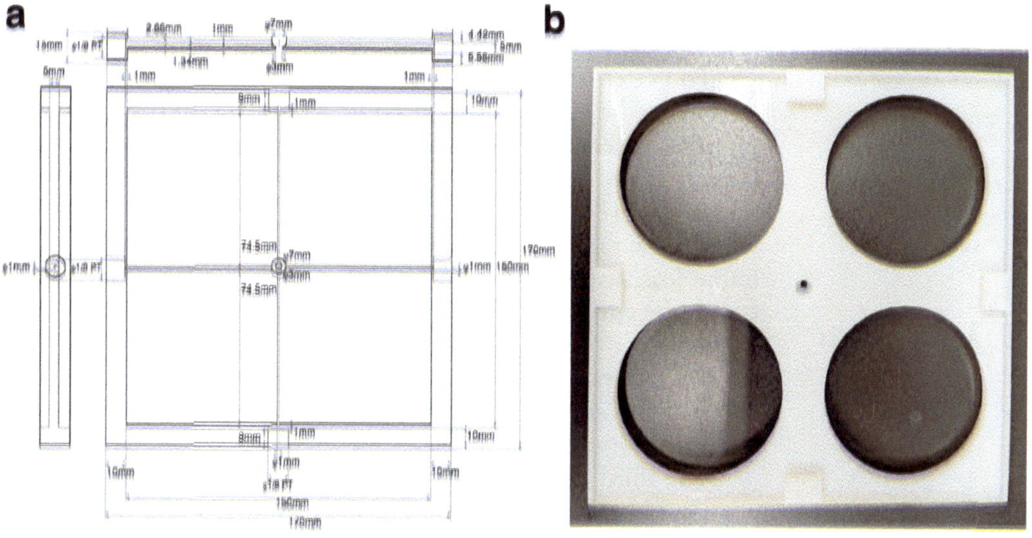

Fig. 13.4 Design of the four-air-inlet flow mount. (**a**) Detailed design information of the four-air-inlet flow mount. (**b**) Photo of the flow mount

design of the flow mount in our system prevents the capture of full images of the ball with a laterally aligned camera. To uniformly illuminate the ball surface and prevent shadows, we used a ring infrared LED illumination system located between the bottom of the ball and the camera. This illumination system can transmit visible or infrared light, which provides the flexibility to switch between visible light for sample preparation and alignment (e.g., applying the tethered fly to the ball) and infrared (850 nm) for image acquisition.

As discussed in Sect. 2.3, the image transfer rate and storage rate heavily influence the performance of the camera-mode treadmill system. Using a high-speed FLIR Grasshopper 3 (GS3-U3-23S6M-C, Point Grey) CCD camera, the maximal resolution was 1920 × 1200 (pixel × pixel), and the maximal pixel depth was 16 bits. A single image with the highest resolution will be 4.6 MB, and images taken at the highest frame rate of 162 frame per second (fps) would require data transfer at 746 MB/s. USB 3.1 Gen 1 is used to transfer data from this camera to the computer, which supports a maximal 625 MB/s transfer rate. This means frame loss will occur during data transfer if frames are taken at the highest camera resolution. Moreover, the storage rate of a standard hard disk drive is 100–250 MB/s (around 520 MB/s of a solid state drive; SSD), which is also smaller than 746 MB/s. Therefore, the image transfer rate and storage rate of the system are limiting factors to determine the ROI size used for ball motion recording. Taking these factors into consideration, we equipped the high-speed CCD camera with a wide-angle lens to magnify the ball and recorded ball motion using a 192 × 200 ROI, 8 bits of pixel depth, and 100 fps capture rate. These settings generated 3.84 MB images in 1 s.

To better understand how the tethered fly drives ball motion, we included a second high-speed CCD camera on the top of flow mount (camera #2, Fig. 13.3 and Table 13.3). This camera supports maximal resolution 2048 × 2048 (pixel × pixel) and maximal frame rate 180 fps. The image was transferred via camera-link, which has a maximal 850 MB/s transfer rate. Yet to avoid potential frame loss, we set the ROI to 200 × 200 (pixel × pixel), which covers the entire fly, and the frame rate was set at 30 fps for storage by computer #2 (equipped with a SSD with read rate of 550 MB/s and write rate of 500 MB/s). To temporally align the frames showing fly movement and ball motion, we synchronized the two computers (right, Fig. 13.3).

The third main component of the treadmill system is the ball motion identification system. We used optic flow (see Sect. 2.3) to quantify the ball motion. Because the ball is stably suspended on the flow mount, its overall position remains the same in each frame (middle, Fig. 13.5a). In the frame, the ball bottom is projected onto a circular area of approximately 1.41 mm in radius (r), which corresponds to about 22% of the ball surface. However, since the two-dimensional image was captured from a curved surface, use of the whole ball bottom area for analysis would misrepresent motion along the curvature and significantly bias the 2-D area calculation of optic flow. Therefore, a small area ($r = 6$ pixel) was further extracted from each frame of the video for analysis.

2.5 Design of the Odor Delivery System

The essential function of the odor delivery system is to provide a controllable and consistent concentration of odor at a stable flow rate. To support this function, we added a mass flow controller between the gas cylinder and the three-way joint to generate a stable air flow with 10 mL/min flow rate, in addition to the built-in control from the gas cylinder (top left, Fig. 13.3 and Table 13.3). The air is split into two airways, each with two normally closed electromagnetic valves. This design allows us to control on and off phases of odor delivery. In one of the airways, a flask with diluted odorant is set between the two electromagnetic valves. A long stainless steel tube (1/8" OD) is inserted into the solution within the flask to bubble air through the odor solution, and a shorter tube with its end above the liquid surface allows the odor-containing air to enter the airway. In the initial odor-off phase, valves #1 and #2 (top, Fig. 13.3) are open, and valves #3 and #4 are closed. When switched to the odor-on phase, valves #1 and #2 are closed, and valves #3 and #4 are open. This setup will allow air to enter the flask and carry the odor out to the nozzle.

The on-off function of the normally closed electromagnetic valves is controlled by an electric signal that triggers a 24-V pulse. To achieve such control, a digital module (NI C-Series 9474) was

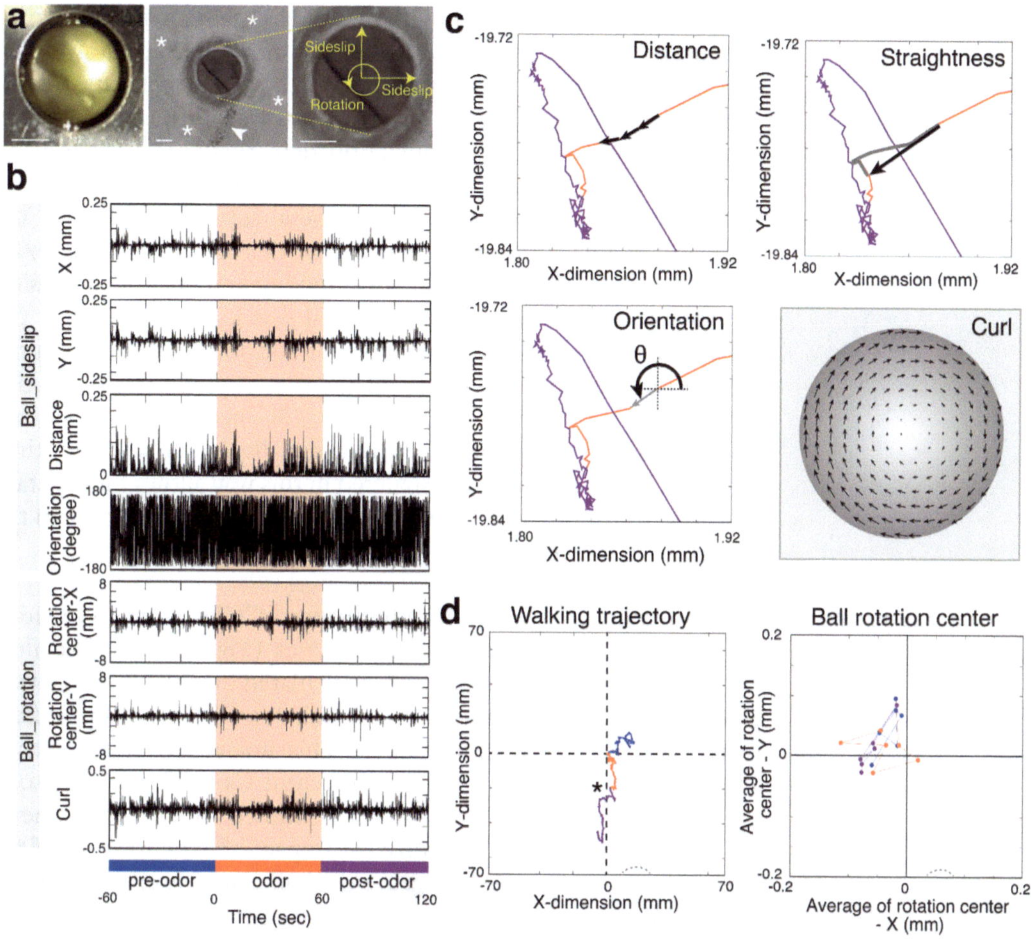

Fig. 13.5 Reconstruction of fly walking trajectories and feature extraction. (**a**) (Left) Bottom view of the ball. (Middle) A single frame of infrared video recording of ball motion. The four air inlets are marked by asterisks. The arrowhead indicates a mark for positioning the nozzle. (Right) The ball bottom image showing three ball motion axes. Scale bars, 1 mm. (**b**) The features derived from ball motion that was driven by a starved fly exposed to a 10^{-1} dilution of pentanoic acid at 17 mL/min. Switching the solenoid valves of the odor delivery system on and then off generated a pre-odor phase (blue), odor phase (orange), and post-odor phase (magenta). The extracted parameters are displacement along X-dimension (X), displacement along Y-dimension (Y), distance, orientation, rotation center at X, rotation center at Y, and curl at every 10 msec interval (sampling time). The speed (mm/10 msec) shows the same pattern as walking distance. (**c**) Four features – distance, straightness, orientation, curl – can be further extracted through the analyses of the parameters in (**b**). See Sect. 2.6 for details. (**d**) (Left) A reconstructed walking trajectory for the same data shown in (**b**). Asterisk indicates the trajectory phase that is enlarged in (**c**). (Right) The position of the ball rotation center derived from the same fly as in (**b**). (Figures were adapted from [78] under CC BY-NC-ND License)

included between computer #1 and the valves to generate an electric signal through LabVIEW in computer #1. The LabVIEW panel is also used to control the image acquisition by high-speed CCD camera #1. Such a design allows for synchronizing the controls of electromagnetic valves and high-speed CCD camera #1. It is worth emphasizing that the delay time between turning on the valves and the arrival of odor at the end of nozzle varies depending on the length and design of the tubing system, as well as the applied odor. Therefore, we used photoionization detectors (see Sect. 3.4) to measure the delay time for each experimental setup. Moreover, with the exception of two stainless metal tubes inserted into the flask and nozzle, all tubing for the odor delivery system was 1/8" OD Teflon to prevent residual odor buildup and possible corrosion caused by the odorant. Stainless metal was used to minimize the possibility of reaction with the solvent and meanwhile maintain a strong seal between the tube and the silicon plug. To avoid odor residue, the nozzle was made from a 23G needle with a 15-mm 1/32" OD polyethylene tube; both are disposable and one-time use. The blunt-end tube constructed in this way allows for constant flow pressure along the nozzle section and limits the air flow to the antennae of the tethered fly (see Sect. 3.3.5).

2.6 Data Acquisition and Feature Extraction

The ball motion captured by a high-speed CCD camera is in video format as a series of consecutive frames (Fig. 13.5a). The flow vector for each pixel of the video is calculated by optic flow analysis, and then the flow vectors are assembled to define an optic field between two consecutive frames. This analysis is done with a custom-made tracking program written in C++. The central zone of the ball bottom in a given frame is aligned to the same zone in the previous frame. To minimize the impact of the curvature of the ball surface, a small ROI covering the ball center is extracted from each frame.

The horizontal and vertical average components of a given optic field are taken as the displacements in the X-dimension and Y-dimension, respectively (top two panels, Fig. 13.5b). The distance between two sampling times is then calculated by the root mean square of the X-dimension and Y-dimension displacement (third panel from the top, Fig. 13.5b). The displacement in a time period is the sum of displacements at each sampling time interval (e.g., each of three black arrows in the top left, Fig. 13.5c). Accumulated distance is the sum of the magnitudes of displacements at each sampling time interval (e.g., the sum of the length of three black arrows in the top left, Fig. 13.5c). Thus, the tendency of a fly to walk along a straight path in a given time period can be quantified as the magnitude of total displacement in that time period divided by the accumulated distance in the same time period (top right panel, Fig. 13.5c).

Orientation, i.e., the direction of displacement, is described as the angular coordinate of a polar coordinate system with the positive-X coordinate equal to 0°; the counterclockwise rotation from this reference is defined as the degree of orientation (fourth panel from the top, Fig. 13.5b, bottom left panel in Fig. 13.5c).

The shifted displacement and the rotation of the ball are considered as ball motion along the X-Y plane and along the rotation axis in any two consecutive frames. The ball motion through the whole recording period can thus be reconstructed by optic flow as a ball motion trajectory, which directly reflects the walking trajectory of the tethered fly (Fig. 13.5d, left).

The estimation of a rotary axis and rotation center is an added capacity that is not achievable with a mouse-mode treadmill system. We found that the ball rotation often switches along the yaw, roll, or pitch axes [78]. This means that reconstructing the trajectories from two orthogonally aligned mouse camera may result in under- or over-estimation of distances. Because we treat the spherical surface of the ROI as a 2-D area, the rotary axis of estimated rotation centers is along the Z-coordinate and not the radial coordinate. Therefore, the shift of rotation center can be described as the X-, Y-, and Z-displacement of rotation center between two consecutive frames, where Z-displacement equals 0 (fifth and sixth panels from the top, Fig. 13.5b, right panel in Fig. 13.5d). Accordingly, the curl is used to describe the magnitude of rotation in the flow field between two consecutive frames (the bottom panel in Fig. 13.5b). Curl is calculated as the sum of all flow vectors in the ROI (bottom right panel in Fig. 13.5c). Therefore, when the estimated rotation center falls in the center of the ROI (X-Y space), the ball exhibits yaw rotation. In cases where the estimated rotation center falls outside the ROI center, the ball is exhibiting a more roll or pitch rotation.

2.7 Individual Fly Walking Trajectories

The walking strategies used by individual flies can be described by four main features extracted from walking trajectories (Fig. 13.5c). The orientation map of a single fly upon different odor phases can be reconstructed as shown in Fig. 13.6a. Distance from the center represents the percentage of time the orientation angle was measured during the odor phase. The orange arc in the figure indicates the positional range of the odor source (~270°–306°). Similarly, the walking patterns of individual flies can be further described and compared through their average of distance, straightness, and average magnitude of absolute curl along time (Fig. 13.6b). It is worth noting that the patterns of the above three parameters may change with different lengths of time bins. In addition, our analyses of these parameters revealed differences between flies on different genetic backgrounds (Fig. 13.7).

3 Methods

Following the design described in Sect. 2.4, a camera-mode fly treadmill system can be built as shown in Fig. 13.8a. To optimize the flow rate and stability of air flow that supports the ball, air

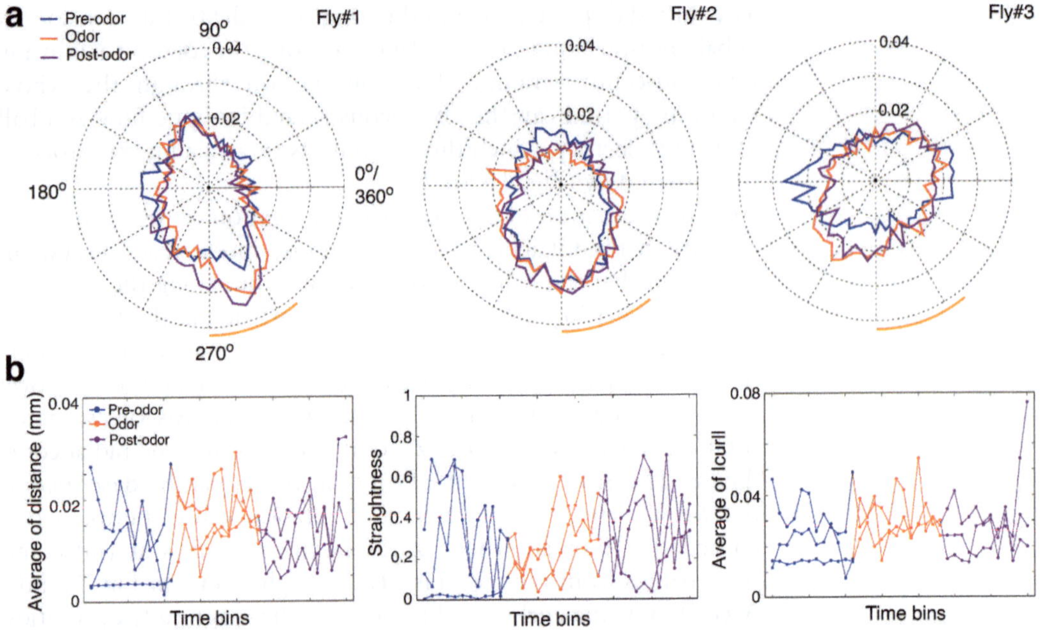

Fig. 13.6 The individuality of fly walking patterns. (**a**) Reconstructed orientation distribution maps for three starved flies during pre-odor (blue), odor (orange), and post-odor (magenta) phases. The odor was a 10^{-1} dilution of pentanoic acid at 17 mL/min. Radius is the percentage of orientation angles (see Sect. 2.6). Orange bracket indicates the position of the odor source (~270°–306°). (**b**) The average of distance (left), dynamics of straightness (middle), and Icurll (right) in a given 5-sec time bin that were derived from the same three starved flies shown in (**a**). Each line represents the average distance traversed by a fly in three odor phases. (Figure 13.6a (left) was adapted from [78] under CC BY-NC-ND License)

3.1 Building the Camera-Mode Treadmill System

pressure of the air compressor and the air storage tank should be controlled. In addition, we have tested different air pressures using a third air regulator (#7, Fig. 13.8a). In our system, an air flow at 0.12 kg/cm^2 pressure can suspend the 0.11 g ball on the float mount, without creating obvious turbulence. In this system, a ring infrared LED was used as a light source (#3, Fig. 13.8a). It is important to adjust the position of the LED ring so that sufficient light can reach the ball but no flare spot will occur on the surface of ball bottom. Flare spots will impede the detection of speckles on the ball surface and thus interfere with the measurement of ball motion.

To compare the ball motion with the walking pattern of the tethered fly and the coordinated motion of the fly's six legs, we used a second high-speed CCD camera in our system (#1, Fig. 13.8a). The video feed from this camera was sent to a second computer (#2 in Fig. 13.3). Therefore, the two computers used for collecting videos from the two CCD cameras needed to be synchronized. We first used the automatic time synchronization in Windows to set the times for the connected computers and CCD cameras. Before each

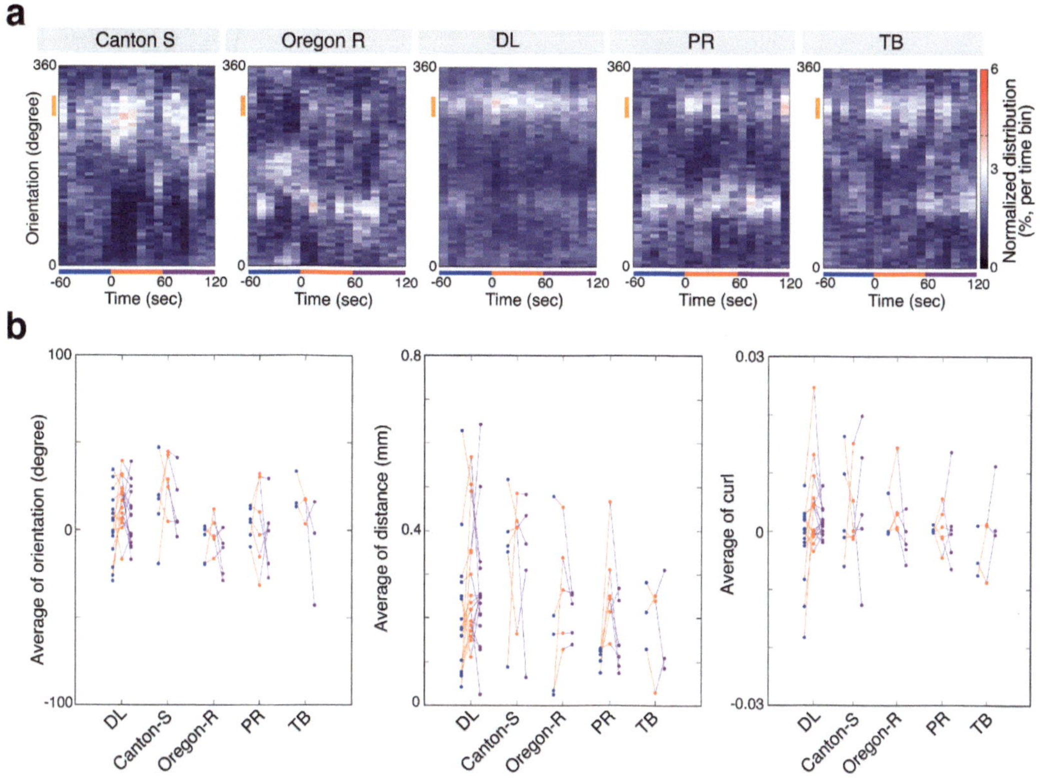

Fig. 13.7 Differential walking strategies of flies on different genetic backgrounds. (**a**) Orientation distribution of five wild-type strains of starved flies in response to a 10^{-1} dilution of pentanoic acid (17 mL/min). The orientations of flies were calculated in 10-sec time bins. The orientation distributions of flies in the same group in a given time bin were normalized and shown in a heat map. Note that DL flies show more consistent orientation dynamics in the three odor phases. The numbers in each group were 5 (Canton S), 5 (Oregon R), 16 (DL), 6 (PR), and 6 (TB) flies. Orange lines indicate the positional range of the odor source (~270°–306°). (**b**) The shift of average of orientation (left), average of distance (middle), and average of curl were derived from individual flies on different genetic backgrounds and under different odor phases. (Figure 13.7a was adapted from [78] under CC BY-NC-ND License)

experiment, we further synchronized the two computers with the Network Time Protocol (NTP) server (e.g., clock.stdtime.gov.tw).

3.2 Building the Odor Delivery System

Following the design described in Sect. 2.5, an odor delivery system can be built and connected to the flow mount of the treadmill system as shown in Fig. 13.8a. We tested different flow rates to carry odor, including 10 mL/min, 17 mL/min, and 22 mL/min. Interestingly, 17 mL/min air flow delivered the strongest stable odor profile [78]. Notably, higher flow rates may trigger wind-sensitive neurons in the antenna and arista [115], which may complicate interpretation of the behavioral readouts. In addition to the flow rate, we also tested different odorants, including absolute EtOH, 10^{-1} pentanoic acid, and distilled water. The delay times of

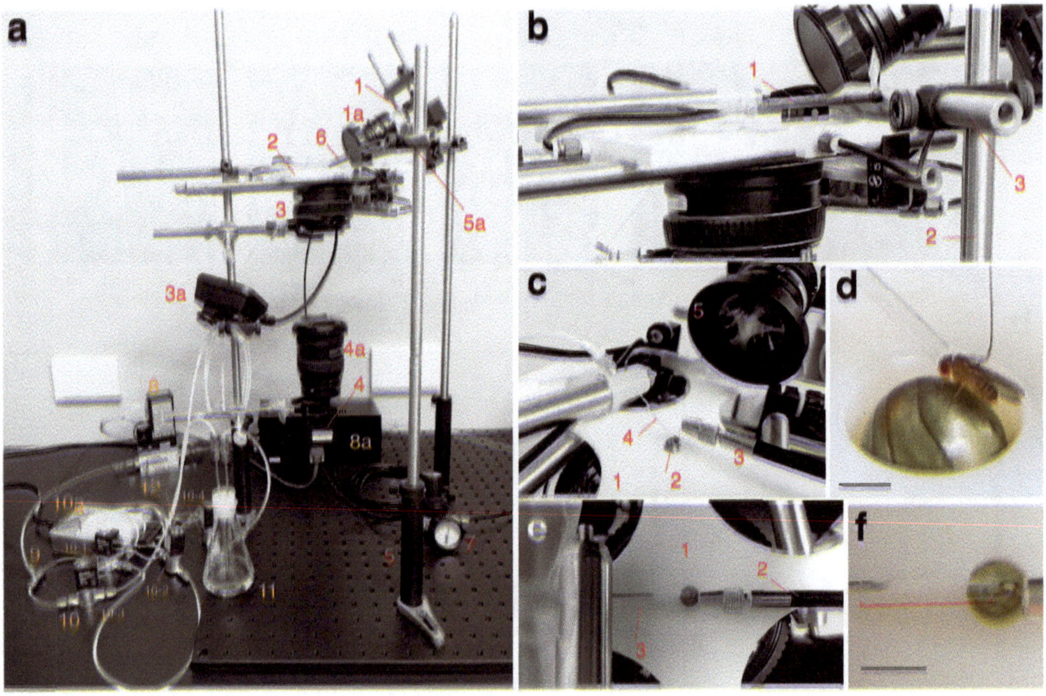

Fig. 13.8 The camera-mode treadmill system. (**a**) Photograph of the camera-mode treadmill system. The parts of treadmill system and odor delivery system are number coded in red and orange, respectively, and follow the items described in Fig. 13.3; 1, high-speed CCD camera #2; 1a, the lens equipped to camera #2; 2, flow mount; 3, ring infrared LED illuminator; 3a, controller for ring infrared LED illuminator; 4, high-speed CCD camera #1); 4a, zoom lens equipped to camera #1; 5, pin holder post; 5a, side arm of the pin holder post; 6, pin holder; 7, air regulator #3 for air pump; 8, mass flow meter; 8a, flow meter power supply; 9, three-way joint #1; 10, four sets of two-way solenoid valves (number coded as in Fig. 13.3); 11, flask with odor solution; 12, three-way joint #2. (**b**) Lateral view of the flow mount showing the relative positions of pin holder (1), pin holder post (2), and the side arm of pin holder post (3). (**c, d**) (**c**) Top lateral view of flow mount shows a tethered fly on the ball. 1, flow mount; 2, ball; 3, pin holder; 4, tip of nozzle; 5, high-speed CCD camera #2. (**d**) The enlarged image shows the relative position of nozzle and the fly in (**c**). Scale bar, 2 mm. (**e, f**) (**e**) Top view of flow mount shows the relative position of the nozzle (needle with a PE tube) and pin holder. 1, flow mount; 2, pin holder; 3, nozzle. (**f**) The enlarged image shows a 1.5-cm disposable PE tube (bracket) fitting to the end of needle. Scale bar, 1 cm. (Figure 13.8a, c, and d were adapted and modified from [78] under CC BY-NC-ND License)

odor arrival for each odor to the nozzle differed. The length of tubing system also contributes to the delay time. Therefore, it is important to measure the delay times for each tested odor in each system (see Sect. 3.4) and calibrate the tethered fly's walking pattern according to the odor arrival time. The efficiency of solenoid control and odor delivery can be further monitored and tested using a photoionization detector (PID; see Sect. 3.4).

Some behavioral experiments were conducted with a vacuum behind or below the fly to quickly eliminate the odor surrounding the fly [116]. This design is optional and dependent on the experiment type. In our system, under the flow rate we used, the residual odor surrounding the fly dropped from ~355 ppb to ~110 ppb within 6 seconds after the solenoids were turned off [78] (also see Sect. 3.4). Therefore, we did not include a vacuum in our system.

3.3 Monitoring the Walking Behavior of a Single Fly

3.3.1 Fly Preparation (Starvation/Agar)

Flies were reared with regular fly cornmeal-saccharose food at 25 °C, 70% humidity, and 12 h–12 h light-dark control incubator (see Sect. 2.2). For starvation experiments, flies were transferred to a 1% agarose-based fly vial at the same environmental conditions for around 16 h. We have tested the locomotor activities of individual flies that were well fed or were starved for 16–23 h or 39–47 h. Starvation for 39–47 h strongly affected the locomotion of flies in the climbing test (see Sect. 3.3.7).

3.3.2 Preparation of the Odor Delivery System

First, 50 mL odorant solution was prepared at the desired dilution and added to the flask. Water, parafilm oil, and mineral oil are commonly used solvents to dilute odors. The flask was then capped with a silicon plug with two stainless tubes, one longer than the other, inserted through the silicon plug. The longer stainless tube was inserted into the odor solution and the shorter one above the solution (Figs. 13.3 and 13.8a). If a silicon plug is used, lubricant may be applied to the contact surface between the stainless tube and hole to prevent possible leakage. The flask was then inserted into the tubing system. The two 3-way joints of the odor delivery system were closed in the initial condition (Fig. 13.3). To ensure the odorant concentration in the tubing and flask remained static, the odorant was prepared and inserted into the apparatus one night ahead of the treadmill experiment. In addition, an air purifier was used to remove the residual odorant in the room. On the second day, electromagnetic valves #1 and #2 (top, Figs. 13.3 and 13.8a) were opened after opening the gas cylinder and adjusting the mass flow meter, as described in Sect. 3.3.3.

3.3.3 Preparation of the Treadmill System

Before the experiment, the pump (bottom left, Fig. 13.3), the gas cylinder and the mass flow meter (upper left, Fig. 13.3), the ring infrared LED (middle, Fig. 13.3), and computers #1 and #2 (right, Fig. 13.3) were sequentially turned on. The clocks of computers #1 and #2 were then synchronized with the Network Time Protocol (NTP) server (e.g., clock.stdtime.gov.tw). The control interface in our system was custom-written with LabVIEW (National Instruments). To establish an appropriately sized ROI, the ring LED was switched to the infrared setting, and then the NI MAX window was opened on computer #1. We usually set the ROI at 192 × 200 pixels for high-speed CCD camera #1 and 200 × 200 for high-speed CCD camera #2 (Fig. 13.3). The destination folders in computers #1 and #2 were then opened for data collection.

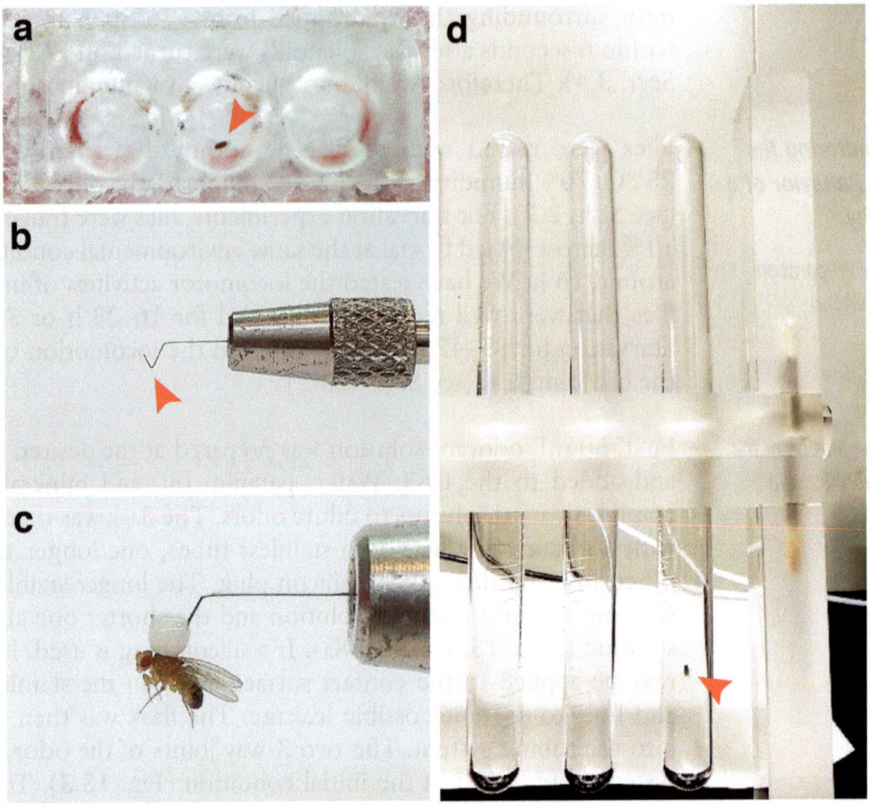

Fig. 13.9 Fly tethering and locomotion test. (**a**) The fly was anesthetized with cold and kept on a pre-cooled cold plate. Arrowhead indicates the fly. (**b**) The insect pin was inserted and fixed to the pin holder, and then it was bent into a Z shape with forceps. Arrowhead indicates the curvature of insect pin to dip with a drop of UV glue. (**c**) Image shows a fly tethered by a drop of UV glue on its notum, which was adhered and secured to the insect pin. (**d**) The climbing assay was used to test each fly's locomotor activity. Arrowhead indicates a fly climbing along the tube after being separated from the UV glue. See Sect. 3.3.7 for detail

3.3.4 Tethering Flies

The food vial with experimental flies was embedded in an ice bath for 3–5 min. A single fly was then transferred with forceps to a cold plate (Fig. 13.9a). To tether the fly, one insect pin was first installed and fixed in the pin holder. The insect pin was then bent into a "Z" shape (Fig. 13.9b). The end of the first curve of the pin should be around 60° to leave space for UV glue adhesive. The curved region of the insect pin was then dipped into a drop of UV glue. The bent insect pin can then be attached with UV glue to the back (notum) of the fly. Contact was maintained between the UV glue, insect pin, and notum, and the UV glue was cured with 20 s low-intensity UV illumination from a UV gun (Fig. 13.9c). (Note: Either an eye protection shield on the UV gun or orange eye protection goggles must be used to prevent eye damage from the UV light.)

3.3.5 Place the Tethered Fly on the Treadmill

For eye protection, it is essential to turn off the infrared illumination before working on the treadmill system. To place the fly onto the treadmill, the side arm of the pin holder post was first slid out of the flow mount. The pin holder hooked a tethered fly was then associated to the side arm, and the holder was fixed in place with 3 M tape. The pin holder should be slightly tilted to prevent collisions with the edge of the float mount in the next adjustment step. The side arm was gently swung to move the pin holder toward the right top of the flow mount and ensure the fly stood on the right top of the ball with a normal posture. The fly was allowed to recover for around 1 h before beginning the experiment. Right before the experiment, we ensured that the fly was standing securely on the ball; then, the nozzle was adjusted to a fixed position and distance relative to the fly (Fig. 13.8c–f). In our setup, we have marked the position of the nozzle on the flow mount to ensure the same nozzle position is used for all flies across trials (Fig. 13.5a, middle). The ring infrared LED was then turned on, and the whole experimental setup was kept in the dark. We used a full blackout curtain to cover the whole system, except for the two computers.

3.3.6 Computer Interface and Program Initiation/Termination

To ensure optimal computer performance during data acquisition, custom-written programs for acquisition of ball motion images and for controlling the solenoid valves in the odor delivery system were installed on computer #1; videos of fly movements from camera #2 were transferred to computer #2 (Fig. 13.3). To ensure that fly movements in different odor phases were fully captured, the control panel in computer #2 was opened first to initiate video recording upon the initiation of a trial. Then, the control panel in computer #1 was opened to initiate the ball motion detection and odor delivery control by the odor-phase switch program. After establishing control of the odor delivery system and image acquisition of ball motion, the monitor was turned off. Subsequently, the light in the behavior room was turned off, and the room was closed. After 65–70 min, the worker reentered the behavior room and turned on the light and computer monitor. The ball motion image acquisition program in computer #1 was confirmed to have ended. Then, video recording by camera #2 was terminated through the control panel on computer #2. Finally, the temperature and humidity shown in the digital monitor around the treadmill system were recorded. (Note: Because flies are sensitive to humidity and temperature, we always record the local humidity and temperature around the treadmill after each experiment.)

3.3.7 Locomotor Activity Test

To ensure that the test fly exhibited normal activity after starvation, the tethered fly was subjected to climbing test after finishing the treadmill test. This was done by releasing the tethered fly from the pin holder and assessing its locomotor activity in a climbing test (Fig. 13.9d). First, the side arm of the pin holder post was moved away from the flow mount and placed above a 14 mL polystyrene round-bottom tube. The insect pin was then gently removed from the fly by inserting the tip of forceps into the space between the UV glue and notum; the UV glue was removed from the fly by slightly opening the forceps. In general, the UV glue was secured to macrochaetes and microchaetes on the notum, which allows one to easily separate the UV glue from the tethered fly. This release of the fly allows it to gently fall into the tube. Then, the tube was gently flipped upside down to shake the fly down to the bottom of the vial. Based on the scale (in mL) along the height of the test tube, the tube was divided to 11 zones: tube bottom to 1 mL was zone 0, 1–2 mL was zone 1, etc. The region from 11 to 12 mL was zone 11. The locomotor activity was scored as the uppermost zone that the fly climbed to before the end of the 15-sec test. The score was 0 when the fly stayed at the bottom of the tube; the maximal score was 11 if the fly climbed to or past zone 11.

3.4 Photoionization Detection of Odor Delay Time and Concentration Dynamics

Several analyses were performed to characterize the solenoid control efficacy, odor delivery efficacy, and odor concentration dynamics in our system, as well as the residual odor clearance in the surrounding environment [78]. To test the efficiency of solenoid valve control and odor delivery, we used absolute EtOH as an odorant. The concentration of EtOH was monitored by a photoionization detector (HAL-HVX501) with a detection range of 0–2000 ppm (Fig. 13.10a). The concentration of EtOH dropped to baseline by the third detection point, demonstrating high efficiency of solenoid control. In addition, based on the dynamics of EtOH in different ON-OFF intervals, no residual odor is expected to accumulate in the tubing system, as the reads in the off intervals are stable at baseline (two right panels, Fig. 13.10a).

To measure the odor delivery delay time (after switching the odor phase from "off" to "on") at the tip of nozzle, we suspended the ball and placed a second PID (PhoCheck TIGER), which has detection range of 0–20,000 ppm and sensitivity of 1 ppb, close to the position of the tethered fly (Fig. 13.10b–d). Our results demonstrated the delay time cannot be ignored; it was as high as 22 ms when the odor was delivered at 17 mL/min (Fig. 13.10e). Slower flow rates and longer tubing systems will cause longer delay times. Therefore, it is recommended to keep the tubing system as short as possible. In addition, the air flow used to support the ball in our system did not appear to affect the odor plume, since no obvious fluctuations in odor concentration were observed in the odor-on phase. The delay time between turning off the solenoid valves and odor concentration drop was also assessed, with the concentration dropping quickly to a low level and then slowly returning to the basal level (Fig. 13.10e).

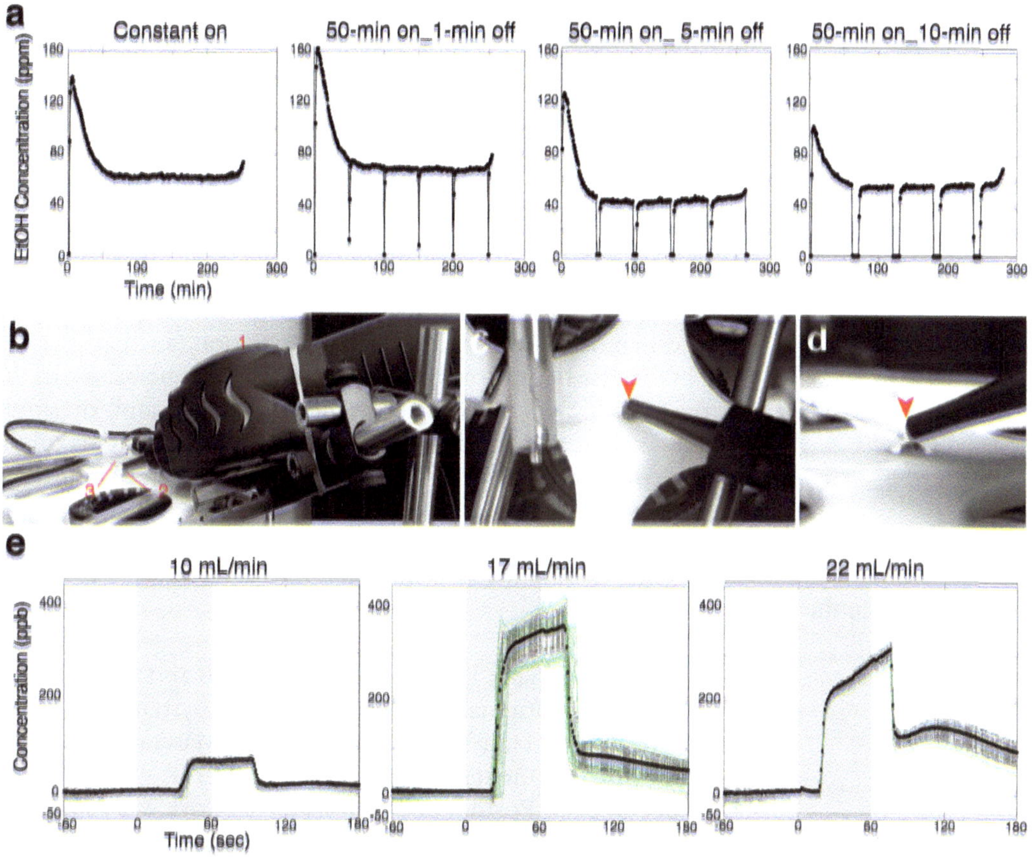

Fig. 13.10 The odor concentration dynamics detected by the photoionization detector. (**a**) (Left) The concentration of absolute EtOH in the air stream at the nozzle with 10 mL/min air flow, monitored real time by a photoionization detector (PID). After a flask with 10 mL absolute EtOH was fit to the odor delivery system, the solenoids were switched on to test the efficiency of our odor delivery system. The odor delivery reached a steady state after 60 min, which is reflected by the accumulation time of EtOH. (Middle to right) The clearance efficiency of residual EtOH was tested by repeatedly switching on and off the solenoid system at three different intervals: 50-min on and 1-min off; 50-min on and 5-min off; or 50-min on and 10-min off (from left to right). Note that the environmental EtOH returned to baseline within 1 min. (**b**-**d**) (**b**) Lateral view of a PID positioned on the top of flow mount with the detector tip in the position of a tethered fly. 1, PID; 2, tip of PID; 3, nozzle. (**c**, **d**) The magnified top view (**c**) and lateral view (**d**) of the posited PID. Arrowheads indicate the detector tip of PID. (**e**) The odor dynamics and delay time in three solenoid on-off phases, as shown in Fig. 13.5b, were measured by a second PID. Gray zones mark the solenoid-on phase; the mean delay times were 34 sec, 22 sec, and 18 sec when odor was delivered at 10 mL/min, 17 mL/min, and 22 mL/min, respectively. Black lines indicate the average (mean ± S.D), and green lines represent individual trials. (Fig. 13.10a and 13.10c–13.10e were adapted from [78] under CC BY-NC-ND License)

4 Conclusion

Video tracking of behaving animals and automatic extraction of embedded features (e.g., EthoVision) have been widely applied to study the responses of animals to treatments and cues [117–

119]. Such unbiased feature extraction is also beneficial when studying individuality. Several recent studies have monitored individual *Drosophila* behaving in singles and demonstrated that flies exhibit individuality in their responses to the same environmental cues. Because social interactions, such as aggressive behaviors among flies of the same sex, inevitably confound the extracted behavioral features of flies in a group, directly linking individuality exhibited in a group setting to circuit connections or activity is a major challenge. Therefore, analysis of a single fly behaving on its own will not be limited by such concerns, but the possibility of minor differences introduced during preparation (such as fly tethering in the case of our treadmill) may introduce some biases. In the future, coupling of optogenetics, chemogenetics, and/or single neuron labeling and manipulation to behavioral paradigms might partially overcome such biases by boosting the readout of individual circuit responses.

Acknowledgments

We would like to thank Drs. Daisuke Yamamoto and Soh Kohatsu for sharing their original designs of treadmill system. We thank Hsuan-Wei Shen, Ting-Wei Liang, and Dr. Marcus Calkins for discussion and comments on the manuscript. K.T.T. was an Academia Sinica Regular Postdoctoral Scholar. This work was supported by Academia Sinica Career Development Award (AS-102-CDA-L02) and Academia Sinica Grand Challenge Seed Program (AS-GC-108-09) to Y.H.C.

References

1. Hiesinger PR, Hassan BA (2018) The evolution of variability and robustness in neural development. Trends Neurosci 41(9): 577–586. https://doi.org/10.1016/j.tins.2018.05.007
2. Honegger K, de Bivort B (2018) Stochasticity, individuality and behavior. Curr Biol 28(1):R8–R12. https://doi.org/10.1016/j.cub.2017.11.058
3. Carbone MA, Yamamoto A, Huang W, Lyman RA, Meadors TB, Yamamoto R, Anholt RR, Mackay TF (2016) Genetic architecture of natural variation in visual senescence in Drosophila. Proc Natl Acad Sci U S A 113(43):E6620–E6629. https://doi.org/10.1073/pnas.1613833113
4. Takagi S, Benton R (2020) Animal behavior: a neural basis of individuality. Curr Biol 30(12): R710–R712. https://doi.org/10.1016/j.cub.2020.04.052
5. Pascual A, Huang KL, Neveu J, Preat T (2004) Neuroanatomy: brain asymmetry and long-term memory. Nature 427(6975): 605–606. https://doi.org/10.1038/427605a
6. Ugur B, Chen K, Bellen HJ (2016) Drosophila tools and assays for the study of human diseases. Dis Model Mech 9(3):235–244. https://doi.org/10.1242/dmm.023762
7. Luo L, Callaway EM, Svoboda K (2018) Genetic dissection of neural circuits: a decade of Progress. Neuron 98(2):256–281. https://doi.org/10.1016/j.neuron.2018.03.040
8. Jenett A, Rubin GM, Ngo TT, Shepherd D, Murphy C, Dionne H, Pfeiffer BD, Cavallaro A, Hall D, Jeter J, Iyer N, Fetter D, Hausenfluck JH, Peng H, Trautman ET, Svirskas RR, Myers EW, Iwinski ZR, Aso Y, DePasquale GM, Enos A, Hulamm P, Lam SC, Li HH, Laverty TR, Long F, Qu L,

Murphy SD, Rokicki K, Safford T, Shaw K, Simpson JH, Sowell A, Tae S, Yu Y, Zugates CT (2012) A GAL4-driver line resource for Drosophila neurobiology. Cell Rep 2(4): 991–1001. https://doi.org/10.1016/j.celrep.2012.09.011

9. Deng B, Li Q, Liu X, Cao Y, Li B, Qian Y, Xu R, Mao R, Zhou E, Zhang W, Huang J, Rao Y (2019) Chemoconnectomics: mapping chemical transmission in Drosophila. Neuron 101(5):876–893. https://doi.org/10.1016/j.neuron.2019.01.045
10. Venken KJ, Schulze KL, Haelterman NA, Pan H, He Y, Evans-Holm M, Carlson JW, Levis RW, Spradling AC, Hoskins RA, Bellen HJ (2011) MiMIC: a highly versatile transposon insertion resource for engineering Drosophila melanogaster genes. Nat Methods 8(9):737–743. https://doi.org/10.1038/nmeth.1662
11. Lee D, Huang TH, De La Cruz A, Callejas A, Lois C (2017) Methods to investigate the structure and connectivity of the nervous system. Fly (Austin) 11(3):224–238. https://doi.org/10.1080/19336934.2017.1295189
12. Scheffer LK, Xu CS, Januszewski M, Lu Z, Takemura SY, Hayworth KJ, Huang GB, Shinomiya K, Maitlin-Shepard J, Berg S, Clements J, Hubbard PM, Katz WT, Umayam L, Zhao T, Ackerman D, Blakely T, Bogovic J, Dolafi T, Kainmueller D, Kawase T, Khairy KA, Leavitt L, Li PH, Lindsey L, Neubarth N, Olbris DJ, Otsuna H, Trautman ET, Ito M, Bates AS, Goldammer J, Wolff T, Svirskas R, Schlegel P, Neace E, Knecht CJ, Alvarado CX, Bailey DA, Ballinger S, Borycz JA, Canino BS, Cheatham N, Cook M, Dreher M, Duclos O, Eubanks B, Fairbanks K, Finley S, Forknall N, Francis A, Hopkins GP, Joyce EM, Kim S, Kirk NA, Kovalyak J, Lauchie SA, Lohff A, Maldonado C, Manley EA, McLin S, Mooney C, Ndama M, Ogundeyi O, Okeoma N, Ordish C, Padilla N, Patrick CM, Paterson T, Phillips EE, Phillips EM, Rampally N, Ribeiro C, Robertson MK, Rymer JT, Ryan SM, Sammons M, Scott AK, Scott AL, Shinomiya A, Smith C, Smith K, Smith NL, Sobeski MA, Suleiman A, Swift J, Takemura S, Talebi I, Tarnogorska D, Tenshaw E, Tokhi T, Walsh JJ, Yang T, Horne JA, Li F, Parekh R, Rivlin PK, Jayaraman V, Costa M, Jefferis GS, Ito K, Saalfeld S, George R, Meinertzhagen IA, Rubin GM, Hess HF, Jain V, Plaza SM (2020) A connectome and analysis of the adult Drosophila central brain. elife 9. https://doi.org/10.7554/eLife.57443
13. Zheng Z, Lauritzen JS, Perlman E, Robinson CG, Nichols M, Milkie D, Torrens O, Price J, Fisher CB, Sharifi N, Calle-Schuler SA, Kmecova L, Ali IJ, Karsh B, Trautman ET, Bogovic JA, Hanslovsky P, Jefferis G, Kazhdan M, Khairy K, Saalfeld S, Fetter RD, Bock DD (2018) A complete electron microscopy volume of the brain of adult Drosophila melanogaster. Cell 174(3):730–743. https://doi.org/10.1016/j.cell.2018.06.019
14. Chiang AS, Lin CY, Chuang CC, Chang HM, Hsieh CH, Yeh CW, Shih CT, Wu JJ, Wang GT, Chen YC, Wu CC, Chen GY, Ching YT, Lee PC, Lin HH, Hsu HW, Huang YA, Chen JY, Chiang HJ, Lu CF, Ni RF, Yeh CY, Hwang JK (2011) Three-dimensional reconstruction of brain-wide wiring networks in Drosophila at single-cell resolution. Curr Biol 21(1):1–11. https://doi.org/10.1016/j.cub.2010.11.056
15. Komiyama T, Luo L (2006) Development of wiring specificity in the olfactory system. Curr Opin Neurobiol 16(1):67–73. https://doi.org/10.1016/j.conb.2005.12.002
16. Winnubst J, Bas E, Ferreira TA, Wu Z, Economo MN, Edson P, Arthur BJ, Bruns C, Rokicki K, Schauder D, Olbris DJ, Murphy SD, Ackerman DG, Arshadi C, Baldwin P, Blake R, Elsayed A, Hasan M, Ramirez D, Dos Santos B, Weldon M, Zafar A, Dudman JT, Gerfen CR, Hantman AW, Korff W, Sternson SM, Spruston N, Svoboda K, Chandrashekar J (2019) Reconstruction of 1,000 projection neurons reveals new cell types and organization of long-range connectivity in the mouse brain. Cell 179(1):268–281. https://doi.org/10.1016/j.cell.2019.07.042
17. Takeuchi H, Sakano H (2014) Neural map formation in the mouse olfactory system. Cell Mol Life Sci CMLS 71(16):3049–3057. https://doi.org/10.1007/s00018-014-1597-0
18. Axel R (1995) The molecular logic of smell. Sci Am 273(4):154–159. https://doi.org/10.1038/scientificamerican1095-154
19. Danchin E, Nobel S, Pocheville A, Dagaeff AC, Demay L, Alphand M, Ranty-Roby S, van Renssen L, Monier M, Gazagne E, Allain M, Isabel G (2018) Cultural flies: conformist social learning in fruitflies predicts long-lasting mate-choice traditions. Science 362(6418):1025–1030. https://doi.org/10.1126/science.aat1590
20. Ramdya P, Schneider J, Levine JD (2017) The neurogenetics of group behavior in Drosophila melanogaster. J Exp Biol 220(Pt 1):35–41. https://doi.org/10.1242/jeb.141457

21. Iyengar A, Imoehl J, Ueda A, Nirschl J, Wu CF (2012) Automated quantification of locomotion, social interaction, and mate preference in Drosophila mutants. J Neurogenet 26(3–4):306–316. https://doi.org/10.3109/01677063.2012.729626
22. Ramdya P, Lichocki P, Cruchet S, Frisch L, Tse W, Floreano D, Benton R (2015) Mechanosensory interactions drive collective behaviour in Drosophila. Nature 519(7542):233–236. https://doi.org/10.1038/nature14024
23. Linneweber GA, Andriatsilavo M, Dutta SB, Bengochea M, Hellbruegge L, Liu G, Ejsmont RK, Straw AD, Wernet M, Hiesinger PR, Hassan BA (2020) A neurodevelopmental origin of behavioral individuality in the Drosophila visual system. Science 367(6482):1112–1119. https://doi.org/10.1126/science.aaw7182
24. Chou YH, Spletter ML, Yaksi E, Leong JC, Wilson RI, Luo L (2010) Diversity and wiring variability of olfactory local interneurons in the Drosophila antennal lobe. Nat Neurosci 13(4):439–449. https://doi.org/10.1038/nn.2489
25. Thum AS, Leisibach B, Gendre N, Selcho M, Stocker RF (2011) Diversity, variability, and suboesophageal connectivity of antennal lobe neurons in D. melanogaster larvae. J Comp Neurol 519(17):3415–3432. https://doi.org/10.1002/cne.22713
26. Chin AL, Lin CY, Fu TF, Dickson BJ, Chiang AS (2014) Diversity and wiring variability of visual local neurons in the Drosophila medulla M6 stratum. J Comp Neurol 522(17):3795–3816. https://doi.org/10.1002/cne.23622
27. Goodman CS (1978) Isogenic grasshoppers: genetic variability in the morphology of identified neurons. J Comp Neurol 182(4):681–705. https://doi.org/10.1002/cne.901820408
28. Zottoli SJ (1978) Comparison of Mauthner cell size in teleosts. J Comp Neurol 178(4):741–757. https://doi.org/10.1002/cne.901780409
29. Tobin WF, Wilson RI, Lee WA (2017) Wiring variations that enable and constrain neural computation in a sensory microcircuit. elife 6. https://doi.org/10.7554/eLife.24838
30. Schulz DJ, Goaillard JM, Marder E (2006) Variable channel expression in identified single and electrically coupled neurons in different animals. Nat Neurosci 9(3):356–362. https://doi.org/10.1038/nn1639
31. Johnston RJ Jr, Desplan C (2014) Interchromosomal communication coordinates intrinsically stochastic expression between alleles. Science 343(6171):661–665. https://doi.org/10.1126/science.1243039
32. Sayin S, De Backer JF, Siju KP, Wosniack ME, Lewis LP, Frisch LM, Gansen B, Schlegel P, Edmondson-Stait A, Sharifi N, Fisher CB, Calle-Schuler SA, Lauritzen JS, Bock DD, Costa M, Jefferis G, Gjorgjieva J, Grunwald Kadow IC (2019) A neural circuit arbitrates between persistence and withdrawal in hungry Drosophila. Neuron 104(3):544–558. https://doi.org/10.1016/j.neuron.2019.07.028
33. Senapati B, Tsao CH, Juan YA, Chiu TH, Wu CL, Waddell S, Lin S (2019) A neural mechanism for deprivation state-specific expression of relevant memories in Drosophila. Nat Neurosci 22(12):2029–2039. https://doi.org/10.1038/s41593-019-0515-z
34. Anderson DJ (2016) Circuit modules linking internal states and social behaviour in flies and mice. Nat Rev Neurosci 17(11):692–704. https://doi.org/10.1038/nrn.2016.125
35. Vogt K, Zimmerman DM, Schlichting M, Hernandez-Nunez L, Qin S, Malacon K, Rosbash M, Pehlevan C, Cardona A, Samuel ADT (2021) Internal state configures olfactory behavior and early sensory processing in Drosophila larvae. Sci Adv 7(1). https://doi.org/10.1126/sciadv.abd6900
36. Honegger KS, Smith MA, Churgin MA, Turner GC, de Bivort BL (2020) Idiosyncratic neural coding and neuromodulation of olfactory individuality in Drosophila. Proc Natl Acad Sci U S A 117(38):23292–23297. https://doi.org/10.1073/pnas.1901623116
37. Kain JS, Stokes C, de Bivort BL (2012) Phototactic personality in fruit flies and its suppression by serotonin and white. Proc Natl Acad Sci U S A 109(48):19834–19839. https://doi.org/10.1073/pnas.1211988109
38. Yuan Q, Joiner WJ, Sehgal A (2006) A sleep-promoting role for the Drosophila serotonin receptor 1A. Curr Biol 16(11):1051–1062. https://doi.org/10.1016/j.cub.2006.04.032
39. Yuan Q, Song Y, Yang CH, Jan LY, Jan YN (2014) Female contact modulates male aggression via a sexually dimorphic GABAergic circuit in Drosophila. Nat Neurosci 17(1):81–88. https://doi.org/10.1038/nn.3581
40. Heisenberg M (2014) The beauty of the network in the brain and the origin of the mind in the control of behavior. J Neurogenet

28(3–4):389–399. https://doi.org/10.3109/01677063.2014.912279

41. Cognigni P, Felsenberg J, Waddell S (2018) Do the right thing: neural network mechanisms of memory formation, expression and update in Drosophila. Curr Opin Neurobiol 49:51–58. https://doi.org/10.1016/j.conb.2017.12.002
42. Gotz KG (1980) Visual guidance in Drosophila. Basic Life Sci 16:391–407. https://doi.org/10.1007/978-1-4684-7968-3_28
43. Buchner E (1976) Elementary movement detectors in an insect visual system. Biol Cybern 24:85–101. https://doi.org/10.1007/BF00360648
44. Bates AS, Schlegel P, Roberts RJV, Drummond N, Tamimi IFM, Turnbull R, Zhao X, Marin EC, Popovici PD, Dhawan S, Jamasb A, Javier A, Serratosa Capdevila L, Li F, Rubin GM, Waddell S, Bock DD, Costa M, Jefferis G (2020) Complete connectomic reconstruction of olfactory projection neurons in the fly brain. Curr Biol 30(16): 3183–3199. https://doi.org/10.1016/j.cub.2020.06.042
45. Shanbhag SR, Muller B, Steinbrecht RA (2000) Atlas of olfactory organs of Drosophila melanogaster 2. Internal organization and cellular architecture of olfactory sensilla. Arthropod Struct Dev 29(3):211–229. https://doi.org/10.1016/s1467-8039(00)00028-1
46. Couto A, Alenius M, Dickson BJ (2005) Molecular, anatomical, and functional organization of the Drosophila olfactory system. Curr Biol 15(17):1535–1547. https://doi.org/10.1016/j.cub.2005.07.034
47. Fishilevich E, Vosshall LB (2005) Genetic and functional subdivision of the Drosophila antennal lobe. Curr Biol 15(17):1548–1553. https://doi.org/10.1016/j.cub.2005.07.066
48. Jefferis GSXE, Marin EC, Stocker RF, Luo L (2001) Target neuron prespecification in the olfactory map of Drosophila. Nature 414(6860):204–208
49. Vosshall LB, Stocker RF (2007) Molecular architecture of smell and taste in Drosophila. Annu Rev Neurosci 30:505–533. https://doi.org/10.1146/annurev.neuro.30.051606.094306
50. Jefferis GS, Potter CJ, Chan AM, Marin EC, Rohlfing T, Maurer CR Jr, Luo L (2007) Comprehensive maps of Drosophila higher olfactory centers: spatially segregated fruit and pheromone representation. Cell 128(6): 1187–1203. https://doi.org/10.1016/j.cell.2007.01.040
51. Caron SJ, Ruta V, Abbott LF, Axel R (2013) Random convergence of olfactory inputs in the Drosophila mushroom body. Nature 497(7447):113–117. https://doi.org/10.1038/nature12063
52. Semmelhack JL, Wang JW (2009) Select Drosophila glomeruli mediate innate olfactory attraction and aversion. Nature 459(7244): 218–223. https://doi.org/10.1038/nature07983
53. Hallem EA, Carlson JR (2006) Coding of odors by a receptor repertoire. Cell 125(1): 143–160. https://doi.org/10.1016/j.cell.2006.01.050
54. Root CM, Semmelhack JL, Wong AM, Flores J, Wang JW (2007) Propagation of olfactory information in Drosophila. Proc Natl Acad Sci U S A 104(28):11826–11831. https://doi.org/10.1073/pnas.0704523104
55. Root CM, Masuyama K, Green DS, Enell LE, Nassel DR, Lee CH, Wang JW (2008) A presynaptic gain control mechanism fine-tunes olfactory behavior. Neuron 59(2):311–321. https://doi.org/10.1016/j.neuron.2008.07.003
56. Olsen SR, Bhandawat V, Wilson RI (2007) Excitatory interactions between olfactory processing channels in the Drosophila antennal lobe. Neuron 54(1):89–103. https://doi.org/10.1016/j.neuron.2007.03.010
57. Liu WW, Wilson RI (2013) Glutamate is an inhibitory neurotransmitter in the Drosophila olfactory system. Proc Natl Acad Sci U S A 110(25):10294–10299. https://doi.org/10.1073/pnas.1220560110
58. Wilson RI (2013) Early olfactory processing in Drosophila: mechanisms and principles. Annu Rev Neurosci 36:217–241. https://doi.org/10.1146/annurev-neuro-062111-150533
59. Huang J, Zhang W, Qiao W, Hu A, Wang Z (2010) Functional connectivity and selective odor responses of excitatory local interneurons in Drosophila antennal lobe. Neuron 67(6):1021–1033. https://doi.org/10.1016/j.neuron.2010.08.025
60. Yaksi E, Wilson RI (2010) Electrical coupling between olfactory glomeruli. Neuron 67(6): 1034–1047. https://doi.org/10.1016/j.neuron.2010.08.041
61. Shang Y, Claridge-Chang A, Sjulson L, Pypaert M, Miesenbock G (2007) Excitatory local circuits and their implications for olfactory processing in the fly antennal lobe. Cell

128(3):601–612. https://doi.org/10.1016/j.cell.2006.12.034

62. Liou NF, Lin SH, Chen YJ, Tsai KT, Yang CJ, Lin TY, Wu TH, Lin HJ, Chen YT, Gohl DM, Silies M, Chou YH (2018) Diverse populations of local interneurons integrate into the Drosophila adult olfactory circuit. Nat Commun 9(1):2232. https://doi.org/10.1038/s41467-018-04675-x
63. Das S, Sadanandappa MK, Dervan A, Larkin A, Lee JA, Sudhakaran IP, Priya R, Heidari R, Holohan EE, Pimentel A, Gandhi A, Ito K, Sanyal S, Wang JW, Rodrigues V, Ramaswami M (2011) Plasticity of local GABAergic interneurons drives olfactory habituation. Proc Natl Acad Sci U S A 108(36):E646–E654. https://doi.org/10.1073/pnas.1106411108
64. Seki Y, Rybak J, Wicher D, Sachse S, Hansson BS (2010) Physiological and morphological characterization of local interneurons in the Drosophila antennal lobe. J Neurophysiol 104(2):1007–1019. https://doi.org/10.1152/jn.00249.2010
65. Sachse S, Rueckert E, Keller A, Okada R, Tanaka NK, Ito K, Vosshall LB (2007) Activity-dependent plasticity in an olfactory circuit. Neuron 56(5):838–850. https://doi.org/10.1016/j.neuron.2007.10.035
66. Liang L, Li Y, Potter CJ, Yizhar O, Deisseroth K, Tsien RW, Luo L (2013) GABAergic projection neurons route selective olfactory inputs to specific higher-order neurons. Neuron 79(5):917–931. https://doi.org/10.1016/j.neuron.2013.06.014
67. Lai SL, Awasaki T, Ito K, Lee T (2008) Clonal analysis of Drosophila antennal lobe neurons: diverse neuronal architectures in the lateral neuroblast lineage. Development 135(17):2883–2893. https://doi.org/10.1242/dev.024380
68. Okada R, Awasaki T, Ito K (2009) Gamma-aminobutyric acid (GABA)-mediated neural connections in the Drosophila antennal lobe. J Comp Neurol 514(1):74–91. https://doi.org/10.1002/cne.21971
69. Hige T, Aso Y, Modi MN, Rubin GM, Turner GC (2015) Heterosynaptic plasticity underlies aversive olfactory learning in Drosophila. Neuron 88(5):985–998. https://doi.org/10.1016/j.neuron.2015.11.003
70. Hige T, Aso Y, Rubin GM, Turner GC (2015) Plasticity-driven individualization of olfactory coding in mushroom body output neurons. Nature 526(7572):258–262. https://doi.org/10.1038/nature15396
71. Chandler CH, Chari S, Dworkin I (2013) Does your gene need a background check? How genetic background impacts the analysis of mutations, genes, and evolution. Trends Genet 29(6):358–366. https://doi.org/10.1016/j.tig.2013.01.009
72. Evangelou A, Ignatiou A, Antoniou C, Kalanidou S, Chatzimatthaiou S, Shianiou G, Ellina S, Athanasiou R, Panagi M, Apidianakis Y, Pitsouli C (2019) Unpredictable effects of the genetic background of transgenic lines in physiological quantitative traits. G3 (Bethesda) 9(11):3877–3890. https://doi.org/10.1534/g3.119.400715
73. Ryder E, Blows F, Ashburner M, Bautista-Llacer R, Coulson D, Drummond J, Webster J, Gubb D, Gunton N, Johnson G, O'Kane CJ, Huen D, Sharma P, Asztalos Z, Baisch H, Schulze J, Kube M, Kittlaus K, Reuter G, Maroy P, Szidonya J, Rasmuson-Lestander A, Ekstrom K, Dickson B, Hugentobler C, Stocker H, Hafen E, Lepesant JA, Pflugfelder G, Heisenberg M, Mechler B, Serras F, Corominas M, Schneuwly S, Preat T, Roote J, Russell S (2004) The DrosDel collection: a set of P-element insertions for generating custom chromosomal aberrations in Drosophila melanogaster. Genetics 167(2):797–813. https://doi.org/10.1534/genetics.104.026658
74. Hoskins RA, Phan AC, Naeemuddin M, Mapa FA, Ruddy DA, Ryan JJ, Young LM, Wells T, Kopczynski C, Ellis MC (2001) Single nucleotide polymorphism markers for genetic mapping in Drosophila melanogaster. Genome Res 11(6):1100–1113. https://doi.org/10.1101/gr.gr-1780r
75. Greenspan RJ (2004) Fly pushing: the theory and practice of Drosophila genetics. Cold Spring Harbor Laboratory Press, Second Edition edn
76. Xiao C, Robertson RM (2015) Locomotion induced by spatial restriction in adult Drosophila. PLoS One 10(9):e0135825. https://doi.org/10.1371/journal.pone.0135825
77. Deshpande SA, Carvalho GB, Amador A, Phillips AM, Hoxha S, Lizotte KJ, Ja WW (2014) Quantifying Drosophila food intake: comparative analysis of current methodology. Nat Methods 11(5):535–540. https://doi.org/10.1038/nmeth.2899
78. Tsai KT, Chou YH (2019) Random walk revisited: quantification and comparative analysis of Drosophila walking trajectories. iScience 19:1145–1159. https://doi.org/10.1016/j.isci.2019.08.054

79. Konopka RJ, Benzer S (1971) Clock mutants of Drosophila melanogaster. Proc Natl Acad Sci U S A 68(9):2112–2116. https://doi.org/10.1073/pnas.68.9.2112
80. Ro J, Harvanek ZM, Pletcher SD (2014) FLIC: high-throughput, continuous analysis of feeding behaviors in Drosophila. PLoS One 9(6):e101107. https://doi.org/10.1371/journal.pone.0101107
81. Sayeed O, Benzer S (1996) Behavioral genetics of thermosensation and hygrosensation in Drosophila. Proc Natl Acad Sci U S A 93(12):6079–6084. https://doi.org/10.1073/pnas.93.12.6079
82. Chen CC, Wu JK, Lin HW, Pai TP, Fu TF, Wu CL, Tully T, Chiang AS (2012) Visualizing long-term memory formation in two neurons of the Drosophila brain. Science 335(6069):678–685. https://doi.org/10.1126/science.1212735
83. Lieber T, Kidd S, Struhl G (2011) DSL-notch signaling in the Drosophila brain in response to olfactory stimulation. Neuron 69(3):468–481. https://doi.org/10.1016/j.neuron.2010.12.015
84. Kasuya J, Iyengar A, Chen HL, Lansdon P, Wu CF, Kitamoto T (2019) Milk-whey diet substantially suppresses seizure-like phenotypes of Para(Shu), a Drosophila voltage-gated sodium channel mutant. J Neurogenet 33(3):164–178. https://doi.org/10.1080/01677063.2019.1597082
85. Liu W, Liang X, Gong J, Yang Z, Zhang YH, Zhang JX, Rao Y (2011) Social regulation of aggression by pheromonal activation of Or65a olfactory neurons in Drosophila. Nat Neurosci 14(7):896–902. https://doi.org/10.1038/nn.2836
86. Chiu H, Hoopfer ED, Coughlan ML, Pavlou HJ, Goodwin SF, Anderson DJ (2021) A circuit logic for sexually shared and dimorphic aggressive behaviors in Drosophila. Cell 184(3):847. https://doi.org/10.1016/j.cell.2021.01.021
87. Schretter CE, Aso Y, Robie AA, Dreher M, Dolan MJ, Chen N, Ito M, Yang T, Parekh R, Branson KM, Rubin GM (2020) Cell types and neuronal circuitry underlying female aggression in Drosophila. elife 9. https://doi.org/10.7554/eLife.58942
88. Yamamoto D, Kohatsu S, Koganezawa M (2013) Insect pheromone behavior: fruit fly. Methods Mol Biol 1068:261–272. https://doi.org/10.1007/978-1-62703-619-1_19
89. Koganezawa M, Kimura K, Yamamoto D (2016) The neural circuitry that functions as a switch for courtship versus aggression in Drosophila males. Curr Biol 26(11):1395–1403. https://doi.org/10.1016/j.cub.2016.04.017
90. Kubli E, Bopp D (2012) Sexual behavior: how sex peptide flips the postmating switch of female flies. Curr Biol 22(13):R520–R522. https://doi.org/10.1016/j.cub.2012.04.058
91. Shao L, Chung P, Wong A, Siwanowicz I, Kent CF, Long X, Heberlein U (2019) A neural circuit encoding the experience of copulation in female Drosophila. Neuron 102(5):1025–1036. https://doi.org/10.1016/j.neuron.2019.04.009
92. Wang K, Wang F, Forknall N, Yang T, Patrick C, Parekh R, Dickson BJ (2021) Neural circuit mechanisms of sexual receptivity in Drosophila females. Nature 589(7843):577–581. https://doi.org/10.1038/s41586-020-2972-7
93. van Breugel F, Huda A, Dickinson MH (2018) Distinct activity-gated pathways mediate attraction and aversion to CO2 in Drosophila. Nature 564(7736):420–424. https://doi.org/10.1038/s41586-018-0732-8
94. Suh GS, Wong AM, Hergarden AC, Wang JW, Simon AF, Benzer S, Axel R, Anderson DJ (2004) A single population of olfactory sensory neurons mediates an innate avoidance behaviour in Drosophila. Nature 431(7010):854–859. https://doi.org/10.1038/nature02980
95. Borst A, Heisenberg M (1982) Osmotropotaxis in Drosophila melanogaster. J Comp Physiol A 147:479–484. https://doi.org/10.1007/BF00612013
96. Takasaki T, Namiki S, Kanzaki R (2012) Use of bilateral information to determine the walking direction during orientation to a pheromone source in the silkmoth Bombyx mori. J Comp Physiol A Neuroethol Sens Neural Behav Physiol 198(4):295–307. https://doi.org/10.1007/s00359-011-0708-8
97. Kain J, Stokes C, Gaudry Q, Song X, Foley J, Wilson R, de Bivort B (2013) Leg-tracking and automated behavioural classification in Drosophila. Nat Commun 4:1910. https://doi.org/10.1038/ncomms2908
98. Gaudry Q, Hong EJ, Kain J, de Bivort BL, Wilson RI (2013) Asymmetric neurotransmitter release enables rapid odour lateralization in Drosophila. Nature 493(7432):424–428. https://doi.org/10.1038/nature11747
99. Kohatsu S, Koganezawa M, Yamamoto D (2011) Female contact activates male-specific

interneurons that trigger stereotypic courtship behavior in Drosophila. Neuron 69(3): 498–508. https://doi.org/10.1016/j.neuron.2010.12.017
100. Clark DA, Bursztyn L, Horowitz MA, Schnitzer MJ, Clandinin TR (2011) Defining the computational structure of the motion detector in Drosophila. Neuron 70(6):1165–1177. https://doi.org/10.1016/j.neuron.2011.05.023
101. Lehnert BP, Baker AE, Gaudry Q, Chiang AS, Wilson RI (2013) Distinct roles of TRP channels in auditory transduction and amplification in Drosophila. Neuron 77(1):115–128. https://doi.org/10.1016/j.neuron.2012.11.030
102. Seelig JD, Chiappe ME, Lott GK, Dutta A, Osborne JE, Reiser MB, Jayaraman V (2010) Two-photon calcium imaging from head-fixed Drosophila during optomotor walking behavior. Nat Methods 7(7):535–540. https://doi.org/10.1038/nmeth.1468
103. Bahl A, Ammer G, Schilling T, Borst A (2013) Object tracking in motion-blind flies. Nat Neurosci 16:730. https://doi.org/10.1038/nn.3386. https://www.nature.com/articles/nn.3386#supplementary-information
104. Ribeiro IMA, Drews M, Bahl A, Machacek C, Borst A, Dickson BJ (2018) Visual projection neurons mediating directed courtship in Drosophila. Cell 174(3):607–621. https://doi.org/10.1016/j.cell.2018.06.020
105. Haberkern H, Basnak MA, Ahanonu B, Schauder D, Cohen JD, Bolstad M, Bruns C, Jayaraman V (2019) Visually guided behavior and Optogenetically induced learning in head-fixed flies exploring a virtual landscape. Curr Biol 29(10):1647–1659. https://doi.org/10.1016/j.cub.2019.04.033
106. Kohatsu S, Yamamoto D (2015) Visually induced initiation of Drosophila innate courtship-like following pursuit is mediated by central excitatory state. Nat Commun 6: 6457. https://doi.org/10.1038/ncomms7457
107. Vishniakou I, Ploger PG, Seelig JD (2019) Virtual reality for animal navigation with camera-based optical flow tracking. J Neurosci Methods 327:108403. https://doi.org/10.1016/j.jneumeth.2019.108403
108. Moore RJ, Taylor GJ, Paulk AC, Pearson T, van Swinderen B, Srinivasan MV (2014) FicTrac: a visual method for tracking spherical motion and generating fictive animal paths. J Neurosci Methods 225:106–119. https://doi.org/10.1016/j.jneumeth.2014.01.010
109. Sancer G, Kind E, Plazaola-Sasieta H, Balke J, Pham T, Hasan A, Munch LO, Courgeon M, Mathejczyk TF, Wernet MF (2019) Modality-specific circuits for skylight orientation in the Fly visual system. Curr Biol 29(17): 2812–2825. https://doi.org/10.1016/j.cub.2019.07.020
110. Handler A, Graham TGW, Cohn R, Morantte I, Siliciano AF, Zeng J, Li Y, Ruta V (2019) Distinct dopamine receptor pathways underlie the temporal sensitivity of associative learning. Cell 178(1):60–75. https://doi.org/10.1016/j.cell.2019.05.040
111. Fisher YE, Lu J, D'Alessandro I, Wilson RI (2019) Sensorimotor experience remaps visual input to a heading-direction network. Nature 576(7785):121–125. https://doi.org/10.1038/s41586-019-1772-4
112. Okubo TS, Patella P, D'Alessandro I, Wilson RI (2020) A neural network for wind-guided compass navigation. Neuron 107(5): 924–940. https://doi.org/10.1016/j.neuron.2020.06.022
113. Pacheco DA, Thiberge SY, Pnevmatikakis E, Murthy M (2021) Auditory activity is diverse and widespread throughout the central brain of Drosophila. Nat Neurosci 24(1):93–104. https://doi.org/10.1038/s41593-020-00743-y
114. Black M (1992) Robust incremental optical flow. Yale University, New Haven, CT
115. Yorozu S, Wong A, Fischer BJ, Dankert H, Kernan MJ, Kamikouchi A, Ito K, Anderson DJ (2009) Distinct sensory representations of wind and near-field sound in the Drosophila brain. Nature 458(7235):201–205. https://doi.org/10.1038/nature07843
116. Chow DM, Frye MA (2008) Context-dependent olfactory enhancement of optomotor flight control in Drosophila. J Exp Biol 211(Pt 15):2478–2485. https://doi.org/10.1242/jeb.018879
117. Branson K, Robie AA, Bender J, Perona P, Dickinson MH (2009) High-throughput ethomics in large groups of Drosophila. Nat Methods 6:451. https://doi.org/10.1038/nmeth.1328. https://www.nature.com/articles/nmeth.1328#supplementary-information
118. Kabra M, Robie AA, Rivera-Alba M, Branson S, Branson K (2013) JAABA: interactive machine learning for automatic annotation of animal behavior. Nat Methods 10(1): 64–67. https://doi.org/10.1038/nmeth.2281
119. Dankert H, Wang L, Hoopfer ED, Anderson DJ, Perona P (2009) Automated monitoring

and analysis of social behavior in Drosophila. Nat Methods 6:297. https://doi.org/10.1038/nmeth.1310. https://www.nature.com/articles/nmeth.1310#supplementary information

120. Zalucki O, Day R, Kottler B, Karunanithi S, van Swinderen B (2015) Behavioral and electrophysiological analysis of general anesthesia in 3 background strains of Drosophila melanogaster. Fly (Austin) 9(1):7–15. https://doi.org/10.1080/19336934.2015.1072663

121. Itskov PM, Moreira JM, Vinnik E, Lopes G, Safarik S, Dickinson MH, Ribeiro C (2014) Automated monitoring and quantitative analysis of feeding behaviour in Drosophila. Nat Commun 5:4560. https://doi.org/10.1038/ncomms5560

122. Alvarez-Salvado E, Licata AM, Connor EG, McHugh MK, King BM, Stavropoulos N, Victor JD, Crimaldi JP, Nagel KI (2018) Elementary sensory-motor transformations underlying olfactory navigation in walking fruit-flies. elife 7. https://doi.org/10.7554/eLife.37815

123. Lints FA, Lints CV, Bullens P, Bourgois M, Delince J (1989) Unexplained variations in life span of the Oregon-R strain of Drosophila melanogaster over a four-year period. Exp Gerontol 24(3):265–271. https://doi.org/10.1016/0531-5565(89)90017-x

124. Bhandawat V, Maimon G, Dickinson MH, Wilson RI (2010) Olfactory modulation of flight in Drosophila is sensitive, selective and rapid. J Exp Biol 213(Pt 21):3625–3635. https://doi.org/10.1242/jeb.040402

125. Frye MA, Dickinson MH (2004) Motor output reflects the linear superposition of visual and olfactory inputs in Drosophila. J Exp Biol 207(Pt 1):123–131. https://doi.org/10.1242/jeb.00725

126. Yin JC, Wallach JS, Del Vecchio M, Wilder EL, Zhou H, Quinn WG, Tully T (1994) Induction of a dominant negative CREB transgene specifically blocks long-term memory in Drosophila. Cell 79(1):49–58. https://doi.org/10.1016/0092-8674(94)90399-9

Chapter 14

Automated Behavior Analysis Using a YOLO-Based Object Detection System

Kai Amino and Takashi Matsuo

Abstract

Object detection methods employing convolutional neural networks have fueled the development of automated behavior analysis, which is traditionally composed of two steps of data processing: tracking and behavior annotation. The former generates the time series of positional data describing the movement of individuals and/or body parts, which is subsequently analyzed in the latter to identify particular behaviors/actions using the mathematically defined criteria. YOLO is a state-of-the-art algorithm for high-speed object detection, which can be used for the tracking of animals in video images. In this section, we extend the use of YOLO for the annotation of various behaviors in a single step. To test the usability of this method, we analyzed the territorial and courtship behavior in a hyper-aggressive fruit fly, *Drosophila prolongata*. Automated behavior analysis using YOLO successfully identified when and where each fight and courtship event occurred, demonstrating that this YOLO-based behavior analysis system can serve as a versatile tool for behavioral studies because it is easy to handle for researchers who are not experts in computer science.

Key words Deep learning, Automated analysis, Behavior annotation, Territoriality

1 Introduction

1.1 Automated Behavior Analysis in Neurogenetics

Automated analysis of animal behavior has been used in various studies [1–4]. In behavior neurogenetics, combinatorial use of optogenetic manipulation of defined neurons and automated detection of particular actions/behaviors resulted in the identification of neural circuits and command neurons controlling various behaviors such as aggression and mating [5, 6]. The power of the high-throughput capacity of this method has been most successfully shown in large-scale screens to find rare and subtle changes in behavioral patterns among many iterations/samples [7, 8]. Automated behavioral analysis has also led to the discovery of a novel behavior that had been otherwise overlooked in human observations [9].

Daisuke Yamamoto (ed.), *Behavioral Neurogenetics*, Neuromethods, vol. 181, https://doi.org/10.1007/978-1-0716-2321-3_14,

1.2 Conventional Methods

Although there are many systems for automated behavior analysis, most of them consist of two parts: detection of individuals and/or body parts (tracking) and detection of behaviors/actions (behavior annotation) [1, 2]. The first part generates a time series of the position of individuals and/or body parts, which is subsequently analyzed in the second part to predict the occurrence possibility of each behavior/action, in some cases using machine learning methods such as a support vector machine. Here, we look at each part closely in ctrax and CADABRA, the most widely used systems for *Drosophila* behavior analysis [10, 11].

The first part of the analysis, tracking, involves three steps: background estimation, foreground extraction, and object detection. Even in controlled laboratory experiments, the background of the video images is often uneven, making the contrast between foreground (animals) and background variable across the recorded area. A simple but widely used method is background subtraction. The background without animals is estimated using multiple frames in the video, in which moving animals are erased by averaging many frames. Subtraction of the estimated background from each frame leaves foreground information. A cluster of foreground pixels often represents an individual animal, but sometimes consists of two or more individuals touching/overlapping each other (occlusion). Ctrax solves this problem by trying multiple hypothetical separations of the pixel cluster into individuals represented as ellipses. The separation hypothesis with the lowest penalty score is taken as the result of object detection. CADABRA circumvents the same problem in a more sophisticated way. It estimates the foreground pixels by fitting a Gaussian mixture model using the expectation-maximization algorithm. This method allows detection of not only the fly body but also their wings simultaneously.

The second part, behavior annotation, is a classifier that assigns the detected objects to the mathematically defined behaviors. It uses the primary parameters (position, orientation of objects) directly derived from the tracking part as well as secondary parameters (velocity calculated between neighboring frames, interval and angle between individuals, etc.) for classification of behavior. Secondary parameters are generated not only from information in a single video frame but also from a sliding window of sequential frames. CADABRA generates 25 parameters for each object in a given frame. These parameters are used to mathematically define each behavior, often aided by machine learning. A frontend helper application provides a visual interface for building a behavior classifier using machine learning [12]. Nevertheless, ctrax and CADABRA mostly adopted intuitively and empirically determined parameter values and ranges for behavior definition because they worked satisfactorily for most behaviors of *Drosophila*.

Although these systems have contributed to numerous studies, there are inevitable problems in the conventional two-step methods. First, object detection in the tracking step is highly vulnerable to the quality of images. Even a faint irregularity in the brightness and contrast in the background, such as a shadow of an animal, results in a false detection. To circumvent this problem, observation chambers need to be smooth and flat with highly coherent illumination. As a consequence, experiments are often performed in the artificial conditions differing substantially from natural conditions. Second, mathematical definitions of behaviors depend on the target species and experimental environment: the whole procedure of empirical optimization of the mathematical behavior definition needs to be repeated for each task. These problems have been an obstacle for the application of the conventional methods to other behaviors of other species.

1.3 Integration of AI

Recently, the development of AI technology based on deep neural networks substantially improved the performance of automated image analysis, particularly the tracking part. Specifically, foreground detection now became highly robust against inconsistent illumination and complex backgrounds. Many free tools are available for tracking tasks [5]. Most of them are quite easy to use and flexibly applicable to various study targets without much effort to adjust the parameter settings.

In contrast, the latter half, behavior annotation, has not yet received the benefit of the AI technology. It remains as a process that requires a lot of trials and optimization of the parameters to obtain satisfactory performance for each task. If integration of AI made the behavior annotation process as easy as the tracking process, such systems would aid in finding otherwise overlooked behavioral phenomena.

1.4 YOLO in Behavioral Analysis

YOLO is a state-of-the-art algorithm developed for real-time object detection using a deep neural network [13]. It detects what objects are located where in a given image. Before YOLO, such calculation was time-consuming (~ 1 frame per sec: fps), whereas YOLO performs object locating and classification simultaneously, enabling processing speeds as high as 30 fps, which is fast enough to analyze video images in real time. Even though YOLO was originally developed for machine vision of automatic vehicle-driving system, its versatility led to its use for many other tasks including tracking animals [14–17]. In such applications, YOLO was particularly useful in generalizing an object's attributes (e.g., objects with different colors can be classified in a single class). However, we thought it can be trained to detect different postures of the same individual as distinct "behavior" classes. In other words, YOLO has the potential to be a single-step behavior detector that can be applied to different tasks flexibly. In fact, there are some examples where YOLO was applied to recognize human facial expressions [18, 19], as well as to detect unusual behaviors of humans, sheep, and chickens [20–22].

In this section, we show that a YOLO-based system is able to detect the aggression and courtship behavior of fruit flies. Our method makes YOLO serve as a reliable behavior detector, opening up a convenient route to automated behavioral analysis.

2 Materials

2.1 Target Species

We demonstrate the usability of a YOLO-based behavior analysis system using *Drosophila prolongata*. Males of *D. prolongata* are hyper-aggressive, and they fight each other over food resources [23–25]. They also show a unique courtship behavior, leg vibration (LV), in which a courting male vibrates their forelegs that are extended along the courted female's body from the front of her. Their aggression and courtship behaviors have been studied independently, but the interaction between the two behaviors has never been studied. For example, it has not been examined quantitatively whether males of *D. prolongata* fight for territoriality to increase mating opportunity or just to keep individual space, which does not necessarily increase mating success [26].

Because of the following reasons, conventional systems are insufficient for automated behavior analysis of *D. prolongata*. First, no mathematical definition is available for *D. prolongata* behavior. Substantial efforts would be required to establish such definitions from scratch. Second, observation of the behavior of *D. prolongata* requires a relatively complex setup of chambers. For instance, a food podium raised from the floor level is necessary to observe their courtship and aggression behavior efficiently. Conventional background extraction methods did not work well in this setup because of the shadow of the podium and individuals. Therefore, a novel method of automated behavior analysis has been awaited.

2.2 Insect Preparation

Two strains of *D. prolongata*, BaVi043 (aggressive) and SaPa014 (unaggressive), were used. These strains had been maintained for more than 100 generations by inbreeding in laboratory conditions with an average population size of around 50 individuals in a culture bottle (Drosophila bottles AS355, Thermo Fisher Scientific, Waltham, MA, USA) [24]. Flies were reared on standard cornmeal medium for *Drosophila* culture [27]. All cultures were maintained at 20 °C, and all experiments were carried out at the same temperature because *D. prolongata* failed to develop at higher temperatures [28]. A 12:12 h light/dark cycle was applied throughout the experiments. In these conditions, the entire life cycle, from an egg to an adult, takes approximately 16 days for females and 18 days for males, and the life span was at least 1 month for both sexes [24, 28]. The rearing density during the developmental stages was controlled by keeping the number of parental flies at 20 males and 20 females in a bottle containing 50 ml of medium for egg laying, and they were transferred to a new bottle every 2 days.

Newly eclosed adults were maintained in glass vials (28.5 mm diameter and 95 mm high) containing culture medium. Because social interaction with other individuals affects male behavior [29], males were maintained individually in separate vials. Females were maintained in a group of up to ten individuals per vial. Color marking of the males was carried out on the fifth day after eclosion using water-based pigment ink (yellow or blue; Posca®, Mitsubishi Pencil Co., Ltd., Tokyo, Japan). A large dot was painted on the dorsal surface of the thorax of males that were anesthetized on ice. Marking of each individual was finished within 90 sec, and very few individuals died after recovery. On the sixth day after eclosion, males were transferred to new vials containing a wet cotton ball for a day of starvation. This treatment increases male locomotor activity and encounter rate during the observation, improving the efficiency of experiments [30].

2.3 Video Recording

Behavioral observation was carried out on the seventh day after eclosion in a glass chamber (50 mm diameter and 70 mm high), the inner wall of which was treated with silicon polish to prevent the flies from climbing. A disc of wet filter paper was placed at the bottom of the chamber, and at the center of it was placed a food podium, consisting of the lid of a disposable conical tube (small podium, a lid of 15 ml tube, 23 mm diameter, and 11 mm high; large podium, a lid of 50 ml tube, 36 mm diameter, and 11 mm high) filled with *Drosophila* instant medium (Formula 4–24 Drosophila Medium, Carolina Biological Supply Co., Burlington, NC, USA). Two sizes of food source were used to examine how the area that is defended by a male is affected by the resource size. Location of the female, as well as the food podium, is an important determinant of the resource value, and the movement of the female is expected to cause temporal and special fluctuation of the resource value. To avoid bringing this unnecessary complexity into the analysis, tethered females were used in accordance with the method described previously [23]. Approximately 3 h before the observation, females were anesthetized on ice, and a small glass bar (8 mm long and 0.56 mm diameter) made from a microcapillary (Microcap 0.5 μl, Drummond Scientific Co., Broomall, PA, USA) was attached to the tip of the abdomen using an instant adhesive (Aron Alpha®, Konishi Co., Osaka, Japan). The hindlegs were also glued on to the bar in order to prevent injuries caused by twisting of the abdomen. The treatment was finished within 3 min for each female. The treated female was tethered at the center of the food podium by inserting the glass bar into the food. The tethered female was able to move her wings and forelegs as well as to turn around but not able to walk or copulate. Up to eight chambers were arranged in two rows, isolated from each other by paper partitions, and covered with glass plates. Two males (yellow and blue) were introduced into each chamber, and their behavior was recorded for 30 min using a digital video camera (HDR-CX630V, Sony, Tokyo, Japan) installed 80 cm above the chambers. For comparison, single male treatments (one male and a tethered female) were also made. All observations were conducted during the last 2 h of the light phase because the highest locomotor activity was observed during either the first 2 h or the last 2 h of the light phase (T. Yoshimizu & T. Matsuo, personal observation).

3 Methods

3.1 Pre-processing

The video was taken at the HD quality (1440×1080 pixels, aspect ratio = 16:9, 30 fps) in the AVCHD (.m2ts) format, which was converted to the MP4 (.mp4) format using the ffmpeg software before further processing (Fig. 14.1). Using MATLAB, the video was down-sampled in resolution and frame rate (960×540 pixels, 16:9, 1 fps) and converted into an uncompressed .avi file, in which the consecutive frame number was imposed. This file was kept as a source video file and used for the following analysis. In training and detection, an image of each chamber (240×240 pixels) was cut out from each frame of the source video.

3.2 Training Datasets

We defined eight object classes: yellow male, blue male, wing vibration (WV) by yellow male, WV by blue male, leg vibration (LV), female, tussling, and side (Fig. 14.2). WV is a courtship behavior commonly observed in many *Drosophila* fruit flies, in which a male performs unilateral wing extension accompanied with a "courtship song" produced by vibration of the extended wing (Fig. 14.2c and d) [26]. LV is a courtship behavior specifically observed in *D. prolongata*, in which a male vibrates its forelegs on the female's body in the final stage of the courtship ritual (Fig. 14.2e) [31–33]. The LV class was not divided by the male identity (marking color) because LV was performed much more rarely than WV and fewer images for the training datasets were available. Tussling is the most intensive male-to-male aggressive interaction in which males stand up on their mid- and hindlegs and thrust their body with their forelegs extended toward the opponent (Fig. 14.2g) [24, 25]. Tussling involves two males by definition. The "side" class was assigned to individuals on the sidewall of the food podium, whose dorsal marking was invisible and their identity could not be determined (Fig. 14.2h).

The images for training datasets were taken from the first 12 source videos. A scene with a particular behavior was identified manually by inspecting the source video using VLC media player software (VideoLan organization), and a snapshot was saved as a

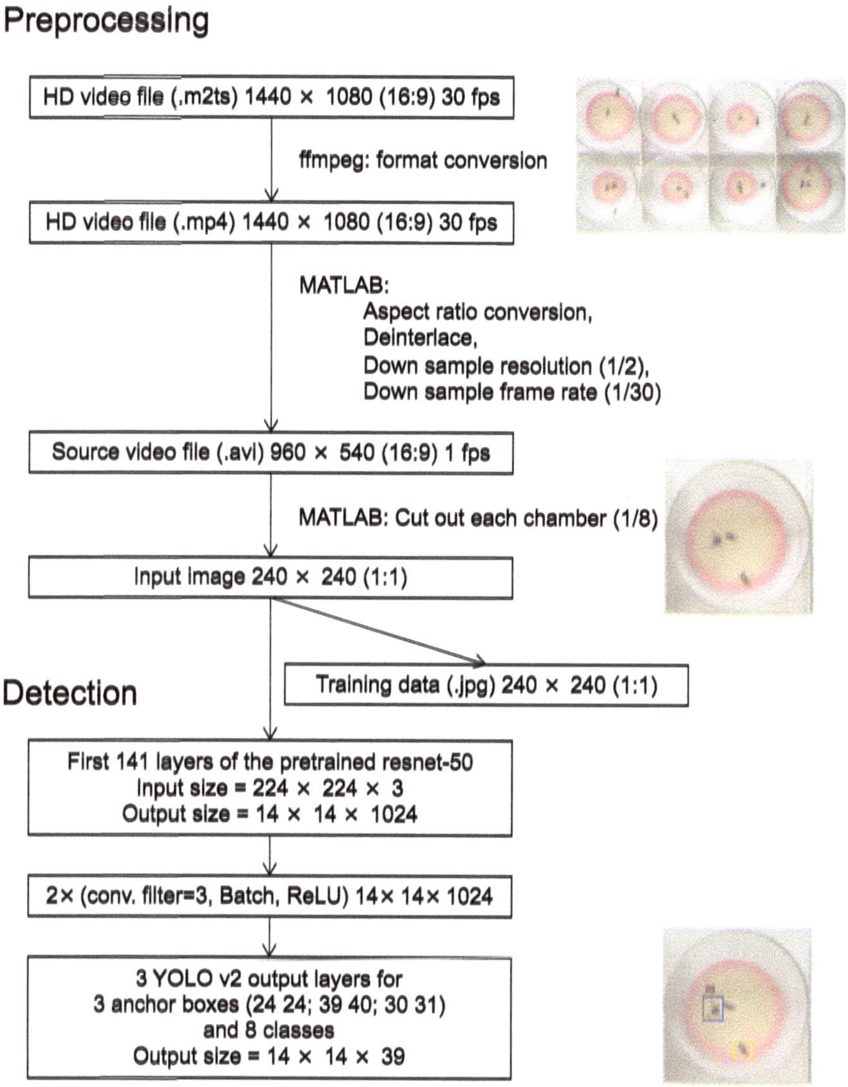

Fig. 14.1 Data processing scheme
The pre-processing section describes the data flow in the preparation of images fed into the detection network. The detection section describes the network layers through which the input image was analyzed

JPEG (.jpg) image file. Then, the image of the corresponding chamber was cut out into a 240 × 240 pixels JPEG file using Gimp software, which was manually annotated using a MATLAB built-in application, Image Labeler. The number of images labeled for training datasets was 3534 at this step.

The original training dataset was augmented three times using the helper functions provided in the MATLAB tutorial document (https://mathworks.com/help/releases/R2019b/vision/ug/train-an-object-detector-using-you-only-look-once.html), which

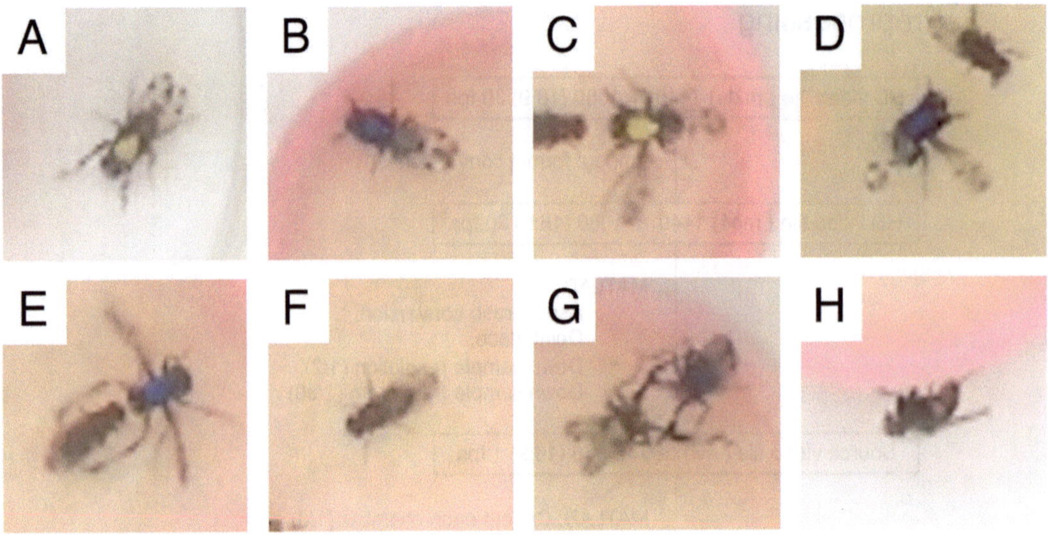

Fig. 14.2 Definition of the object classes
(**a**) A yellow-marked male. (**b**) A blue-marked male. (**c**) Wing vibration (WV) performed by a yellow male. (**d**) WV performed by a blue male. (**e**) Leg vibration (LV). (**f**) A female. (**g**) Tussling. (**h**) A male on the sidewall of the food podium

involved a random XY flip, rotation (within ±15°), and color jittering (hue, contrast, saturation, and brightness).

3.3 Training of the YOLO v2 Network and Detection

The whole process of transfer learning and detection was performed on MATLAB R2019b in accordance with the MATLAB tutorial document. Briefly, the last 36 layers of the ResNet-50 convolutional neural networks pre-trained with the ImageNet datasets were replaced with the 9 layers designed for YOLO detection. The resulting network was trained with 14,136 training images prepared as described above for 200 epochs on a Windows PC (CPU, Core i5 9400F 6 core 2.9 GHz; GPU, NVIDIA TITAN RTX × 1). Prior to detection, the position of the food podium in each chamber of the source video was specified manually using a custom-made MATLAB interface. A class-wise non-maximal suppression was applied to the detected bounding boxes because in each chamber the number of objects belonging to each class should not be more than one in this experiment. A CSV file of a table containing the positions of all bounding boxes in every frame was generated for each chamber.

3.4 Reliability: Accuracy of Detection and Correction of Errors

To confirm the accuracy of the YOLO behavior detection, we re-examined the results manually. The number of video frames in which tussling or LV were detected was not large (4943 for tussling and 832 for LV), and all the frames were re-examined manually whether the detections were correct. The false detections were excluded from the subsequent analyses. Among 4943 "tussling"

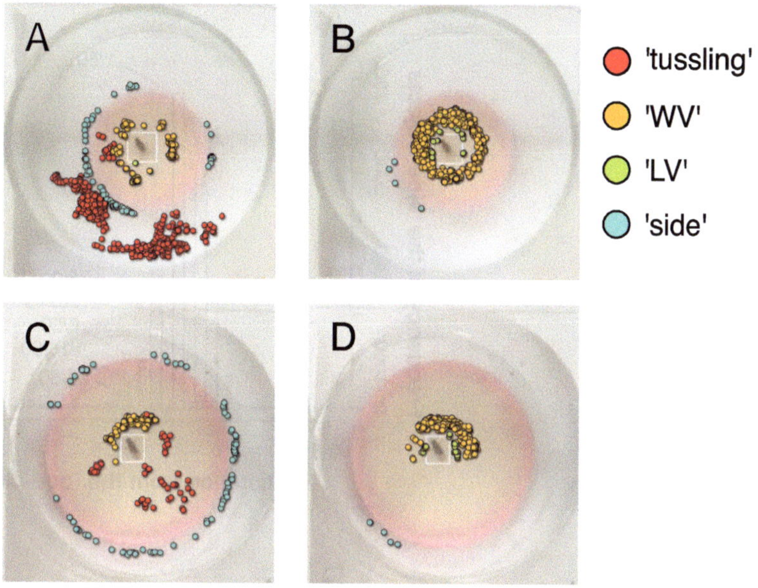

Fig. 14.3 Distribution of the detected events
Positions of the objects (events) detected during 30 min of observation were plotted on the background image. A female was tethered at the center of the food podium. The "side" class represents the individuals on the sidewall of the podium. (**a**, **b**) Small food podium. (**c**, **d**) Large food podium. (**a**, **c**) Two males. (**b**, **d**) Single male

class objects detected, 4170 (84.4%) were confirmed as tussling, and 390 (7.9%) were antagonistic interactions that were not exactly tussling. Only 383 (7.7%) were erroneous detections that were not aggressive interactions. Among 832 "LV" class objects detected, 683 (82.1%) were accurately recognized as LV, and 28 (3.4%) were courtship behaviors other than LV. Among erroneous detections, 91 (10.9%) were non-courtship male-female interactions, and 31 (3.7%) were other than male-female interactions.

Some classes of objects were expected to be found at certain locations (Fig. 14.3). For example, "side" is a class dedicated for the individuals on the sidewall of the food podium, whose color markings on their back were invisible. By definition, "side" class objects should only be detected near the edge of the podium. This feature is used to test the level of accuracy of our detection system. In fact, the positions of the "side" class objects were narrowly distributed proximately outside the edge of the podium, with quite a few outliers that should be erroneous detections (Fig. 14.4a). We sampled 50 frames from each bin (of 0.25 mm width) to manually count false detections, such as flies on the arena floor instead of on the sidewall. Within 1.25 mm from the edge of the food podium, more than 95% of the detected "side" class objects were accurate, suggesting that 1.25 mm corresponds to

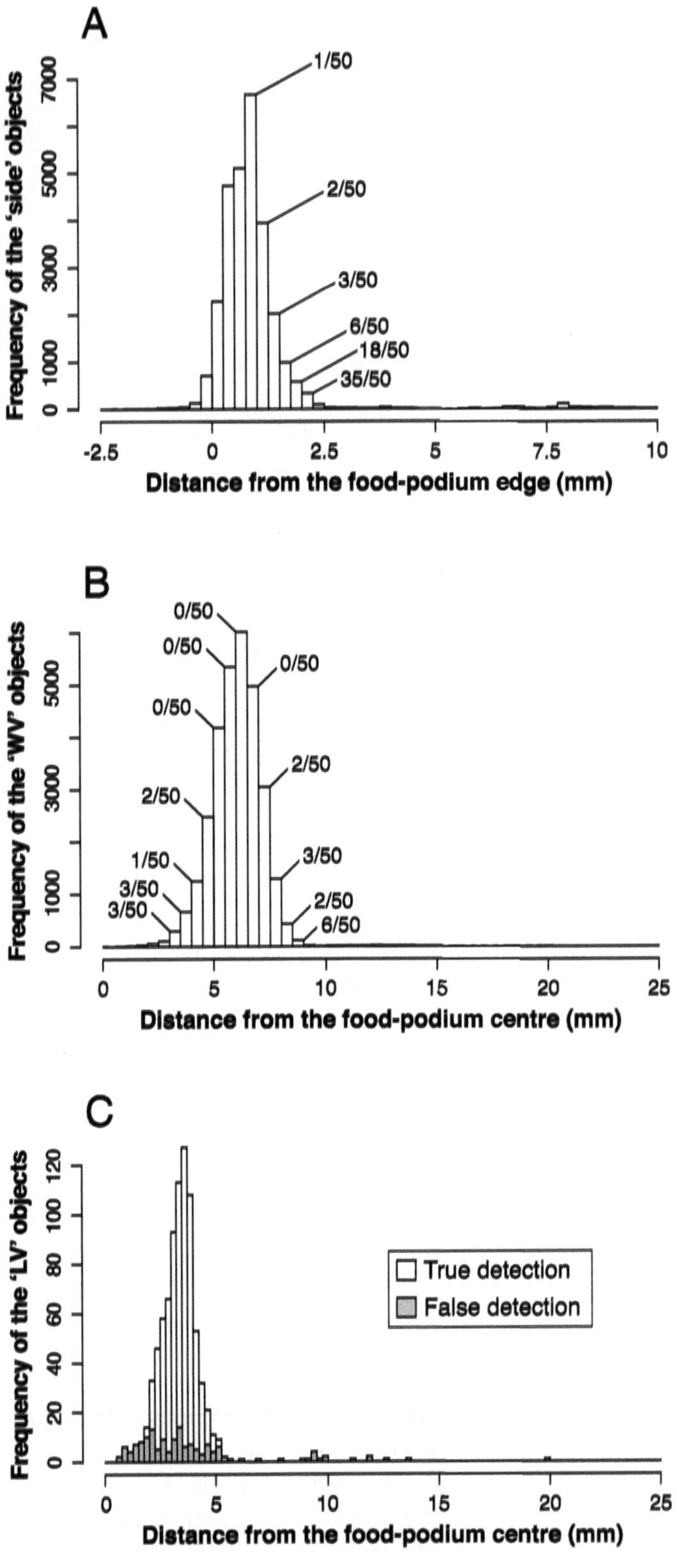

Fig. 14.4 Accuracy of the detected objects
(**a**) "Side" class. For each bin, 50 detections were sampled for manual

the height of a body extruding from the podium wall. Therefore, we set this value as the operational boundary between the food podium and arena. We manually re-examined part of the WV detections. All WV objects were sorted according to the distance from the podium center. From each bin of 0.5 mm width, 50 detections were randomly sampled and manually re-examined.

Because females were tethered at the center of the podium in this study, courtship behavior should also be observed near the center area of the food podium. Detected WV class objects were distributed in a short range from the center of the podium (Fig. 14.3, Fig. 14.4b). Manual re-examination of sampled frames showed that more than 95% of detections were accurate within a range of 4.0–7.5 mm from the podium center. The WV class objects detected out of this range were excluded from the subsequent analyses. Another type of courtship, LV, which was usually performed at the last stage of the courtship ritual, was also detected around the tethered female, a bit closer to the center than WV because LV involves physical contact with the female (Fig. 14.2c). LV class objects detected at positions farther than 5 mm from the center were mostly false detections, but the number of such detections was quite small (Fig. 14.4c).

3.5 Analysis of the Relationship Between Aggression and Courtship

To demonstrate the capacity of the YOLO-based system, we analyzed the aggression and courtship behavior among three individuals (2 males and 1 female). Location of each male, distances between individuals, and occurrence of tussling or courtship (WV and LV) can be detected simultaneously and recorded every second during 30 min of the observation (Fig. 14.5). Such data allow examination of the relationship between tussling events and the male-to-male distance in a sequential context (Fig. 14.6), as well as the interaction between the male-to-male distance and the male-to-female distance (Fig. 14.7).

3.6 Weaknesses

The disadvantage of the YOLO-based system is the absence of time factors from the detection process. Because YOLO uses information within a single video frame and does not refer to the previous or subsequent frames, the results do not reflect any information about sequential changes in the appearance or velocity of movements, which are important for accurate discrimination of behaviors resembling each other. Appropriate post-processing of the

Fig. 14.4 (continued) re-examination. The proportion of false detections among 50 randomly selected detections for manual re-examination is indicated above each bar. (**b**) "WV" class. For each bin, 50 detections were sampled for manual re-examination. (**c**) "LV" class. All the detected objects were manually re-examined for this class. False detections (excluded from the subsequent analyses) are shaded

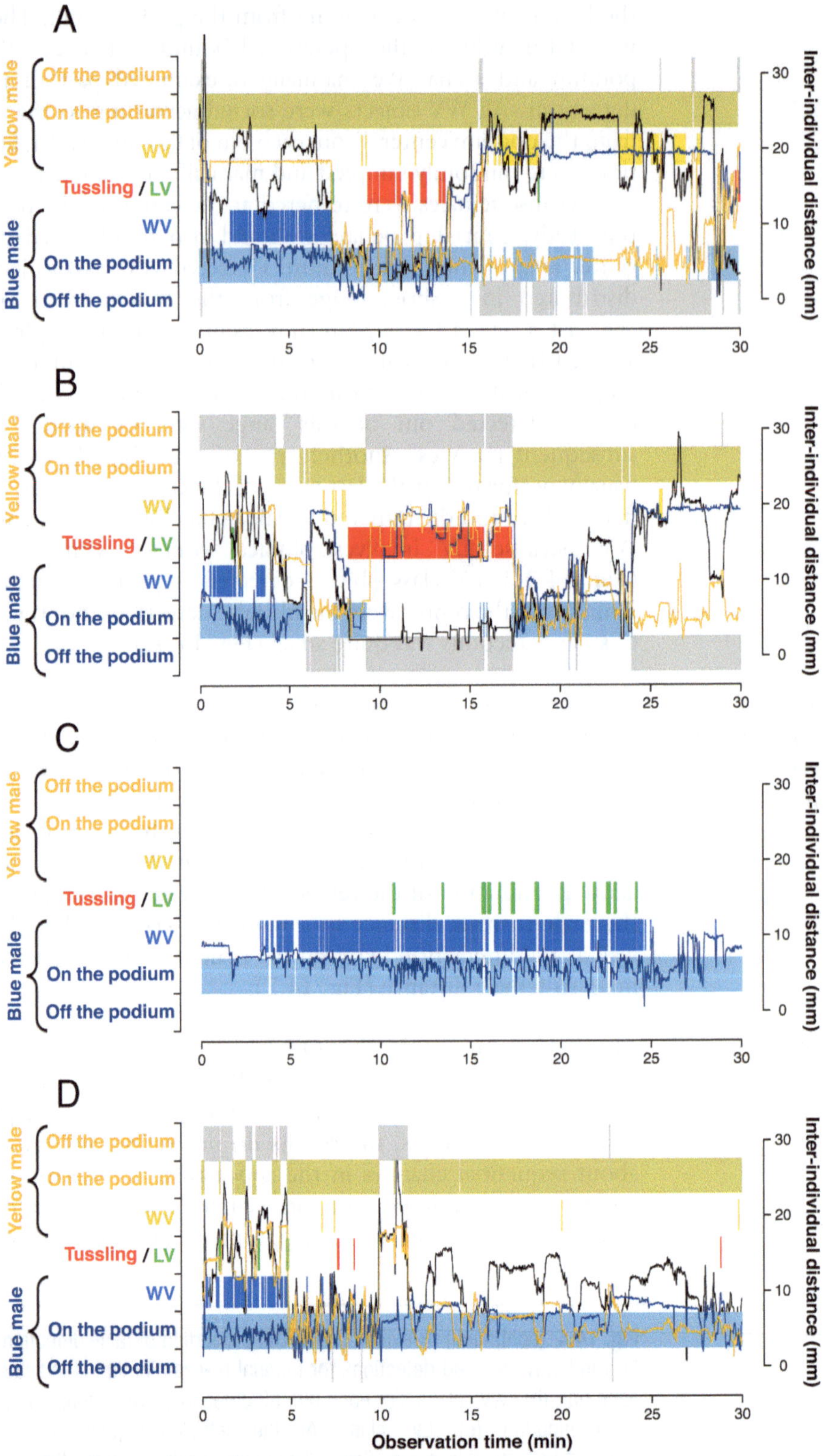

Fig. 14.5 Sequential behavior annotation during 30 min of observation
Each panel represents the results from a single observation chamber. Location of each male, distances

obtained data may overcome this weakness. For example, tracing an individual identity through a time course provides an opportunity to assign a higher-level behavior annotation to a particular sequence of posture detections. Nevertheless, we need to be aware of the consequence of too much elaboration in this direction, which leads the whole system to return to the conventional two-step method.

4 Future Directions

4.1 Use of Free-Moving Females

Although the tethered female method was helpful to simplify the ecological context in the example shown in this chapter, it is obviously preferred to use free-moving females for experiments in more realistic situations. We tried to analyze the videos from experiments using free-moving females with the YOLO network trained using videos with tethered females. Such a network could detect only females on the food surface and failed to recognize those on the arena or sidewall of the podium, probably because the training data did not contain such images. Independent training of a YOLO network with a dedicated training dataset will make it possible to analyze video data of free-moving females.

4.2 Multiple Class Detection

Although we used version 2 of YOLO in this study primarily because it was implemented in MATLAB as a built-in function, version 3 is used more widely. Among a few differences between the two versions, one has a potential to improve the performance of the system. In YOLO v2, object class prediction is exclusive; if an object is predicted to be class A, it cannot be class B at the same time. This does not perfectly fit our needs in behavioral annotation, where individual A makes action B. Because "identity" and "action" are represented by different classes, they are exclusive in YOLO v2. This is problematic in two points.

First, it leads to the fragmentation of tracks. Not only the behavioral annotation but also the reliable tracking of each individual is an indispensable basis for any kind of automated behavior analysis. In YOLO v2, however, the "identity" class was not assigned to the object when an "action" class had a higher prediction score, resulting in the fragmentation of the tracks. We circumvented this problem by making classes have information about both

Fig. 14.5 (continued) between individuals (lines; black, male-male; yellow, yellow male-female; blue, blue male-female), and occurrence of tussling or courtship (LV and WV) were detected simultaneously. (**a**) The blue male courted the female initially, but the yellow male took over the position after tussling events. (**b**) Two males competed for a long time, and finally the blue male retreated. (**c**) A single male performed intensive courtship throughout the observation without a rival. (**d**) Males of the unaggressive strain co-existed on the food podium and rarely performed tussling

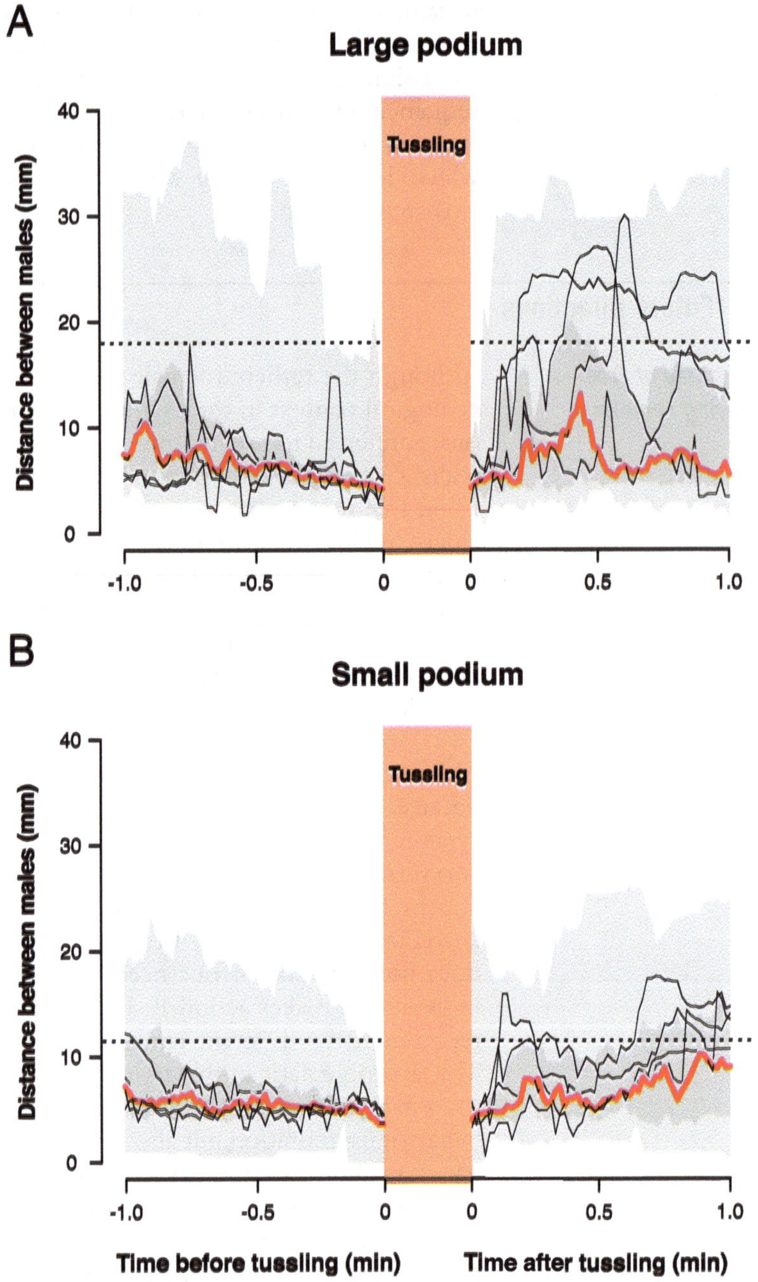

Fig. 14.6 Time course of the male-to-male distance before/after tussling events Shading represents the maximum/minimum values (pale gray) and the quartiles (dark gray). Lines represent the traces of representative pairs (black), the mean of 30 cases (red), and the podium radius (dashed). Note that the upper quartile was beyond the podium radius after tussling events, suggesting that more males were expelled after tussling

identity and action at the same time. For example, for the frequently observed courtship behavior WV, two distinct classes

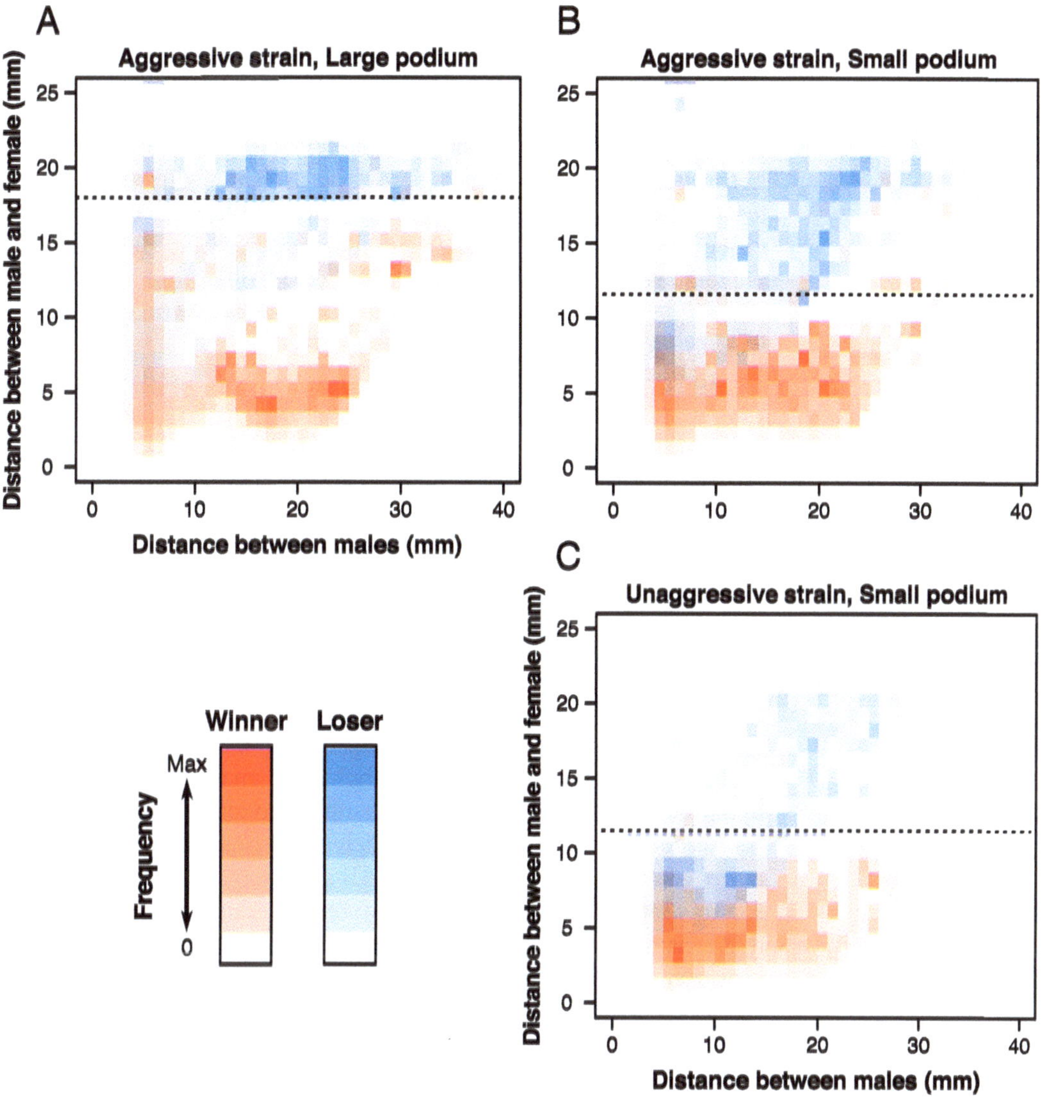

Fig. 14.7 Relationship between the male-to-male distance and the male-to-female distance "Winner" represents the male that was closer to the female at that moment. The dashed line indicates the podium radius. (**a**) The aggressive strain with the large podium. Losers were mostly found outside of the podium. (**b**) The aggressive strain with the small podium. The podium size defined the boundary of the male distribution. (**c**) The unaggressive strain with the small podium. Most males closely located around the female

were defined according to the performer ("WV by yellow male" and "WV by blue male"). This allowed the reconstruction of the track by post-processing of the data. However, it is obviously preferable to obtain the complete tracks in the first place.

Second, single class prediction of YOLO v2 leads to interference between classes during training. In our case, detection of "action" classes was interfered by the prediction score of "identity" classes. If the training data contain many "identity" class images and the network learned "identity" better than "action," the prediction scores for a given object tend to be higher for "identity" than for "action" classes, suppressing the detection of "action" classes. This is problematic because better learning may result in poorer detection, and such interference is difficult to predict beforehand (e.g., we cannot know beforehand when to stop the learning of "identity" classes). As described above, defining the "identity + action" classes partially circumvented this problem. Nevertheless, "tussling" and "LV" were still mutually exclusive with "identity," and the identity of males under tussling or LV was not detected.

In YOLO v3, multiple class prediction is used during the training stage to avoid unwanted suppression of learning one class by others. Unfortunately, in the detection stage, YOLO v3 still outputs only one class with the highest prediction score (non-maximal suppression). Further modification of the YOLO v3 output section is necessary to obtain multiple class predictions for each object, which would substantially improve the performance in the behavioral annotation tasks. It may be possible to make a system to detect three layers of classes (e.g., action + identity + body parts) simultaneously.

4.3 Real-Time Analysis

Although we analyzed the recorded video images in this section, there are demands of real-time behavior analysis. For example, dynamic activation of neurons using optogenetics according to the target position/action is a widely used paradigm in behavioral neurogenetics. The use of MATLAB as a platform is advantageous for this kind of extension of the system. Because the USB camera driver is available in MATLAB, only a few lines of the code are required to be modified to implement real-time analysis. Similarly, various hardware interfaces are provided, being ready to be integrated with the YOLO detection system to control peripheral instruments depending on the result of behavior detection.

5 Conclusions

Our analysis demonstrated that behavioral annotation using YOLO has several advantages over conventional methods.

1. Simple and quick: YOLO can perform tracking and behavioral annotation simultaneously, which is clearly advantageous over conventional two-step systems composed of independent frameworks for each task. Because YOLO was originally designed for real-time use with relatively high frame rate movies, future applications may include real-time behavior analysis using this method.

2. Easy to train: Users can control the system just by preparing training datasets. In other words, we do not need to be concerned with numerous parameters for optimization of the system. Instead, we just need to focus on preparing good training datasets, which should be an accomplishable task even for researchers who are not familiar with computer sciences. Furthermore, transfer learning reduced the amount of required training data, making the whole process easier. In fact, only a few thousand pictures were manually annotated for successful training in this study.
3. Robust and flexible: Conventional object detection methods are comprised of background subtraction and foreground extraction using binarization of images, which makes the process highly sensitive to disturbance such as shadows, collisions of objects, and inconsistent lighting conditions among videos taken on different occasions. This constrained the design of experimental setups (chambers). For example, a flat, counter-illuminated floor without any other objects such as food sources has been preferentially used. In contrast, object detection using a convolutional neural network (including YOLO) is highly robust on uneven backgrounds in terms of brightness, contrast, and color balance. Therefore, it is widely applicable to the various behavioral experiments using more realistic setups involving non-flat floors, food sources, shelters, etc.

These advantages make YOLO-based behavioral analysis unparalleled among various methods that have been used for the same purpose. Because of its simplicity, easy operation, and versatility to different tasks, we expect that this method can be applied to a wide range of behavioral analyses.

References

1. Dell AI, Bender JA, Branson K et al (2014) Automated image-based tracking and its application in ecology. Trends Ecol Evol 29:417–428. https://doi.org/10.1016/j.tree.2014.05.004
2. Egnor SER, Branson K (2016) Computational analysis of behavior. Annu Rev Neurosci 39: 217–236. https://doi.org/10.1146/annurev-neuro-070815-013845
3. Robie AA, Seagraves KM, Egnor SER, Branson K (2017) Machine vision methods for analyzing social interactions. J Exp Biol 220:25–34. https://doi.org/10.1242/jeb.142281
4. Valletta JJ, Torney C, Kings M, Thornton A, Madden J (2017) Applications of machine learning in animal behavior studies. Anim Behav 124:203–220. https://doi.org/10.1016/j.anbehav.2016.12.005
5. Abbas W, Rodo DM (2019) Computer methods for automatic locomotion and gesture tracking in mice and small animals for neuroscience applications: a survey. Sensors 19:3274. https://doi.org/10.3390/s19153274
6. Moulin TC, Covill LE, Itskov PM, Williams MJ, Sciöth HB (2021) Rodent and fly models in behavioral neuroscience: an evaluation of methodological advances, comparative research, and future perspectives. Neurosci Biobehav Rev 120:1–12. https://doi.org/10.1016/j.neubiorev.2020.11.014
7. Hoopfer ED, Jung Y, Inagaki HK, Rubin GM, Anderson DJ (2015) P1 interneurons promote a persistent internal state that enhances inter-male aggression in *Drosophila*. eLife 4:e11346. https://doi.org/10.7554/eLife.11346

8. Scaplen KM, Mei NJ, Bounds HA, Song SL, Azanchi R, Kaun KR (2019) Automated real-time quantification of group locomotor activity in Drosophila melanogaster. Sci Rep 9:4427. https://doi.org/10.1038/s41598-019-40952-5
9. Clemens J, Coen P, Roemschied FA, Pereira T, Mazumder D, Aldarondo D, Pacheco D, Murthy M (2018) Discovery of a new song mode in *drosophila* reveals hidden structure in the sensory and neural drivers of behavior. Curr Biol 28:2400–2412
10. Branson K, Robie AA, Bender J, Perona P, Dickinson MH (2009) High-throughput ethomics in large groups of *drosophila*. Nat Methods 6:451–457
11. Dankert H, Wang L, Hoopfer ED, Anderson DJ, Perona P (2009) Automated monitoring and analysis of social behavior in *drosophila*. Nat Methods 6:297–303
12. Kabra M, Robie AA, Rivera-Alba M, Branson S, Branson K (2013) JAABA: interactive machine learning for automatic annotation of animal behavior. Nat Methods 10:64–67
13. Redmon J, Divvala S, Girshick R, Farhadi A (2015) You only look once: unified, real-time object detection arXiv:1506.02640
14. Hong S-J, Han Y, Kim S-Y, Lee A-Y, Kim G (2019) Application of deep-learning methods to bird detection using unmanned aerial vehicle imagery. Sensors 19:1651. https://doi.org/10.3390/s19071651
15. Lu H, Uemura T, Wang D, Zhu J, Huang Z, Kim H (2020) Deep-sea organisms tracking using dehazing and deep learning. Mobile Networks Appl 25:1008–1015. https://doi.org/10.1007/s11036-018-1117-9
16. Silvirianti, Shin SV (2019) UAV based search and rescue with honeybee flight behavior in forest. In: Proceedings of 5th international conference on mechatronics and robotics engineering (ICMRE 2019). Italy, Rome. https://doi.org/10.1145/3314493.3314497
17. Wageeh Y, Mohamed HE-D, Fadl A, Anas O, ElMasry N, Nabil A, Atia A (2021) YOLO fish detection with Euclidean tracking in fish farms. J Ambient Intell Humaniz Comput 12:5–12. https://doi.org/10.1007/s12652-020-02847-6
18. Luh G-C, Wu H-B, Yong Y-T, Lai Y-J, Chen Y-H (2019) Facial expression based emotion recognition employing YOLOv3 deep neural networks. Int Conf Machine Learning Cybernetics (ICMLC). https://doi.org/10.1109/ICMLC48188.2019.8949236
19. Nguyen CC, Tran GS, Nghiem TP, Burie JC, Luong CM (2019) Real-time smile detection using deep learning. J Comput Sci Cybernetics 35:135–145. https://doi.org/10.15625/1813-9663/35/2/13315
20. Qian H, Zhou X, Zheng M (2020) Abnormal behavior detection and recognition method based on improved ResNet model. Comput Mat Continua 65:2153–2167. https://doi.org/10.32604/cmc.2020.011843
21. Thenmozhi M, Saravanan M, Kumar KPM, Suseela S, Deepan S (2020) Improving the prediction rate of unusual behaviors of animal in a poultry using deep learning technique. Soft Comput 24:14491–14502. https://doi.org/10.1007/s00500-020-04801-2
22. Wang J, Wang N, Li L, Ren Z (2019) Real-time behavior detection and judgement of egg breeders based on YOLO v3. Neural Comput & Applic 32:5471–5481. https://doi.org/10.1007/s00521-019-04645-4
23. Amino K, Matsuo T (2020) Intra- versus intersexual selection on sexually dimorphic traits in *Drosophila prolongata*. Zool Sci 37:210–216. https://doi.org/10.2108/zs200010
24. Kudo A, Takamori H, Watabe H, Ishikawa Y, Matsuo T (2015) Variation in morphological and behavioral traits among isofemale strains of *Drosophila prolongata* (Diptera: Drosophilidae). Entomol Sci 18:221–229. https://doi.org/10.1111/ens.12116
25. Kudo A, Shigenobu S, Kadota K, Nozawa M, Shibata TF, Ishikawa Y, Matsuo T (2017) Comparative analysis of the brain transcriptome in a hyper-aggressive fruit fly, *Drosophila prolongata*. Insect Biochem Mol Biol 82:11–20. https://doi.org/10.1016/j.ibmb.2017.01.006
26. Kaufmann JH (1983) On the definitions and functions of dominance and territoriality. Biol Rev 58:1–20. https://doi.org/10.1111/j.1469-185X.1983.tb00379.x
27. Setoguchi S, Takamori H, Aotsuka T, Sese J, Ishikawa Y, Matsuo T (2014) Sexual dimorphism and courtship behavior in *Drosophila prolongata*. J Ethol 32:91–102. https://doi.org/10.1007/s10164-014-0399-z
28. Hitoshi Y, Ishikawa Y, Matsuo T (2016) Intraspecific variation in heat tolerance of *Drosophila prolongata* (Diptera: Drosophilidae). Appl Entomol Zool 51:515–520. https://doi.org/10.1007/s13355-016-0425-4
29. Matsuo T (2018) Effect of social condition on behavioral development during early adult phase in *Drosophila prolongata*. J Ethol 36:15–22. https://doi.org/10.1007/s10164-017-0524-x
30. Ando Y, Yoshimizu T, Matsuo T (2020) Food availability reverses the effect of hunger state on

copulation rate in *Drosophila prolongata* females. Anim Behav 166:51–59. https://doi.org/10.1016/j.anbehav.2020.06.003

31. Setoguchi S, Kudo A, Takanashi T, Ishikawa Y, Matsuo T (2015) Social context-dependent modification of courtship behavior in *Drosophila prolongata*. Proc R Soc B 282:20151377. https://doi.org/10.1098/rspb.2015.1377
32. Minekawa K, Miyatake T, Ishikawa Y, Matsuo T (2018) The adaptive role of a species-specific courtship behavior in coping with remating suppression of mated females. Anim Behav 140:29–37. https://doi.org/10.1016/j.anbehav.2018.04.002
33. Minekawa K, Amino K, Matsuo T (2020) A courtship behavior that makes monandrous females polyandrous. Evolution 74:2483–2493. https://doi.org/10.1111/evo.14098

Chapter 15

Proteomic Analysis of *C. Elegans* Neurons Using TurboID-Based Proximity Labeling

Murat Artan and Mario de Bono

Abstract

The proteomes of specialized structures, and the interactomes of proteins of interest, provide entry points to elucidate the functions of molecular machines. Here, we review a proximity-labeling strategy that uses the improved *E. coli* biotin ligase TurboID to characterize *C. elegans* protein complexes. Although the focus is on *C. elegans* neurons, the method is applicable regardless of cell type. We describe detailed extraction procedures that solubilize the bulk of *C. elegans* proteins and highlight the importance of tagging endogenous genes, to ensure physiological expression levels. We review issues associated with non-specific background noise and the importance of appropriate controls. As proof of principle, we review our analysis of the interactome of a presynaptic active zone protein, ELKS-1. Our aim is to provide a detailed protocol for TurboID-based proximity labeling in *C. elegans* and to highlight its potential and its limitations to characterize protein complexes and subcellular compartments in this animal.

Key words TurboID, Proximity-dependent protein labeling, *C. elegans*, Proteomics, Neurons, Synapses, Interactome, Protein complexes

1 Introduction

Characterizing the proteomes of cells and subcellular structures, and identifying the interactome of specific proteins, can provide insights into the molecular functions of these structures or proteins. Conventional methods utilized to identify protein-protein interactions (PPIs) include yeast-two-hybrid (Y2H), affinity purification (AP), and phage display. These methods have proven useful, but have limitations. Y2H suffers from high false positive rates and, like for phage display, detects interactions in a non-physiological context. AP is poor at detecting transient or weak PPIs and usually requires gentle solubilization and careful washing to keep complexes intact, making it a challenging approach for studying membrane proteins or insoluble protein complexes that require aggressive solubilization. AP is also prone to artifacts as non-physiological complexes can form once cells/tissues are

Daisuke Yamamoto (ed.), *Behavioral Neurogenetics*, Neuromethods, vol. 181, https://doi.org/10.1007/978-1-0716-2321-3_15,

broken open. Nevertheless, AP can be successfully used to identify protein interactors in *C. elegans* neurons even when the protein of interest is expressed in relatively few neurons [1]. However, reliably detecting PPIs in specific cell types, tissues, or subcellular compartments in vivo remains a challenge [2–6]. Proximity-dependent protein labeling provides a way to overcome many of the challenges associated with defining PPIs [2, 7–9]. In proximity labeling, the protein of interest is fused to an enzyme that can add a covalently linked tag, typically biotin, to proteins in close proximity. Biotin-tagged proteins can be purified using their femtomolar affinity for streptavidin and then identified using mass spectrometry (MS) (reviewed in *2*, *3*, *8*). Currently, the most widely used enzymes for proximity-dependent protein labeling in living cells and organisms are variants of the *E. coli* biotin ligase (BirA), ascorbate peroxidase (APEX), and horseradish peroxidase (HRP) [2]. Among the BirA derivatives, TurboID and miniTurbo, which are catalytically more active forms of BioID obtained by directed evolution, have proven most versatile [10]. TurboID and miniTurbo work well in cultured mammalian cells, plants, flies, and worms [10–15].

Obtaining enough materials can be a challenge when performing biochemistry in neurons. Neurons are terminally differentiated, non-dividing cells, and the heterogeneity of the nervous system makes the problem of studying protein complexes in specific neural types more difficult. Nevertheless, the Ting group has successfully identified novel excitatory and inhibitory synaptic cleft proteins by tagging post-synaptic proteins (Lrrtm1, Lrrtm2, Slitrk3, and Nlgn2) with horseradish peroxidase (HRP) in embryonic rat cortical cultures [16]. Cultured rat cortical neurons transduced with rAAV, and transfected cultured rat hippocampal neurons, have also been used to identify novel synaptic cleft proteins by using HRP-tagged SynCAM1, a synaptic cleft protein found in excitatory neurons, as a bait [17]. A LAMP1-APEX2 fusion protein has been used to screen for lysosomal proximal proteins in human iPSC-derived neurons [18]. Lentiviral-mediated transduction of APEX2 tagged with α-synuclein (αSyn) has revealed previously uncharacterized interactors of αSyn in cultured rat primary cortical neurons [19]. Li et al. [20] analyzed the surface proteomic profile of fly olfactory projection neurons at different developmental stages by expressing HRP on their cell surface. Recently, the nuclear, cytoplasmic, and cell membrane proteomes of Drd1 receptor-positive mouse striatal neurons have been characterized using APEX2 fusions transduced via AAV [21].

The nematode *Caenorhabditis elegans* offers many advantages for proximity labeling. Gram quantities of worms can be grown quickly and cheaply. CRISPR/Cas9 allows efficient and precise knockin of the required enzyme tags, and the functionality of the resulting fusion proteins can usually be tested by examining the phenotype of knockin animals. Neuron-specific promoters allow targeted expression in desired neuron types. Conversely, because *C. elegans* neurons are small, optimizing signal-to-noise ratio is important, and there are challenges associated with effective protein extraction. Here, we review in detail a TurboID proximity-labeling protocol that can be used to characterize the interactome of specific proteins or the proteome of specific cells or subcellular structures [15]. We have successfully used the protocol with TurboID fusion proteins expressed at the active zones of synapses, the endoplasmic reticulum, endosomes, and the plasma membrane [[15]; unpublished]. As an example, we focus on work with a knockin strain that fuses TurboID to the presynaptic active zone protein ELKS-1 (see Fig. 15.1a for a simple schematic of the method). Following affinity purification, we observe enrichment of the known synaptic proteins UNC-10/RIM, SYD-1/SYDE1, SYD-2/liprin-alpha, SAD-1/BRSK1, RIM-binding protein RIMB-1, C16E9.2/Sentryn, and previously uncharacterized proteins, some of which we show localize to the synapses. Our protocol is applicable to different kinds of proteins and allows non-specialists to apply TurboID-mediated proximity labeling to identify bona fide interactors of their protein of interest in vivo.

2 Materials and Methods

2.1 TurboID-Based Enzymatic Protein Labeling and Extraction of Biotinylated Proteins from C. Elegans

2.1.1 Materials

-Sodium hypochlorite 11–14% available chlorine (Alfa Aesar, #L14709).

-Biotin (Sigma Aldrich, #B4501).

-cOmplete protease inhibitor cocktail (Roche Applied Science, #11836145001).

-Ultracentrifuge tubes (Beckman, 26.3 ml polycarbonate, #P25 837 or 10 ml polycarbonate, Lot No: P90404MPI).

-Ultracentrifuge (e.g., Beckman Optima XPN-100 with Beckman 70Ti rotor or Beckman Optima Max XP with Beckman MLA-80 rotor).

-*C. elegans* M9 buffer ($Na_2HPO_4.2H_2O$, KH_2PO_4, NaCl, $MgSO_4$).

-Liquid nitrogen.

-Cryogenic mill (Spex 6875D).

-Sonicator (Qsonica 700 with 1.6 mm microtip).

–20% SDS solution.

-1 M DTT (1,4-dithiothreitol powder, Roche Applied Science, #10197777001).

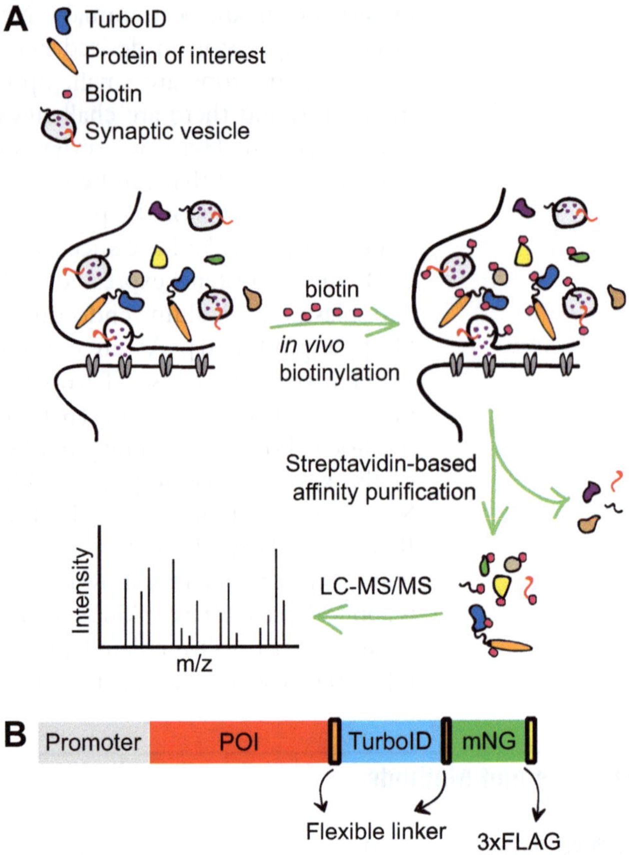

Fig. 15.1 TurboID-mediated proximity labeling of *C. elegans* synapses
(**a**) Schematic of TurboID-mediated proximity labeling in worm synapses. TurboID fused to an active zone protein, such as ELKS-1, localizes to presynaptic active zones and biotinylates proteins in its vicinity. Biotinylation is favored by exogenous addition of biotin. Biotin-labeled proteins are affinity purified using streptavidin beads, whereas unlabeled proteins are washed away. Biotinylated proteins are digested by trypsin and identified by LC-MS/MS. (**b**) Schematic of a basic bait::TurboID construct. mNG, mNeongreen; POI, protein of interest

-Zeba™ spin desalting columns, 7 K MWCO, 10 ml (Thermo Scientific, #89893).

-8 M urea buffer (8 M urea, 1% SDS, 50 mM Tris-HCl, 150 mM NaCl).

-RIPA buffer (1% Triton X-100, 1 mM EDTA, 0.5% Na deoxycholate, 0.1% SDS, 150 mM NaCl, 50 mM Tris-HCl, pH 7.4).

-High SDS and urea RIPA buffer (1% Triton X-100, 1 mM EDTA, 0.5% Na deoxycholate, 1% SDS, 2 M urea, 150 mM NaCl, 50 mM Tris-HCl, pH 7.4).

–100 mM PMSF in ethanol (phenylmethylsulfonyl fluoride powder, Roche Applied Science, #39212625).

-Pierce™ 660 nm protein assay kit (Thermo Scientific, #22662).

-Ionic Detergent Compatibility Reagent (IDCR) for Pierce™ 660 nm protein assay reagent (Thermo Scientific, #22663).

–2X TBS buffer (300 mM NaCl, 100 mM Tris-HCl, pH 7.6).

-The biotin-auxotrophic *E. coli* strain MG1655bioB:kan.

-*C. elegans* strains:

N2, the wild-type Bristol strain.

AX7647 *dbIs37[rab-3p::mNeongreen::3XFLAG]*.

The *rab-3* promoter (*rab-3p*) drives pan-neuronal expression.

AX7526 *dbIs24[rab-3p::CeTurboID::mNeongreen::3XFLAG.*

*]*PHX1710 *elks-1(syb1710)*. The *syb1710* allele is a knockin that expresses ELKS-1 tagged C-terminally with CeTurboID:: mNeongreen::3XFLAG (see Fig. 15.1b).

2.2 Methods

2.2.1 Worm Husbandry

Worms can be grown either on solid nematode growth medium (NGM) plates or in liquid culture using S Medium (see Wormbook). Liquid culture facilitates growth of large worm populations but is more prone to contamination. Care needs to be taken throughout the protocol to minimize contamination with other organisms or keratin from human skin or hair. To minimize free biotin prior to the addition of exogenous biotin, animals can be grown on *E. coli* MG1655, a biotin auxotroph. This provides some control over the time interval when most labeling occurs. *C. elegans* husbandry otherwise follows standard protocols [22]. We prefer to use staged animals rather than mixed populations, to ensure differences between experimental and control samples do not reflect differences in population structure. To obtain synchronized populations of animals, gravid adult *C. elegans* are bleached and the eggs transferred to NGM plates seeded with *E. coli* MG1655. 4–5 large (90 mm) NGM plates containing gravid adult worms are bleached, and the eggs are transferred to 12–15 large plates to obtain 10–15 mg of total worm protein. The number of plates can be increased if one desires to grow the worms on less crowded plates. We usually harvest animals at the L4 or young adult stage. We have also harvested younger larvae, e.g., L2, but these are harder to collect as they tend not to settle under gravity. Animals are washed three times in M9 buffer and incubated for 2 h at room temperature (22 °C) in M9 buffer supplemented with 1 mM biotin and *E. coli* MG1655, to allow proximity labeling. Worms are then washed three times in M9 buffer and allowed to settle on ice. After carefully aspirating as much of the M9 buffer as possible without removing worms, we add an equal volume of 2X TBS buffer supplemented with 1 mM PMSF and cOmplete EDTA-free protease inhibitor cocktail (Roche Applied Science) on top of the packed animals. After thorough mixing, the animals are again allowed to settle on ice and then added dropwise to liquid N_2 in a Styrofoam box, to obtain frozen worm "popcorn." The worm popcorn can then be transferred to a 50 ml Falcon tube. At this stage, samples can be stored at –80 °C for later processing.

2.2.2 Protein Extraction

We use a Spex 6875D cryogenic mill (Fig. 15.2a) to grind the frozen *C. elegans* to a fine powder under liquid N_2 (see Fig. 15.2a for freeze mill settings). Different mill inserts allow processing of different amounts of material, from <0.5 g to 10 g; eight samples can be processed simultaneously if worm volumes are small (Fig. 15.2b). After grinding, the worm powder is thawed by distributing it evenly along the length of a 15 ml Falcon tube and rolling it on a tube roller at 4 °C. The completely thawed sample is then briefly centrifuged (1000 rpm, 1 min, 4 °C) to collect the sample at the bottom of the tube, and SDS and DTT are added to final concentrations of 1% and 10 mM, respectively, from stock solutions of 20% SDS and 1 M DTT. The tubes are inverted several times and immediately incubated at 90 °C for 5 min in a heat block. After heating, the samples are sonicated continuously for 1 min twice, with a 1 min cooling at room temperature between the sonication steps (note: SDS precipitates below 15 °C). For sonication, we use a probe sonicator (Qsonica 700) with a 1.6 mm microtip and an amplitude setting of 16/max. Sonication helps fragment DNA, making samples less viscous. Following sonication, the samples are cooled to room temperature and adjusted to 2 M urea using a stock solution (8 M urea, 1% SDS, 50 mM Tris-HCl, 150 mM NaCl) that leaves the concentrations of other buffer components unchanged. Samples are then transferred to ultracentrifuge tubes (Beckman, 26.3 ml polycarbonate, for larger volumes, or 10 ml polycarbonate, for smaller volumes), centrifuged at 35000 rpm for 45 min at 22 °C using an ultracentrifuge, e.g., Beckman Optima XPN-100 with Beckman 70Ti rotor, for larger volumes, or at 50000 rpm for 45 min at 22 °C using a benchtop ultracentrifuge, e.g., Beckman Optima Max XP with Beckman MLA-80 rotor, for smaller volumes. The clear supernatant between the pellet and surface lipid layer is transferred to a new 15 ml Falcon tube. If worms have been extracted effectively, both the pellet and the lipid layer should be small (see Fig. 15.2c). Ideally, only the supernatant between the pellet and the lipid layer is transferred; in reality, it is difficult not to contaminate the samples with the lipid layer. This is not problematic – we find that lipid transferred with the supernatant is lost during the desalting step, leaving a clear supernatant.

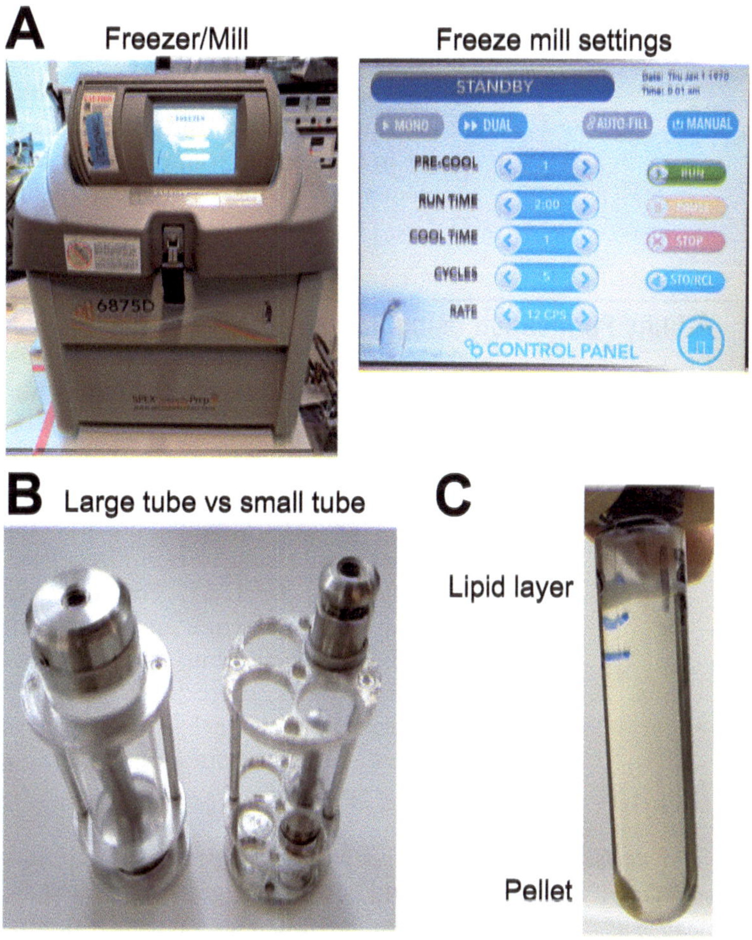

Fig. 15.2 Protein extraction from *C. elegans*
(**a**) Worms are ground to a fine powder in a Spex 6875D freezer/mill under liquid nitrogen at the settings shown. (**b**) Two large tubes or eight small tubes can be processed simultaneously. The tube size used depends on the amount of the frozen sample. (**c**) Extracted worm samples are clarified by ultracentrifugation. This separates the sample into a lipid layer at the top and a pellet, which should be small if solubilization was effective. Tube capacity in the image is 10.4 ml

2.2.3 Desalting to Remove Free Biotin

Before samples can be affinity purified on streptavidin beads, the copious amount of free biotin carried over from the proximity-labeling step needs to be removed. To do this, we equilibrate Zeba™ spin desalting columns (7 K MWCO) (Thermo Fisher) three times with 5 ml RIPA buffer containing 1% SDS and 2 M urea and freshly supplemented with protease inhibitors (Roche cOmplete EDTA-free protease inhibitor cocktail 1 tablet/25 ml; PMSF 1 mM) by centrifuging them at 1000 g for 5 min (or until the buffer is completely eluted from resin). Around 4 ml of clarified sample is then loaded onto the equilibrated spin column and desalted by centrifugation at 1000 g for 5 min (or until the sample is completely eluted from the resin). Equilibrated desalting columns should be used only once and loaded with no more than 4 ml of clarified samples. We repeat the desalting step once using freshly equilibrated columns. We then measure the protein concentration in the samples using Pierce™ 660 nm protein assay reagent supplemented with Ionic Detergent Compatibility Reagent (IDCR) (Thermo Fisher Scientific).

3 Affinity Purification Using Streptavidin Beads

We have successfully used two affinity purification protocols. In the "elution-digestion" protocol, proteins bound to streptavidin beads are eluted from the beads and run on an SDS/PAGE gel, and the gel is cut into slices that are each treated with trypsin for mass spectrometry. The second "on-bead digestion" protocol digests bound proteins using trypsin directly on beads. Each method has advantages (e.g., see [23]). The "elution-digestion" approach separates the protein mixture and thus reduces the complexity of the peptide mix injected into the mass spectrometer. However, this method requires more mass spectrometer time, as peptides extracted from each gel slice are analyzed separately. "On-bead digestion" results in one pot of peptides that can be analyzed in one mass spec run.

3.1 Streptavidin-Based Affinity Purification of Biotinylated Proteins and Streptavidin Magnetic Bead Washing

Note: this part of the protocol is common for the "elution-digestion" and "on-bead digestion" protocols, except that the "on-bead digestion" protocol uses Pierce™ streptavidin magnetic beads (see below), whereas the "elution-digestion" protocol uses Dynabeads™ MyOne™ streptavidin C1 (Invitrogen) which, in our experience, are easier to handle.

3.1.1 Materials

- Neodymium magnets.
- Pierce™ streptavidin magnetic beads (Thermo Scientific, #88817) (on-bead digestion) or Dynabeads™ MyOne™ streptavidin C1 (Invitrogen, #65002) (elution-digestion).
- MagRack™ 6 (GE Healthcare, #GE28–9489-64).
- 2% SDS wash buffer (150 mM NaCl, 1 mM EDTA, 2% SDS, 50 mM Tris-HCl).
- KCl wash buffer (1 M KCl, 1 mM EDTA, 50 mM Tris-HCl, pH 7.4).
- Na_2CO_3 buffer (0.1 M Na_2CO_3, pH 11.5).
- 2 M urea buffer (2 M urea, 10 mM Tris-HCl, pH 8.0).

- TBS buffer (150 mM NaCl, 50 mM Tris-HCl, pH 7.6).
- Protein LoBind tubes (Eppendorf).

3.1.2 Method

Desalted, clarified samples (10 mg of total protein) are mixed with magnetic beads (200 μl of bead slurry per 10 mg of proteins) in a 5 ml Protein LoBind tube (Eppendorf) and incubated overnight on a tube roller at room temperature. To collect beads, a neodymium magnet is taped to the tube and incubated on a rocking platform for 1 h. Unbound lysate is removed, and the beads are transferred to a 2 ml Protein LoBind tube (Eppendorf) using 2% SDS wash buffer (150 mM NaCl, 1 mM EDTA, 2% SDS, 50 mM Tris-HCl, pH 7.4). Beads are washed twice with 2% SDS wash buffer, once with TBS buffer (150 mM NaCl, 50 mM Tris-HCl, pH 7.6), twice with 1 M KCl wash buffer (1 M KCl, 1 mM EDTA, 50 mM Tris-HCl, pH 7.4), twice with 0.1 M Na_2CO_3 buffer (0.1 M Na_2CO_3, pH 11.5), twice with 2 M urea buffer (2 M urea, 10 mM Tris-HCl, pH 8.0), and five times with TBS buffer (150 mM NaCl, 50 mM Tris-HCl, pH 7.6). During each wash, beads are incubated on a rocking platform for 10–15 min and transferred to new tubes after each wash or as often as possible. At the end of this step, beads are kept in TBS buffer.

3.2 "Elution-Digestion" Protocol

3.2.1 Materials

- 4X Bolt™ LDS sample buffer (Thermo Scientific, #B0007).
- Bolt™ 4–12% Bis-Tris mini protein gel (Thermo Scientific, #NW04120BOX).
- 10X Bolt™ sample reducing agent (Thermo Scientific, #B0009).
- Biotin (Sigma Aldrich, #B4501).
- Protein LoBind tubes (Eppendorf).
- InstantBlue™ (Expedeon, #190718010).

3.2.2 Method

Elution Elution sample buffer is prepared by dissolving free biotin (Sigma) to saturation in sample buffer (NuPAGE LDS sample buffer 4X) containing reducing agent (NuPAGE sample reducing agent 10X). The elution sample buffer is then centrifuged at 13000 rpm for 5 min to remove undissolved biotin, and the biotin-saturated elution sample transferred to a new tube. 100 μl of elution buffer is applied to the Dynabeads and, after gentle vortexing, heated for 5 min at 90 °C; the vortexing and heating steps are repeated once. Dynabeads are then collected using the neodymium magnet, and the buffer containing the eluted biotinylated proteins is transferred to a fresh 1.5 ml Protein LoBind tube (Eppendorf). 70 μl of sample is then electrophoresed on a NuPAGE™ 4–12% Bis-Tris protein gel (Invitrogen) and stained with InstantBlue (Expedeon). We typically slice the gel lane into 20 slices for further processing.

3.3 Digestion and Mass Spectrometry (Performed by Mass Spectrometry Facility)

The polyacrylamide gel slices (1–2 mm wide) containing the purified proteins can be prepared for mass spectrometric analysis using the Janus liquid handling system (PerkinElmer, UK). Briefly, each excised protein gel slice is placed in a well of a 96-well microtiter plate, destained with 50% v/v acetonitrile and 50 mM ammonium bicarbonate, reduced with 10 mM DTT, and alkylated with 55 mM iodoacetamide. After alkylation, proteins are digested with 6 ng/μL trypsin (Promega, UK) overnight at 37 °C. The resulting peptides are extracted in 2% v/v formic acid and 2% v/v acetonitrile. The digest is then analyzed by nano-scale capillary LC-MS/MS using an Ultimate U3000 HPLC (Thermo Scientific Dionex, San Jose, USA) to deliver a flow of approximately 300 nL/min. A C18 Acclaim PepMap100 5 μm, 100 μm x 20 mm nanoViper (Thermo Scientific Dionex, San Jose, USA), traps the peptides prior to separation on an EASY-Spray PepMap RSLC 2 μm, 100 Å, 75 μm x 250 mm nanoViper column (Thermo Scientific Dionex, San Jose, USA). Peptides are eluted using a 60 min gradient of acetonitrile (2% to 80%). The analytical column outlet is directly interfaced via a nano-flow electrospray ionization source with a hybrid quadrupole-orbitrap mass spectrometer (Q Exactive Orbitrap, Thermo Scientific, San Jose, USA). Data collection is performed in data-dependent acquisition (DDA) mode with an r = 70,000 (@ *m/z* 200) full MS scan from *m/z* 380 to 1600 with a target AGC value of 1e6 ions followed by 15 MS/MS scans at r = 17,500 (@ *m/z* 200) at a target AGC value of 1e5 ions. MS/MS scans are collected using a threshold energy of 27 for higher-energy collisional dissociation (HCD) and a 30 s dynamic exclusion employed to increase depth of coverage. LC-MS/MS data are then searched against a protein database (UniProt KB) using the Mascot search engine program (Matrix Science, UK) [24]. Database search parameters are set with a precursor tolerance of 10 ppm and a fragment ion mass tolerance of 0.8 Da. One missed enzyme cleavage is allowed, and variable modifications for oxidized methionine and N-terminal acetylation are included. Carbamidomethylation of cysteine is specified as a fixed modification. MS/MS data are validated using the Scaffold program (Proteome Software Inc., USA) [25]. All data are additionally interrogated manually.

3.4 "On-Bead Digestion" Protocol

3.4.1 Streptavidin Magnetic Bead Acetylation

On bead digestion of streptavidin-bound proteins using trypsin can release significant amounts of streptavidin peptides. This reduces signal to noise in the mass spectrometer analysis. Trypsin cleaves proteins on the C-terminal side of lysine and arginine residues. To minimize trypsin digestion of streptavidin, beads can be acetylated before use, to block the amine groups of these basic residues [26].

Materials

- Buffer 1 (50 mM HEPES-NaOH, pH 7.8, 0.2% Tween-20)
- Buffer 2 (50 mM ammonium bicarbonate, 0.2% Tween-20)
- PierceTM Sulfo-NHS-Acetate (Thermo Scientific, #26777, dissolved in DMSO to make 100 mM stock solution)
- MagRackTM 6 (GE Healtcare, #GE28-9489-64)
- PierceTM Streptavidin magnetic beads (Thermo Scientific, #88817)

Method

PierceTM Streptavidin magnetic beads (Thermo Scientific, catalog # 88817, 200 μl bead slurry per 10 mg of C. elegans total protein lysate) are first washed three times with 1 ml of Buffer 1 (50 mM HEPES-NaOH, pH 7.8, 0.2% Tween-20) by briefly incubating them in the buffer and using magnetic separation to retain beads while discarding buffer. Beads are then resuspended in a mix of 190 μl of Buffer 1 and 10 μl of 100 mM PierceTM Sulfo-NHS-Acetate (Thermo Scientific, catalog #26777, dissolved in DMSO), and incubated for 1 h at room temperature to acetylate free amines. Beads are then washed three times with 1 ml of Buffer 2 (50 mM ammonium bicarbonate, 0.2% Tween-20) to stop the reaction.

3.4.2 Digestion and Mass Spectrometry (Performed by Mass Spectrometry Facility)

Following the affinity purification step, beads are resuspended in 50 μl 1 M urea and 50 mM ammonium bicarbonate. Disulfide bonds are reduced with 2 μl of 250 mM dithiothreitol (DTT) for 30 min at room temperature before adding 2 μl of 500 mM iodoacetamide and incubating for 30 min at room temperature in the dark. The remaining iodoacetamide is quenched with 1 μl of 250 mM DTT for 10 min. Proteins are digested with 150 ng trypsin (Trypsin Gold, Promega) in 1.5 μL 50 mM ammonium bicarbonate at 37 °C overnight. The supernatant without beads is then transferred to a new tube, the digest is stopped by the addition of trifluoroacetic acid (TFA) to a final concentration of 0.5%, and the peptides are desalted using C18 StageTips [26].

Peptides are separated on an Ultimate 3000 RSLC nano-flow chromatography system (Thermo Fisher), using a pre-column for sample loading (Acclaim PepMap C18, 2 cm × 0.1 mm, 5 μm, Thermo Fisher) and a C18 analytical column (Acclaim PepMap C18, 50 cm × 0.75 mm, 2 μm, Thermo Fisher), applying a segmented linear gradient from 2% to 35% and finally 80% solvent B (80% acetonitrile, 0.1% formic acid; solvent A 0.1% formic acid) at a flow rate of 230 nL/min over 120 min. Eluting peptides are analyzed on a Q Exactive HF-X Orbitrap mass spectrometer (Thermo Fisher), coupled to the column with a nanospray ion source using coated emitter tips (PepSep, MSWil).

The mass spectrometer is operated in data-dependent acquisition (DDA) mode, and survey scans are obtained in a mass range of 375–1500 m/z with lock mass activated, at a resolution of 120 k at

200 m/z and an AGC target value of 3E6. The 20 most intense ions are selected with an isolation width of 1.4 m/z, fragmented in the HCD cell at 28% collision energy and the spectra recorded for max. 120 ms at a target value of 1E5 and a resolution of 30 k. Peptides with a charge of +2 to +6 are included for fragmentation, the peptide match and the exclude isotopes features enabled, and selected precursors are dynamically excluded from repeated sampling for 30 seconds.

Raw data are processed using the MaxQuant software package (version 1.6.17.0, [27]) and the UniProt *Caenorhabditis elegans* reference proteome (www.uniprot.org), as well as a database of most common contaminants. The search is performed with full trypsin specificity and a maximum of two missed cleavages at a protein and peptide spectrum match false discovery rate of 1%. Carbamidomethylation of cysteine residues is set as fixed, and oxidation of methionine and N-terminal acetylation are set as variable modifications. For label-free quantification, the "match between runs" feature and the LFQ function are activated – all other parameters are left at default.

3.4.3 Data Analysis Using R Scripts

MaxQuant output tables are further processed in R (R Core Team, 2018). Reverse database identifications, contaminant proteins, protein groups identified only by a modified peptide, protein groups with less than three quantitative values in one experimental group, and protein groups with less than two razor peptides are removed for further analysis. Missing values are replaced by randomly drawing data points from a normal distribution modeled on the whole dataset (data mean shifted by −1.8 standard deviations, width of distribution of 0.3 standard deviations). Differences between groups are statistically evaluated using the LIMMA package [28] at 5% FDR (Benjamini-Hochberg).

3.4.4 Proteomics Data Deposition

The mass spectrometry proteomics data have been deposited to the ProteomeXchange Consortium via the PRIDE partner repository [29] with the dataset identifier PXD029060.

4 Analytical SDS-PAGE and Immunoblotting

Analytical gels can be used to evaluate efficiency of different steps in the protocol, for example, the in vivo proximity labeling.

4.1 Materials

- 4X Bolt™ LDS sample buffer (Thermo Scientific, #B0007).
- Immobilon™-FL transfer membrane (Merck Millipore, #IPFL00010, pore size 0.45 μm).
- Bolt™ 4–12% Bis-Tris mini protein gel (Thermo Scientific, #NW04120BOX).

- 10X Bolt™ sample reducing agent (Thermo Scientific, #B0009).
- Bovine serum albumin (Fisher Scientific, #BP9703–100).
- IRDye® 800CW streptavidin (1:7000) (LI-COR Biosciences, #926–32,230) or streptavidin-HRP (1:5000) (Cell Signaling, #3999S).

4.2 Method

Synchronized populations of *C. elegans* grown on *E. coli* MG1655 are harvested at L4 or young adult stage, washed three times in M9 buffer, and incubated at room temperature (22 °C) in M9 buffer supplemented with 1 mM biotin and *E. coli* MG1655 for 2 h. Two hours later, the worms are washed three times in M9 buffer, flash-frozen after the M9 buffer is completely aspirated and 4X Bolt™ LDS sample buffer supplemented with fresh DTT added. The samples are then thawed, boiled for 10 min at 90 ° C, vortexed for 10 min, and centrifuged for 30 min at 15000 rpm at 4 ° C and the supernatant collected. Proteins are electrophoresed using a Bolt 4–12% Bis-Tris Plus gel (Thermo Fisher Scientific) and transferred to a PVDF membrane (Thermo Fisher Scientific). Membranes are blocked for 1 h at room temperature with TBS-T buffer containing 5% BSA and incubated for 2 h at room temperature with IRDye® 800CW streptavidin (1:7000) (LI-COR Biosciences). Membranes are then washed three times with TBS-T and three times with TBS before being imaged using a ChemiDoc Imaging System (Model MP, Bio-Rad).

5 Experimental Design and Analysis of Mass Spectrometry Data

5.1 Ensuring Tagged Proteins Are Functional and Appropriately Localized

An advantage of using *C. elegans* is that the functionality of TurboID-tagged proteins can be explicitly tested if a mutant phenotype is available. Proteins tagged at their N- or C-terminus often, but not always, retain function. We recommend making knockins for proximity-labeling experiments, rather than integrated multicopy arrays. Knockins help ensure physiological levels of expression of the tagged protein. Overexpression associated with multicopy arrays is likely to increase noise, especially for proteins that enter the secretory pathway, as the tagged protein can accumulate in abnormal subcellular locations [30]. TurboID-tagged proteins that accumulate in inappropriate locations will highlight a misleading set of proteins. We also recommend adding a fluorescent reporter, so that the subcellular localization of the fusion protein can be monitored in the knockin animals. This is especially useful for proteins known to localize at specific subcellular locations, e.g., synapses, cilia. Confirming that the fusion protein localizes as expected provides a reality check early on, especially if phenotypic assays are not available to confirm functionality of the TurboID knockin. In general, it is advisable to place the TurboID domain proximal to the protein of interest and the fluorescent reporter distal. The labeling radius of TurboID inside cells is reported to be ~10 nm; the long axis of GFP is ~4.2 nm. Note that tagging large proteins at different sequence locations can result in different proteins being labeled.

5.1.1 Background Noise

The femtomolar affinity of streptavidin for biotin makes it possible to extract and affinity purify biotinylated proteins under strongly denaturing conditions and to wash streptavidin beads extensively following capture. Despite this, background noise is a significant confound in these experiments. The most prominent proteins detected by the mass spectrometer are four endogenously biotinylated proteins, the carboxylases PCCA-1/PCCA (propionyl coenzyme A carboxylase alpha subunit), PYC-1/PC (pyruvate carboxylase), MCCC-1/MCCC (methylcrotonyl coenzyme A carboxylase), and POD-2/ACACA (acetyl coenzyme A carboxylase). These carboxylases can easily be seen by Western blot, including in wild-type controls that have no TurboID expression. Depending on how widely and strongly TurboID is expressed in the worm, these carboxylases may comprise 8%–85% of all the peptides identified in the MS analysis. Depleting these endogenously biotinylated proteins, e.g., by affinity purification, prior to streptavidin capture, should increase the sensitivity and specificity of mass spec experiments substantially. While the carboxylases dominate the background noise, there are also scores of other proteins that repeatedly come through to the mass spectrometer, even when extracts are made from control worms that do not express TurboID. This "crapome" likely consists of sticky proteins that bind non-specifically to the streptavidin beads. Finding wash conditions that can disrupt this non-specific binding to beads would also increase sensitivity.

5.1.2 Controls

Designing good controls is key to interpreting TurboID experiments. For our ELKS-1-TurboID experiments, our controls included extracts made from wild-type animals and from transgenic animals expressing soluble, free TurboID throughout the nervous system [15] (data not shown). As expected, mass spectrometry analysis of the ELKS-1-TurboID samples revealed enrichment of synaptic proteins compared to both control samples. Future control experiments could include analysis of extracts made from animals in which TurboID is targeted to the plasma membrane or of other synaptic or peri-synaptic proteins tagged with TurboID. Matching the expression levels of the control and experimental TurboID fusion proteins can make comparisons more straightforward. For this purpose, control TurboID constructs can be expressed in an operon with the gene of interest (see Fig. 15.1b). However, even when this precaution is taken, different turnover rates of control and experimental TurboID fusion proteins can lead

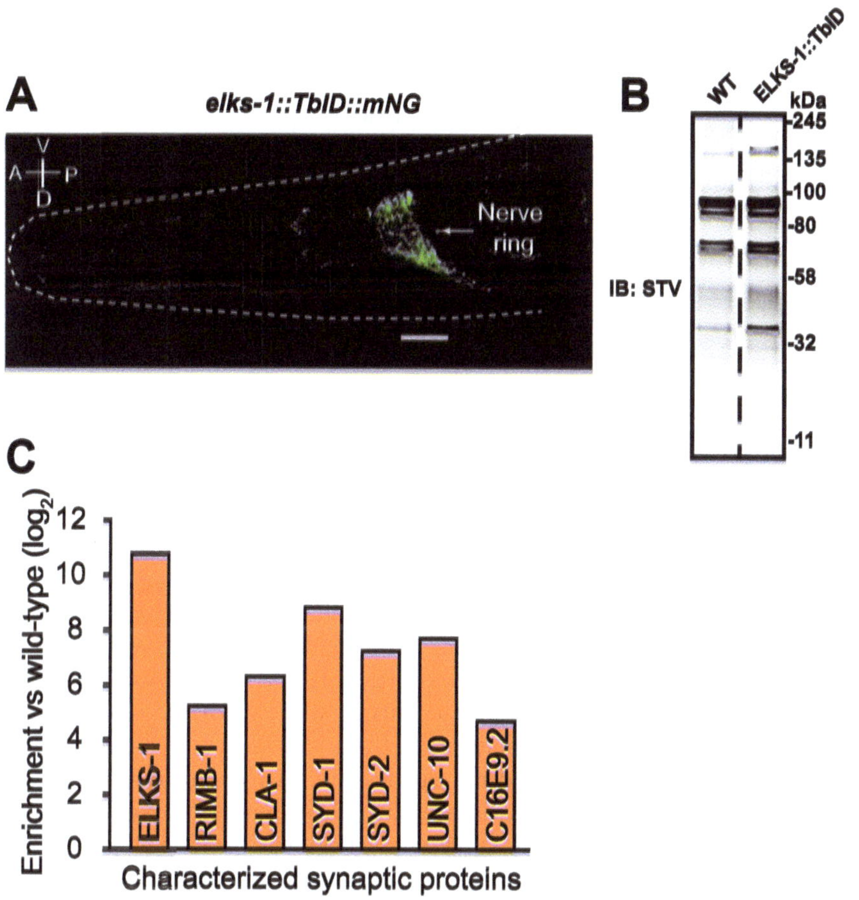

Fig. 15.3 Interactome analysis of the presynaptic active zone protein ELKS-1
(**a**) Confocal microscopy image of *C. elegans* harboring an *elks-1::TurboID::mNeongreen* knockin allele generated by CRISPR/Cas9-mediated genome editing. (**b**) Western blot analysis of biotinylated proteins obtained from animals expressing the ELKS-1::TurboID::mNeongreen fusion protein. (**c**) The synaptic proteins ELKS-1/ELKS, UNC-10/RIM, SYD-1/SYDE1, SYD-2/liprin-alpha, SAD-1/BRSK1, CLA-1/CLArinet, C16E9.2/Sentryn, and the RIM-binding protein RIMB-1 are highly enriched following proximity labeling using ELKS-1::TurboID samples. Scale bar: 5 μm

to different expression levels. What is encouraging is that analyzing a battery of protein-TurboID fusion proteins allows each fusion to act as a control for others.

6 Conclusion

Here, we review a detailed protocol for TurboID-based proximity labeling in nematodes. TurboID works well in all *C. elegans* tissues tested [15], and we have successfully used the protocol with TurboID fusion proteins that have different subcellular localizations, including the cytoplasm, the endoplasmic reticulum, other parts of the secretory pathway, and the plasma membrane (unpublished data). We have successfully studied soluble, membrane-associated, and transmembrane proteins (unpublished data). We observe efficient labeling with TurboID on both the cytoplasmic and the luminal sides of endosomal compartments, as has been observed in mammalian tissue culture cells [2, 10]. Our success in identifying many active zone proteins when we knocked in *TurboID::mNeongreen* into the endogenous *elks-1* locus (Fig. 15.3a-c), which encodes an active zone protein [31], suggests our extraction conditions can solubilize even difficult structures such as active zones. Our data suggest the protocol is versatile.

Material is rarely limiting in these experiments: growing grams of *C. elegans* is straightforward. Non-specific noise is, however, a significant confounding factor, particularly when the TurboID fusion protein is expressed in a small number of cells or at low levels. Reducing this background would substantially increase the utility of the method. A different issue is the effective range of proximity labeling. This is reported to be about 10 nm intracellularly [2], where the high protein concentration means the reactive biotin generated by TurboID rapidly reacts with an amine group. The effective range of TurboID proximity labeling may be significantly greater extracellularly, where protein concentration is much lower. This may make the method useful for describing the cell surface proteome, but less useful to identify extracellular interactomes of specific proteins [32]. Being able to tune the half-life of the proximity-labeling reaction may provide ways to alter the labeling diameter to suit the question being studied [32]. A related problem is temporal control of the labeling reaction. In cultured mammalian cells, incubation times of as low as 10 min yielded robust biotinylation with TurboID or miniTurbo [10]. Feeding biotin-auxotroph *E. coli* to worms permits a semi-controlled start for biotinylation by TurboID. In our experiments, a 2 h incubation with exogenous biotin was sufficient for biotinylation in worm neurons [15]. For time-sensitive experiments, shorter incubation time may, however, also be sufficient.

In summary, TurboID promises to be a reliable and useful tool by which the *C. elegans* community can map the proteome of cells and subcellular structures and characterize the interactomes of specific proteins. It will be interesting to test whether by further decreasing background noise, the interactomes of specific proteins can be characterized even at single neuron resolution. This would complement and extend the power of RNA Seq experiments to characterize the biology of single neurons. Reducing background will also make it easier to characterize post-translational modifications and for quantitative comparisons across genotypes and after animals have had different experiences.

Acknowledgments

We thank de Bono lab members for the helpful comments on the manuscript. The biotin-auxotrophic *E. coli* strain MG1655bioB: kan was a generous gift from J. Cronan (University of Illinois) and was kindly sent to us by Jessica Feldman and Ariana Sanchez (Stanford University). dg398 pEntryslot2_mNeongreen::3X-FLAG::stop and dg397 pEntryslot3_mNeongreen::3XFLAG::stop::unc-54 3'UTR entry vector were kindly sent by Dr. Dominique Glauser (University of Fribourg). This work was supported by an Advanced ERC Grant (269058 ACMO) and a Wellcome Investigator Award (209504/Z/17/Z) to MdB and an ISTplus Fellowship to MA (Marie Sklodowska-Curie agreement No 754411).

References

1. Beets I, Zhang G, Fenk LA, Chen C, Nelson GM, Félix MA, de Bono M (2020) Natural variation in a dendritic scaffold protein remodels experience-dependent plasticity by altering neuropeptide expression. Neuron 105:106–121
2. Qin W, Cho KF, Cavanagh PE, Ting AY (2021) Deciphering molecular interactions by proximity labeling. Nat Methods. https://doi.org/10.1038/s41592-020-01010-5
3. Kim DI, Roux KJ (2016) Filling the void: proximity-based labeling of proteins in living cells. Trends Cell Biol 26:804–817
4. Varnaitė R, MacNeill SA (2016) Meet the neighbors: mapping local protein interactomes by proximity-dependent labeling with BioID. Proteomics 16:2503–2518
5. Trinkle-Mulcahy L (2019) Recent advances in proximity-based labeling methods for interactome mapping [version 1; referees: 2 approved]. F1000Research. https://doi.org/10.12688/f1000research.16903.1
6. Rees JS, Li XW, Perrett S, Lilley KS, Jackson AP (2015) Protein neighbors and proximity proteomics. Mol Cell Proteomics 14:2848–2856
7. Xu Y, Fan X, Hu Y (2021) In vivo interactome profiling by enzyme-catalyzed proximity labeling. Cell Biosci 11:1–9
8. Samavarchi-Tehrani P, Samson R, Gingras AC (2020) Proximity dependent biotinylation: key enzymes and adaptation to proteomics approaches. Mol Cell Proteomics 19:757–773
9. Gingras AC, Abe KT, Raught B (2019) Getting to know the neighborhood: using proximity-dependent biotinylation to characterize protein complexes and map organelles. Curr Opin Chem Biol 48:44–54
10. Branon TC, Bosch JA, Sanchez AD, Udeshi ND, Svinkina T, Carr SA, Feldman JL, Perrimon N, Ting AY (2018) Efficient proximity labeling in living cells and organisms with TurboID. Nat Biotechnol 36:880–898
11. Zhang Y, Song G, Lal NK, Nagalakshmi U, Li Y, Zheng W, Huang PJ, Branon TC, Ting AY, Walley JW, Dinesh-Kumar SP (2019) TurboID-based proximity labeling reveals that UBR7 is a regulator of N NLR immune receptor-mediated immunity. Nat Commun. https://doi.org/10.1038/s41467-019-11202-z
12. Mair A, Xu SL, Branon TC, Ting AY, Bergmann DC (2019) Proximity labeling of protein complexes and cell type specific organellar proteomes in Arabidopsis enabled by TurboID. elife 8:1–46
13. Shinoda N, Hanawa N, Chihara T, Koto A, Miura M (2019) Dronc-independent basal executioner caspase activity sustains drosophila imaginal tissue growth. Proc Natl Acad Sci U S A 116:20539–20544
14. Sanchez AD, Branon TC, Cote LE, Papagiannakis A, Liang X, Pickett MA, Shen K, Jacobs-Wagner C, Ting AY, Feldman JL (2021) Proximity labeling reveals non-centrosomal microtubule-organizing center components required for microtubule growth and localization. Curr Biol 31:3586–3600
15. Artan M, Barratt S, Flynn SM, Begum F, Skehel M, Nicolas A, de Bono M (2021) Interactome analysis of C. elegans synapses by TurboID-based proximity labeling. J Biol

Chem. https://doi.org/10.1016/j.jbc.2021.101094

16. Loh KH, Stawski PS, Draycott AS, Stevens B, Carr SA, Ting AY, Loh KH, Stawski PS, Draycott AS, Udeshi ND, Lehrman EK, Wilton DK (2016) Proteomic analysis of unbounded cellular compartments : synaptic clefts resource proteomic analysis of unbounded cellular compartments: synaptic clefts. Cell 166:1295–1307
17. Cijsouw T, Ramsey AM, Lam TKT, Carbone BE, Blanpied TA, Biederer T (2018) Mapping the proteome of the synaptic cleft through proximity labeling reveals new cleft proteins. Proteomes. https://doi.org/10.3390/proteomes6040048
18. Liao Y, Fernandopulle MS, Wang G, George-hyslop PS, Lippincott-schwartz J, Ward ME, Liao Y, Fernandopulle MS, Wang G, Choi H, Hao L, Drerup CM (2019) Article RNA granules hitchhike on lysosomes for long- distance transport , using Annexin A11 as a Article RNA granules hitchhike on lysosomes for long-distance transport, using Annexin A11 as a molecular tether. Cell 179:147–164
19. Chung CY, Khurana V, Yi S, Vidal M, Ting AY (2017) In situ peroxidase labeling and mass-spectrometry connects alpha-Synuclein directly to endocytic trafficking and mRNA metabolism in neurons report in situ peroxidase labeling and mass-spectrometry connects alpha-Synuclein directly to endocytic trafficking. Cell Syst 4:242–250
20. Li J, Han S, Li H, Carr SA, Ting AY, Luo L (2020) Cell-surface proteomic profiling in the Fly brain cell-surface proteomic profiling in the Fly brain uncovers wiring regulators. Cell 180: 373–386
21. Dumrongprechachan V, Salisbury RB, Soto G, Kumar M, MacDonald ML, Kozorovitskiy Y (2021) Cell-type and subcellular compartment-specific APEX2 proximity labeling reveals activity-dependent nuclear proteome dynamics in the striatum. Nat Commun 12:1–16
22. Brenner S (1974) The genetics of *Caenorhabditis elegans*. Genetics 77(1):71–94
23. Zhang Y, Sun H, Zhang J, Brasier AR, Zhao Y (2017) Quantitative assessment of the effects of trypsin digestion methods on affinity purification-mass spectrometry-based protein-protein interaction analysis. J Proteome Res 16:3068–3082
24. Perkins DN, Pappin DJC, Creasy DM, Cottrell JS (1999) Probability-based protein identification by searching sequence databases using mass spectrometry data. Electrophoresis 20: 3551–3567
25. Keller A, Nesvizhskii AI, Kolker E, Aebersold R (2002) Empirical statistical model to estimate the accuracy of peptide identifications made by MS/MS and database search. Anal Chem 74: 5383–5392
26. Rappsilber J, Mann M, Ishihama Y (2007) Protocol for micro-purification, enrichment, pre-fractionation and storage of peptides for proteomics using StageTips. Nat Protoc 2: 1896–1906
27. Tyanova S, Temu T, Cox J (2016) The Max-Quant computational platform for mass spectrometry-based shotgun proteomics. Nat Protoc 11:2301–2319
28. Ritchie ME, Phipson B, Wu D, Hu Y, Law CW, Shi W, Smyth GK (2015) Limma powers differential expression analyses for RNA-sequencing and microarray studies. Nucleic Acids Res 43:e47
29. Perez-Riverol Y, Csordas A, Bai J, Bernal-Llinares M, Hewapathirana S, Kundu DJ, Inuganti A, Griss J, Mayer G, Eisenacher M, Pérez E, Uszkoreit J, Pfeuffer J, Sachsenberg T, Yilmaz Ş, Tiwary S, Cox J, Audain E, Walzer M, Jarnuczak AF, Ternent T, Brazma A, Vizcaíno JA (2019) The PRIDE database and related tools and resources in 2019: improving support for quantification data. Nucleic Acids Res 47:D442–D450
30. Grote E, Kelly RB (1996) Endocytosis of VAMP is facilitated by a synaptic vesicle targeting signal. J Cell Biol 132:537–547
31. Südhof TC (2012) The presynaptic active zone. Neuron 75:11–25
32. Geri JB, Oakley JV, Reyes-Robles T, Wang T, McCarver SJ, White CH, Rodriguez-Rivera FP, Parker DL, Hett EC, Fadeyi OO, Oslund RC, MacMillan DWC (2020) Microenvironment mapping via Dexter energy transfer on immune cells. Science (80-) 367:1091–1097

Index

Daisuke Yamamoto (ed.), *Behavioral Neurogenetics*, Neuromethods, vol. 181, https://doi.org/10.1007/978-1-0716-2321-3,

MIX
Papier aus verantwortungsvollen Quellen
Paper from responsible sources
FSC® C105338

If you have any concerns about our products,
you can contact us on
ProductSafety@springernature.com

In case Publisher is established outside the EU,
the EU authorized representative is:
Springer Nature Customer Service Center GmbH
Europaplatz 3, 69115 Heidelberg, Germany

Printed by Libri Plureos GmbH
in Hamburg, Germany